Pitman Research Notes in Mathematics Series

Submission of proposals for consideration
Suggestions for publication, in the form of outlines and representative samples, are invited by the Editorial Board for assessment. Intending authors should approach one of the main editors or another member of the Editorial Board, citing the relevant AMS subject classifications. Alternatively, outlines may be sent directly to the publisher's offices. Refereeing is by members of the board and other mathematical authorities in the topic concerned, throughout the world.

Preparation of accepted manuscripts
On acceptance of a proposal, the publisher will supply full instructions for the preparation of manuscripts in a form suitable for direct photo-lithographic reproduction. Specially printed grid sheets are provided and a contribution is offered by the publisher towards the cost of typing. Word processor output, subject to the publisher's approval, is also acceptable.

Illustrations should be prepared by the authors, ready for direct reproduction without further improvement. The use of hand-drawn symbols should be avoided wherever possible, in order to maintain maximum clarity of the text.

The publisher will be pleased to give any guidance necessary during the preparation of a typescript, and will be happy to answer any queries.

Important note
In order to avoid later retyping, intending authors are strongly urged not to begin final preparation of a typescript before receiving the publisher's guidelines and special paper. In this way it is hoped to preserve the uniform appearance of the series.

Longman Scientific & Technical
Longman House
Burnt Mill
Harlow, Essex, UK
(tel (0279) 26721)

Titles in this series

1 Improperly posed boundary value problems
A Carasso and A P Stone
2 Lie algebras generated by finite dimensional ideals
I N Stewart
3 Bifurcation problems in nonlinear elasticity
R W Dickey
4 Partial differential equations in the complex domain
D L Colton
5 Quasilinear hyperbolic systems and waves
A Jeffrey
6 Solution of boundary value problems by the method of integral operators
D L Colton
7 Taylor expansions and catastrophes
T Poston and I N Stewart
8 Function theoretic methods in differential equations
R P Gilbert and R J Weinacht
9 Differential topology with a view to applications
D R J Chillingworth
10 Characteristic classes of foliations
H V Pittie
11 Stochastic integration and generalized martingales
A U Kussmaul
12 Zeta-functions: An introduction to algebraic geometry
A D Thomas
13 Explicit *a priori* inequalities with applications to boundary value problems
V G Sigillito
14 Nonlinear diffusion
W E Fitzgibbon III and H F Walker
15 Unsolved problems concerning lattice points
J Hammer
16 Edge-colourings of graphs
S Fiorini and R J Wilson
17 Nonlinear analysis and mechanics: Heriot-Watt Symposium Volume I
R J Knops
18 Actions of fine abelian groups
C Kosniowski
19 Closed graph theorems and webbed spaces
M De Wilde
20 Singular perturbation techniques applied to integro-differential equations
H Grabmüller
21 Retarded functional differential equations: A global point of view
S E A Mohammed
22 Multiparameter spectral theory in Hilbert space
B D Sleeman
24 Mathematical modelling techniques
R Aris
25 Singular points of smooth mappings
C G Gibson
26 Nonlinear evolution equations solvable by the spectral transform
F Calogero
27 Nonlinear analysis and mechanics: Heriot-Watt Symposium Volume II
R J Knops
28 Constructive functional analysis
D S Bridges
29 Elongational flows: Aspects of the behaviour of model elasticoviscous fluids
C J S Petrie
30 Nonlinear analysis and mechanics: Heriot-Watt Symposium Volume III
R J Knops
31 Fractional calculus and integral transforms of generalized functions
A C McBride
32 Complex manifold techniques in theoretical physics
D E Lerner and P D Sommers
33 Hilbert's third problem: scissors congruence
C-H Sah
34 Graph theory and combinatorics
R J Wilson
35 The Tricomi equation with applications to the theory of plane transonic flow
A R Manwell
36 Abstract differential equations
S D Zaidman
37 Advances in twistor theory
L P Hughston and R S Ward
38 Operator theory and functional analysis
I Erdelyi
39 Nonlinear analysis and mechanics: Heriot-Watt Symposium Volume IV
R J Knops
40 Singular systems of differential equations
S L Campbell
41 N-dimensional crystallography
R L E Schwarzenberger
42 Nonlinear partial differential equations in physical problems
D Graffi
43 Shifts and periodicity for right invertible operators
D Przeworska-Rolewicz
44 Rings with chain conditions
A W Chatters and C R Hajarnavis
45 Moduli, deformations and classifications of compact complex manifolds
D Sundararaman
46 Nonlinear problems of analysis in geometry and mechanics
M Atteia, D Bancel and I Gumowski
47 Algorithmic methods in optimal control
W A Gruver and E Sachs
48 Abstract Cauchy problems and functional differential equations
F Kappel and W Schappacher
49 Sequence spaces
W H Ruckle
50 Recent contributions to nonlinear partial differential equations
H Berestycki and H Brezis
51 Subnormal operators
J B Conway

52 Wave propagation in viscoelastic media
F Mainardi
53 Nonlinear partial differential equations and their applications: Collège de France Seminar. Volume I
H Brezis and J L Lions
54 Geometry of Coxeter groups
H Hiller
55 Cusps of Gauss mappings
T Banchoff, T Gaffney and C McCrory
56 An approach to algebraic K-theory
A J Berrick
57 Convex analysis and optimization
J-P Aubin and R B Vintner
58 Convex analysis with applications in the differentiation of convex functions
J R Giles
59 Weak and variational methods for moving boundary problems
C M Elliott and J R Ockendon
60 Nonlinear partial differential equations and their applications: Collège de France Seminar. Volume II
H Brezis and J L Lions
61 Singular systems of differential equations II
S L Campbell
62 Rates of convergence in the central limit theorem
Peter Hall
63 Solution of differential equations by means of one-parameter groups
J M Hill
64 Hankel operators on Hilbert space
S C Power
65 Schrödinger-type operators with continuous spectra
M S P Eastham and H Kalf
66 Recent applications of generalized inverses
S L Campbell
67 Riesz and Fredholm theory in Banach algebra
B A Barnes, G J Murphy, M R F Smyth and T T West
68 Evolution equations and their applications
F Kappel and W Schappacher
69 Generalized solutions of Hamilton-Jacobi equations
P L Lions
70 Nonlinear partial differential equations and their applications: Collège de France Seminar. Volume III
H Brezis and J L Lions
71 Spectral theory and wave operators for the Schrödinger equation
A M Berthier
72 Approximation of Hilbert space operators I
D A Herrero
73 Vector valued Nevanlinna Theory
H J W Ziegler
74 Instability, nonexistence and weighted energy methods in fluid dynamics and related theories
B Straughan
75 Local bifurcation and symmetry
A Vanderbauwhede
76 Clifford analysis
F Brackx, R Delanghe and F Sommen
77 Nonlinear equivalence, reduction of PDEs to ODEs and fast convergent numerical methods
E E Rosinger
78 Free boundary problems, theory and applications. Volume I
A Fasano and M Primicerio
79 Free boundary problems, theory and applications. Volume II
A Fasano and M Primicerio
80 Symplectic geometry
A Crumeyrolle and J Grifone
81 An algorithmic analysis of a communication model with retransmission of flawed messages
D M Lucantoni
82 Geometric games and their applications
W H Ruckle
83 Additive groups of rings
S Feigelstock
84 Nonlinear partial differential equations and their applications: Collège de France Seminar. Volume IV
H Brezis and J L Lions
85 Multiplicative functionals on topological algebras
T Husain
86 Hamilton-Jacobi equations in Hilbert spaces
V Barbu and G Da Prato
87 Harmonic maps with symmetry, harmonic morphisms and deformations of metrics
P Baird
88 Similarity solutions of nonlinear partial differential equations
L Dresner
89 Contributions to nonlinear partial differential equations
C Bardos, A Damlamian, J I Díaz and J Hernández
90 Banach and Hilbert spaces of vector-valued functions
J Burbea and P Masani
91 Control and observation of neutral systems
D Salamon
92 Banach bundles, Banach modules and automorphisms of C*-algebras
M J Dupré and R M Gillette
93 Nonlinear partial differential equations and their applications: Collège de France Seminar. Volume V
H Brezis and J L Lions
94 Computer algebra in applied mathematics: an introduction to MACSYMA
R H Rand
95 Advances in nonlinear waves. Volume I
L Debnath
96 FC-groups
M J Tomkinson
97 Topics in relaxation and ellipsoidal methods
M Akgül
98 Analogue of the group algebra for topological semigroups
H Dzinotyiweyi
99 Stochastic functional differential equations
S E A Mohammed

151 A stochastic maximum principle for optimal control of diffusions
U G Haussmann
152 Semigroups, theory and applications. Volume II
H Brezis, M G Crandall and F Kappel
153 A general theory of integration in function spaces
P Muldowney
154 Oakland Conference on partial differential equations and applied mathematics
L R Bragg and J W Dettman
155 Contributions to nonlinear partial differential equations. Volume II
J I Díaz and P L Lions
156 Semigroups of linear operators: an introduction
A C McBride
157 Ordinary and partial differential equations
B D Sleeman and R J Jarvis
158 Hyperbolic equations
F Colombini and M K V Murthy
159 Linear topologies on a ring: an overview
J S Golan
160 Dynamical systems and bifurcation theory
M I Camacho, M J Pacifico and F Takens
161 Branched coverings and algebraic functions
M Namba
162 Perturbation bounds for matrix eigenvalues
R Bhatia
163 Defect minimization in operator equations: theory and applications
R Reemtsen
164 Multidimensional Brownian excursions and potential theory
K Burdzy
165 Viscosity solutions and optimal control
R J Elliott
166 Nonlinear partial differential equations and their applications. Collège de France Seminar. Volume VIII
H Brezis and J L Lions
167 Theory and applications of inverse problems
H Haario
168 Energy stability and convection
G P Galdi and B Straughan
169 Additive groups of rings. Volume II
S Feigelstock
170 Numerical analysis 1987
D F Griffiths and G A Watson
171 Surveys of some recent results in operator theory. Volume I
J B Conway and B B Morrel
172 Amenable Banach algebras
J-P Pier
173 Pseudo-orbits of contact forms
A Bahri
174 Poisson algebras and Poisson manifolds
K H Bhaskara and K Viswanath
175 Maximum principles and eigenvalue problems in partial differential equations
P W Schaefer
176 Mathematical analysis of nonlinear, dynamic processes
K U Grusa
177 Cordes' two-parameter spectral representation theory
D F McGhee and R H Picard
178 Equivariant K-theory for proper actions
N C Phillips
179 Elliptic operators, topology and asymptotic methods
J Roe
180 Nonlinear evolution equations
J K Engelbrecht, V E Fridman and E N Pelinovski
181 Nonlinear partial differential equations and their applications. Collège de France Seminar. Volume IX
H Brezis and J L Lions
182 Critical points at infinity in some variational problems
A Bahri
183 Recent developments in hyperbolic equations
L Cattabriga, F Colombini, M K V Murthy and S Spagnolo
184 Optimization and identification of systems governed by evolution equations on Banach space
N U Ahmed
185 Free boundary problems: theory and applications. Volume I
K H Hoffmann and J Sprekels
186 Free boundary problems: theory and applications. Volume II
K H Hoffmann and J Sprekels
187 An introduction to intersection homology theory
F Kirwan
188 Derivatives, nuclei and dimensions on the frame of torsion theories
J S Golan and H Simmons
189 Theory of reproducing kernels and its applications
S Saitoh
190 Volterra integrodifferential equations in Banach spaces and applications
G Da Prato and M Iannelli
191 Nest algebras
K R Davidson
192 Surveys of some recent results in operator theory. Volume II
J B Conway and B B Morrel
193 Nonlinear variational problems. Volume II
A Marino and M K Murthy
194 Stochastic processes with multidimensional parameter
M E Dozzi
195 Prestressed bodies
D Iesan
196 Hilbert space approach to some classical transforms
R H Picard
197 Stochastic calculus in application
J R Norris
198 Radical theory
B J Gardner
199 The C* – algebras of a class of solvable Lie groups
X Wang

200 Stochastic analysis, path integration and dynamics
D Elworthy
201 Riemannian geometry and holonomy groups
S Salamon
202 Strong asymptotics for extremal errors and polynomials associated with Erdös type weights
D S Lubinsky
203 Optimal control of diffusion processes
V S Borkar
204 Rings, modules and radicals
B J Gardner
205 Numerical studies for nonlinear Schrödinger equations
B M Herbst and J A C Weideman
206 Distributions and analytic functions
R D Carmichael and D Mitrović
207 Semicontinuity, relaxation and integral representation in the calculus of variations
G Buttazzo
208 Recent advances in nonlinear elliptic and parabolic problems
P Bénilan, M Chipot, L Evans and M Pierre
209 Model completions, ring representations and the topology of the Pierce sheaf
A Carson
210 Retarded dynamical systems
G Stepan
211 Function spaces, differential operators and nonlinear analysis
L Paivarinta
212 Analytic function theory of one complex variable
C C Yang, Y Komatu and K Niino
213 Elements of stability of visco-elastic fluids
J Dunwoody
214 Jordan decompositions of generalised vector measures
K D Schmidt
215 A mathematical analysis of bending of plates with transverse shear deformation
C Constanda
216 Ordinary and partial differential equations Vol II
B D Sleeman and R J Jarvis
217 Hilbert modules over function algebras
R G Douglas and V I Paulsen
218 Graph colourings
R Wilson and R Nelson
219 Hardy-type inequalities
A Kufner and B Opic
220 Nonlinear partial differential equations and their applications. College de France Seminar Volume X
H Brezis and J L Lions

Recent advances in nonlinear elliptic and parabolic problems

P Bénilan, M Chipot, L C Evans, M Pierre (Editors)
Université de Franche-Comté/Université de Metz/University of Maryland/
Université de Nancy I

Recent advances in nonlinear elliptic and parabolic problems

Proceedings of an international conference, Nancy, France, March 1988

Longman Scientific & Technical
Copublished in the United States with
John Wiley & Sons, Inc., New York

Longman Scientific & Technical,
Longman Group UK Limited,
Longman House, Burnt Mill, Harlow
Essex CM20 2JE, England
and Associated Companies throughout the world.

Copublished in the United States with
John Wiley & Sons, Inc., 605 Third Avenue, New York, NY 10158

First published 1989

AMS Subject Classification: (main) 35KXX, 35JXX, 35BXX
(subsidiary) 49A29, 76S05, 92XX

ISSN 0269-3674

British Library Cataloguing in Publication Data

Recent advances in nonlinear elliptic and parabolic problems
1. Nonlinear partial differential equations
I. Bénilan, P.
515.3′53

ISBN 0-582-03538-4

Library of Congress Cataloging-in-Publication Data
(Pitman research notes in mathematics series, 208)
1. Differential equations, Elliptic – Congresses.
2. Differential equations, Parabolic – Congresses.
I. Bénilan, P. (Philippe), 1940– . II. Series.
QA372.R38 1989 515.3′53 88-27296
ISBN 0-470-21344-2

Printed and bound in Great Britain
by Biddles Ltd, Guildford and King's Lynn

C Stack
SCIMON

Contents

Preface

This volume collects most of the lectures and communications presented to the International Conference which took place in Nancy in March 1988. The main issues addressed were: nonlinear elliptic equations and systems, parabolic equations time-dependent systems, and the calculus of variations.

The meeting was part of the Special Year on Mathematical Study of Nonlinear Phenomena supported by the "Centre National de la Recherche Scientifique" and the "Ministère de la Recherche et des Enseigements Supérieurs". Additional support was also provided by the "Région Lorraine", the City of Nancy, the District of Nancy, the "Société Nancéienne Varin-Bernier" and the Universities of Besançon, Metz and Nancy. It is our pleasure to thank all these institutions for their help.

We would also like to express our special thanks to Longman for publishing these proceedings.

The Editors

List of conference participants

J. Alvarez-Contreras (Madrid)
A. Alvino (Naples)
H. Amann (Zurich)
S. Amraoui (Besançon)
D.G. Aronson (Minneapolis)
G. Aronsson (Linkoping)
M. Artola (Bordeaux)
Y. Atik (Alger)
M. Badii (Rome)
C. Bandle (Basel)
P. Baras (Grenoble)
M. Barcelo Conesa (Barcelona)
M. Bardi (Padova)
D. Barlet (Nancy)
L. Barthelemy (Besançon)
L. Beaud (Pont à Mousson)
S. Benachour (Alger)
P. Bénilan (Besançon)
N. Benkafadar (Constantine)
L. Berard Bergery (Nancy)
M. Bernardi (Brescia)
P. Biler (Orsay)
L. Boccardo (Rome)
G. Bohnke (Nancy)
M. Boudourides (Xanthi)
P. Bougerol (Nancy)
C.M. Brauner (Bordeaux)
A. Brillard (Mullhouse)
S. Campanato (Pisa)
J. Carrillo (Madrid)
V. Caselles (Besançon)
F. Catte (Paris)
M. Chipot (Metz)
P. Clément (Delft)
J.L. Clerc (Nancy)
F. Conrad (Nancy)
O. Coulaud (Nancy)
R. Dal Passo (Rome)
A. Damlamian (Palaiseau)
T. Del Vecchio (Naples)
P. Deuring (Bayreuth)
R. Deville (Besançon)
E. Di Benedetto (Evanston)
J.I. Diaz (Madrid)
G. Dubois (Lyon)
H. Ducauquis Lanchon (Nancy)
L.C. Evans (College Park)
A. Feggous (Metz)
I. Fonseca (Pittsburgh)
M. Freidlin (College Park)
A. Friedman (Minneapolis)
G. Gagneux (Pau)
C. Gardin (Nancy)
R. Gariepy (Lexington)
L. Ghannam (Toulouse)
M. Giaquinta (Florence)
G. Gilardi (Pavia)
J. Goncerzewicz (Wroclaw)
M. Grinfeld (Edinburgh)
L. Hedberg (Linkoping)
A. Henrot (Nancy)
P. Hess (Zurich)

D. Hilhorst (Orsay)
A. Huard (Besançon)
R. Jensen (Chicago)
S. Kamin (Tel-Aviv)
O. Kavian (Nancy)
N. Kenmochi (Chiba-Shi)
D. Kinderlehrer (Minneapolis)
M. Kirane (Rome)
E. Laamri (Nancy)
H. Labani (Besançon)
M.T. Lacroix (Besançon)
M. Langlais (Bordeaux)
P. Lesaint (Besançon)
P.L. Lions (Paris)
M. Madaune-Tort (Pau)
M. Marjani (Besançon)
R.H. Martin (Raleigh)
G. Michaille (Metz)
G. Modica (Florence)
A. Mokrane (Alger)
J.M. Morel (Paris)
M. Moulay (Nancy)
F. Murat (Paris)
S. Nicaise (Mons)
A. Nikoudes (Xanthi)
P. Noverraz (Nancy)
L. Persson (Umea)
C. Picard (Amiens)
M. Pierre (Nancy)
R. Pinsky (Haifa)
M. Porzio (Rome)
M.A. Pozio (Rome)
G. Pozzi (Pavia)
J.P. Puel (Orléans)
M. Rihani (Cassablanca)
J.R. Roche (Nancy)
J.F. Rodrigues (Lisbon)
P. Ryckelinck (Nancy)
B. Sadallah (Alger)
K. Schilling (Gottingen)
C. Schmidt-Lainé (Ecully)
A. Sili (Alger)
J. Spruck (Amherst)
B. Stoth (Bonn)
G. Sweers (Delft)
B. Terreni (Milan)
A. Tesei (Rome)
G. Tissier (Nancy)
G. Trombetti (Naples)
P. Trouve (Orsay)
R.E.L. Turner (Madison)
J.L. Vazquez (Madrid)
A. Visintin (Trento)
W. von Wahl (Bayreuth)
F. Weissler (Paris)
N. Yebari (Saint Etienne)

List of unpublished communications

J. Alvarez-Contreras
Comparison of the solutions of a nonlinear elliptic equation in hydrodynamic lubrication.

H. Amann
Feedback stabilization of semilinear parabolic systems.

D.G. Aronson
Regularity of flows in porous media.

G. Aronsson
Some aspect of p-harmonic functions in the plane.

M. Bardi
Singular perturbation problems of large deviations type for elliptic equations.

S. Benachour
Analyticity of the solutions of Vlassov-Poisson equations.

P. Bertrand, A. Ghizzo, E. Fijalkow, T.W. Johnston, M.M. Shoucri
Modelization and simulation of the acceleration of particules by laser-plasma interaction.

M.A. Boudourides
Local solvability of an initial free-boundary value problem for the Navier-Stokes equations.

C.M. Brauner, C. Schmidt-Lainé
A combustion problem with singular travelling waves.

J. Carrillo
Uniqueness and asymptotic behaviour of the solution of the problem of the flow through a porous medium of arbitrary shape.

E. Di Benedetto
Initial traces revisted.

M. Freidlin
Reaction-diffusion equations with small parameter.

A. Friedman
Nonlinear problems arising in industry.

R. Gariepy
Caccioppoli-type inequalities and the partial regularity of minimizers in the calculus of variations.

M. Giaquinta
Existence of minimizers in nonlinear elasticity.

P. Hess
On asymptotic periodic initial boundary value problem.

J. Jensen
Nonlinear partial differential equations and viscosity solutions.

S. Kamin
On the system of degenerate parabolic equations arising in plasma diffusion.

A. Kolpakov
Homogenization in a problem of thermochemistry.

D. Kinderlehrer
Variational principles associated to elastic crystals.

P.L. Lions
Viscosity solutions for second-order equations and boundary conditions.

R.H. Martin
Asymptotic behaviour of solutions to quasi-monotone parabolic systems.

A. Myslinski
Shape optimization problem for the von Kármán system.

A. Nikoudes
Existence of solutions of the Bénard convection equations with temperature-dependent viscosity.

J. Simon
On nonhomogeneous viscous incompressible fluids.

J. Spruck

Dirichlet problem for positive data and critical Sobolev growth.

J.L. Vazquez

Nonuniqueness of solutions of a singular diffusion equation.

W. von Wahl

Singular initial values and global solvability for semilinear parbolic equations.

F. Weissler

Blow up for a parabolic equation with a first-order damping term.

1. PARABOLIC EQUATIONS

S.N. ANTONTSEV AND J.I. DIAZ

On space or time localization of solutions of nonlinear elliptic or parabolic equations via energy methods

1. INTRODUCTION

In this article we present some new results on space or time localization of solutions of nonlinear elliptic or parabolic equations with "sources", i.e. with a prescribed right-hand part.

The results will be obtained by suitable energy methods (previously suggested and justified by the authors in [2], [3] and [13]), where homogeneous nonlinear equations were investigated. Here the keystone is the study of several nonhomogeneous nonlinear ordinary differential inequalities satisfied by the corresponding energy functions.

Some of the qualitative properties obtained here seem to be new in the literature. This is the case of the instantaneous extinction time or the nondiffusion of the support properties for nonhomogeneous parabolic equations. Other results of this work generalize to very general formulations. Some qualitative properties only well-known before for some special formulations. This is the case for the waiting time (or metastable localization) of parabolic equations and the nondiffusion of the support for elliptic equations. Both properties were earlier investigated by several authors by other methods (see the review expositions in [12] and [14]).

Energy methods are also applied to systems of combined-type equations. Applications to several nonlinear systems in continuum mechanics will be given in [6].

A first announcement of part of the result of this work was made in [5].

1. Parabolic equations

We consider a general class of nonlinear parabolic equations of the form

$$\frac{\partial\psi(u)}{\partial t} - \operatorname{div} \vec{A}(t,x,u,\nabla u) + B(t,x,u) = f(t,x), \qquad (1)$$

where ψ is a continuous nondecreasing real function such that

$$C_1|r|^{\beta-1}r \leq \psi(r) \leq C_2|r|^{\beta-1}r \tag{2}$$

for some constants $C_2 \geq C_1 > 0$ and $\beta > 0$ and for every $r \in \mathbb{R}$. We will also assume the following structural assumptions on $\vec{A}$ and B:

$$|\vec{A}(t,x,u,\xi)| \leq C_3|\xi|^{p-1}, \tag{3}$$

$$\vec{A}(t,x,u,\xi)\xi \geq C_4|\xi|^{p-2}, \tag{4}$$

$$B(t,x,u)u \geq C_5|u|^{q+1}, \tag{5}$$

for some positive constants C_3, C_4, $p > 1$ and $C_5 \geq 0$, $q \geq 0$. Here $u \in \mathbb{R}$ and $\xi \in \mathbb{R}^N$.

Notice that (1) contains, as a particular case, the following generalization of the porous media equation:

$$\frac{\partial u^{1/m}}{\partial t} - \Delta_p u + \lambda u^q = 0, \tag{6}$$

where $\Delta_p u = \text{div}(|\nabla u|^{p-2}\nabla u)$, $\lambda \geq 0$, $m > 0$ and the expressions $u^{1/m}$ and u^q must be substituted by $|u|^{1/m-1}u$ and $|u|^{q-1}u$ if u changes sign.

1.1 Localization in time: Instantaneous extinction time

Let us begin by studying the vanishing, in a finite time, of global solutions of (1) satisfying the following initial and boundary conditions:

$$u(0,x) = u_0(x) \text{ in } \Omega, \tag{7}$$

$$u(t,x) = 0 \quad \text{on } \Sigma = (0,T) \times \partial\Omega \quad , \tag{8}$$

where Ω is a bounded regular open set of $\mathbb{R}^N$ and $T > 0$. The existence and uniqueness of weak solutions of problems (1), (7), (8) have been considered by many authors (see, e.g. [1], [8] and [9]). In particular, it is well known that if $u_0 \in L^{\beta+k}(\Omega)$ and $f \in L^{p'}(0, T: L^{(\beta+k)/\beta}(\Omega))$ then there exists a unique weak solution $u \in L^p(0, T: W_0^{1,p}(\Omega)) \cap L^1(0, T: L^{\beta+k}(\Omega))$ for any $k \geq 0$.

The following result is peculiar to two alternative phenomena: a "fast diffusion", which corresponds to the range of parameters $\beta > (p-1)$ and $C_3 \geqq 0$; or a "strong absorption with respect to the accumulation term" which corresponds to the assumption $C_3 > 0$ and $q < \beta$.

THEOREM 1: Assume that one of the following conditions holds:

$$\beta > (p-1) \text{ and } C_3 \geqq 0, \tag{9}$$

or

$$C_3 > 0 \text{ and } \beta > q. \tag{10}$$

Let u_0 and f as mentioned with K large enough. We also assume that f vanishes after a finite time $T_f < T$ and that

$$\int_\Omega |f(t,x)|^{(\beta+k)/\beta} dx \leqq C_6 (T_f - t)_+^{\alpha/(1-\alpha)} \quad \text{for a.e. } t \in (t_1, T), \tag{11}$$

where $h_+ = \max(h,0)$, $0 \leqq t_1 < T_f$, α and C_6 are suitable positive constants $\alpha \in (0,1)$ and C_6 small. Then there exists a constant C_7 (depending on C_6, $\|u_0\|_{\beta+k}$, and α) such that if

$$\int_\Omega |u(t_1,x)|^{\beta+k} dx \leqq C_7 (T_f - t_1)^\alpha, \tag{12}$$

then $u(t,\cdot)$ vanishes on Ω for any $t \geqq T_f$. More precisely

$$\int_\Omega |u(t,x)|^{\beta+k} dx \leqq C_7 (T_f - t)_+^\alpha \tag{13}$$

for any $t \in (t_1, T)$.

PROOF: We first consider the case of assumption (9). Multiplying the equation by $|u|^{K-1}u$, integrating by parts, and using Sobolev-Poincaré, Hölder and Young inequalities it is not difficult to show that the function

$$y(t) = \int_\Omega |u(t,x)|^{\beta+k} dx \tag{14}$$

satisfies the ordinary differential inequality

$$\frac{dy}{dt}(t) + C_8\, y^{\alpha}(t) \leqq C_9(T_f-t)^{\alpha/(1-\alpha)} \text{ for } t \in (t_1,T), \tag{15}$$

where $\alpha = (p-1+k)/(\beta+k)$ and C_8 and C_9 are given in the following way: First, we fix $\varepsilon > 0$ and then

$$C_9 = \frac{\beta+k}{\beta p'} C_9 \varepsilon^{-p'}, \quad C_8 = \frac{\beta+k}{\beta} \left(C_4C_{10} - \frac{\varepsilon^p}{p} (1 + C_4^2 C_{10})\right)$$

and C_{10} is the constant of the Sobolev-Poincaré inequality in $W_0^{1,p}(\Omega)$ (when assuming (9), K must be any positive real number such that $K \geqq (N(\beta-p+1)/p)-\beta$ if $p < N$ and $K \geqq 1$ arbitrary if $p \geqq N$).

In the case of assumption (10) the inequality (15) is also obtained for another exponent $\alpha \in (0,1)$ and positive constants C_8 and C_9. To do that an interpolation inequality of the form

$$\|v\|^{a}_{\beta+k(\Omega)} \leqq \varepsilon^{-\mu} \|v\|^{q+k}_{q+k(\Omega)} + C\varepsilon^{\mu'} \int_\Omega |v|^{k-1} |\nabla v|^p dx$$

must be used (see [14] or [15] for the case of homogeneous equations), where now $k \geqq 1$, $q > 0$, $p > 1$ and $\varepsilon > 0$ are arbitrary and the constants a, μ and C are suitably chosen.

The conclusion of Theorem 1 comes from the investigation of inequality (15) which is made in the following lemma.

<u>LEMMA 1</u>: Let $y(t) \geqq 0$ be such that $y'(t) + \phi(y(t)) \leqq F((T_f-t)_+)$ a.e. on (t_1,T), where ϕ is a nondecreasing function such that $\phi(0) = 0$ and $(1/\phi(\cdot)) \in L^1(0,1)$. For any $\mu > 0$ and $\tau > 0$ we define

$$\theta_\mu(\tau) = \int_0^\tau \frac{ds}{\mu\phi(s)} \tag{16}$$

and $\eta_\mu(s) = \theta_\mu^{-1}(s)$. Assume

$$\exists \bar{\mu} < 1 \text{ such that } F(s) \leqq (1-\bar{\mu})\phi(\eta_{\bar{\mu}}(s))$$

and

$$(T_f-t_1) \geqq \theta_{\bar{\mu}}(y(t_1)).$$

Then $y(t) \equiv 0$ for any $t \in [T_f,T]$.

PROOF OF LEMMA 1: The function $\bar{y}(t) = \eta_{\bar{\mu}}((T_f-t)_+)$ satisfies $\bar{y}' + \phi(\bar{y}) = (1-\bar{\mu})\phi(\eta_{\bar{\mu}}((T_f-t)_+)$, and so it is a supersolution for the ordinary differential inequality. □

REMARK 1: The conclusion of Theorem 1 can be interpreted as an instantaneous extinction time (the solution vanishes from the time in which f vanishes). When $t_1 = 0$, condition (12) only affects the initial datum u_0. If $t_1 > 0$ (12) can be obtained throughout u_0 and f by using the well-known a priori estimate

$$\|u(t_1)\|_{\beta+k(\Omega)} \leqq \|u_0\|_{\beta+k(\Omega)} + \int_0^{t_1} \|f(s,\cdot)\|_{L^{(\beta+k)/\beta}(\Omega)}.$$

REMARK 2: The instantaneous extinction time also holds for other boundary conditions. The case of the Cauchy problem, $\Omega = R^N$ can also be considered under assumptions stronger than (9) or (10). So, if, for instance, we consider the case of "fast diffusion", (9) must be replaced by

$$\beta > \max p-1 < \frac{N(p-1)}{N-p} \quad . \tag{17}$$

REMARK 3: Inequality (15) is also useful to show the "nondegeneracy" of the energy y(t) near its first zero T_0. Applications to the continuous dependence of T_0 with respect to u_0 and f can be obtained from this property. Those results will be published elsewhere.

REMARK 4: Lemma 1 is inspired in Theorem 1.15 of [12]. The particular case of $\phi(s) = CS^{\alpha}$ with $\alpha \in (0,1)$ can be investigated by other methods.

1.2. Space localization: Waiting time and nondiffusion of the support

In this section we will study the local vanishing of solutions of (1). The local nature of this property will allow us to work merely with local solutions of (1), i.e. functions satisfying (1) on sets of the form $6 \times (0,T)$ with $\sigma \subset \Omega$ but without any information on the values of u or of $\nabla u\cdot\vec{n}$ on $\partial\Omega$.

The main conclusion of this section is, again, peculiar to two alternative phenomena: a "slow diffusion", which corresponds to the assumption $\beta < (p-1)$ and $C_3 \geqq 0$; or a "strong absorption with respect to the diffusion", which corresponds to conditions $C_3 > 0$ and $q < (p-1)$. Concerning the first case, we have:

THEOREM 2: Assume $\beta < (p-1)$ and $C_3 \geqq 0$. Let $B_\rho(x_0) = \{x \in \Omega : |x-x_0| < \rho\}$ and assume that there exists $\rho_0 > 0$, $t_f > 0$, $\delta > 0$ and $\varepsilon > 0$ (ε small enough) such that

$$\int_{B_\rho(x_0)} |u(0,x)|^{\beta+1} dx + \left(\int_0^{t_f} \|f(s,\cdot)\|_{L^{r(r-1)}(B_\rho(x_0))}^{p/(p-\theta)} ds\right)^{(p-\theta)/p(1-\lambda)}$$

$$\leqq \varepsilon(\rho-\rho_0)_+^{1/(1-\sigma)}$$

for a.e. $\rho \in (0,\rho_0+\delta)$, where

$$\sigma = \frac{p}{p-1}\left(1 - \left(\frac{\theta}{p} + \frac{1-\theta}{\beta+1}\right)\right), \quad \theta = \left(\frac{1}{\beta+1} - \frac{1}{r}\right)\Big/\left(\frac{1}{\beta+1} - \frac{N-p}{Np}\right),$$

$$\lambda = \left(\frac{1-\theta}{\beta+1} + \frac{\theta}{p}\right), \qquad \beta + 1 \leqq r \leqq \frac{Np}{N-p}$$

Then there exists $t^* > 0$, $t^* \leqq t_f$, such that $u(t,x) = 0$ a.e. $x \in B_{\rho_0}(x_0)$ and $t \in [0,t^*]$.

PROOF: We introduce the energy functions ([13])

$$E(t,\rho) \equiv \int_0^t \int_{B_\rho} \vec{A}(s,x,u,\nabla u)\cdot\nabla u \, dx \, dt, \; b(t,\rho) = \operatorname*{ess\,sup}_{\tau\in[0,t]} \int_{B_\rho} |u(\tau,x)|^{\beta+1} dx.$$

Integrating by parts, using the interpolation inequality

$$\|u(t,\cdot)\|_{L^r} \leqq C \, \|u(t,\cdot)\|_{L^{\beta+1}}^{1-\theta} \left(\|\nabla u(t,\cdot)\|_{L^p} + \|u(t,\cdot)\|_{L^{\beta+1}}\right)^{\theta}$$

with $r \in [\beta+1, N_p/(N-p)]$, $\theta = (1/(\beta+1) - 1/r)/(1/(\beta+1) - (N-p)/N_p)$, and using the interpolation-trace lemma (see [13]), we conclude that

$$(E + b) \leqq Ct^{\frac{(1-\theta)H}{p}} \left(\frac{\partial E}{\partial \rho}\right)^{\frac{(p-1)H}{p}} + \varepsilon(\rho-\rho_0)_+^{\frac{1}{1-\sigma}} \tag{19}$$

for any $\rho > 0$ and $t \leq t_f$, where

$$H = \frac{1}{1 - (\frac{\theta}{p} + \frac{(1-\theta)}{\beta+1})} \,. \tag{20}$$

The proof of Theorem 2 ends with the following lemma:

<u>LEMMA 2</u>: Let $y \in C^0[0,t_1] \times [0,\rho_0+\rho])$, $y \geq 0$, be such that for any $t \leq t_1$ and for some $\omega > 0$ and $\delta > 0$

$$\phi(y(t,\rho)) \leq Ct^{\omega} \frac{\partial y}{\partial \rho}(t,\rho) + G((\rho-\rho_0)_+), \quad \text{a.e. } \rho \in (0,\rho_0+\delta), \tag{21}$$

where ϕ is a nondecreasing continuous function such that $\phi(0) = 0$ and $1/\phi(\cdot) \in L^1(0,1)$. As in Lemma 1, given $\mu > 0$, we define $\theta_\mu(\tau)$ and $\eta_\mu(s)$. We suppose that

$$\exists \bar{\mu} > 0 \text{ and } \varepsilon < 1 \text{ such that } G(s) < \varepsilon\phi(\eta_{\bar{\mu}}(s)) \text{ a.e. } s \in (0,\delta). \tag{22}$$

Then there exists $t^* \leq t_1$ such that $y(t,\rho) = 0$ for any $0 \leq \rho \leq \rho_0$ and $t \in [0,t^*]$.

<u>PROOF OF LEMMA 2</u>: It is easy to see that function $\bar{y}(\rho) = \eta_\mu((\rho-\rho_0)_+)$ satisfies

$$-Ct^{\omega} \frac{\partial \bar{y}}{\partial \rho} + \phi(\bar{y}) = (-Ct^{\omega}\mu + 1)\phi(\eta_\mu((\rho-\rho_0)_+)).$$

Then, taking $\mu \geq \bar{\mu}$ and $t \leq t^*$ with t^* such that

$$t^* \leq (\frac{1-\varepsilon}{C\mu})^{1/\omega}$$

we have that $\bar{y}$ is a supersolution of the differential inequality. In order to conclude that $y(t,\rho) \leq \bar{y}(\rho)$ for $\rho \in (0,\rho_0+\delta)$ we only need to have $y(t,\rho_1) \leq \bar{y}(\rho_1)$. This last condition holds if we take μ large enough such that

$$\frac{1}{\rho_0+\delta} \int_0^M \frac{ds}{\phi(s)} \leq \mu$$

where $M = \sup \{y(t,\rho_0 +\delta): t \in [0,t_1]\}$. □

REMARK 5: In the special case $f \equiv 0$ the time t^* is called the waiting time. The existence of t^* was previously shown, for particular formulations of (1), by different authors (see, e.g. [14]). It is not difficult to check that our condition (18) coincides with the one in the literature for the one-dimensional porous media equation. The case of f 0 was previously treated by an energy method in [4], where a different proof of Lemma 2 (for $\phi(s) = s^\alpha$, $\alpha \in (0,1)$) was given.

REMARK 6: The conclusion of Theorem 2 also holds when $C_3 > 0$ and we replace the condition $\beta < (p-1)$ by the assumption $q < (p-1)$. In that case the exponents in (18) are different and in fact $t^* = t_f$ (*nondiffusion of the support of the solution*). These results, applications to suitable systems, the treatment of the case in which B also depends on ∇u and $f(t,\cdot) \in W^{-1,p'}(\Omega)$, etc. will be given in [7].

REMARK 7: The existence of a waiting time for higher order nonlinear parabolic equations is the object of [11] (see [10] for the proof of the finite speed of propagation for higher order equations by an energy method).

2. ELLIPTIC EQUATIONS

The local energy method used in Section 1.2 can also be applied to the study of nonlinear elliptic equations of the form

$$- \operatorname{div} \vec{A}(x,u,\nabla u) + B(x,u) = f(x), \tag{23}$$

where $\vec{A}$ and B satisfy conditions (3), (4) and (5).

THEOREM 3: Assume $C_3 > 0$ and $q < p - 1$. Assume that there exists $\rho_0 > 0$, $\delta > 0$ and $\varepsilon > 0$ (ε small enough), such that

$$\|f(x)\|^{1/(1-\lambda)}_{L^{r/(r-1)}(B_\rho(x_0))} \leq \varepsilon(\rho-\rho_0)_+^{1/(1-\sigma)} \quad \text{a.e. } \rho \in (0,\rho_0 + \delta), \tag{24}$$

where

$$q+1 \leq r \leq \frac{Np}{N-p}, \quad \theta = \left(\frac{1}{q+1} - \frac{1}{r}\right) / \left(\frac{1}{q+1} - \frac{N-p}{Np}\right),$$

$$\lambda = \left(\frac{1-\theta}{q+1} + \frac{\theta}{p}\right), \quad \sigma = \frac{p}{(p-1)}\left(\frac{\theta}{p} + \frac{(1-\theta)}{q+1}\right)).$$

Let u be any local weak solution of (23) and assume that the following a priori estimate holds

$$\|\nabla u\|_{L^p(B_{\rho+\delta}(x_0))} \leq \varepsilon^*$$

for some $\varepsilon^* > 0$ small enough. Then $u(x) = 0$ a.e. $x \in B_{\rho_0}(x_0)$.

<u>PROOF</u>: We introduce the energy functions

$$E(\rho) = \int_{B_\rho} \vec{A}(x,u,\nabla u)\cdot\nabla u \, dx, \quad b(\rho) = \int_{B_\rho} |u|^{q+1} \, dx.$$

Using the interpolation inequality

$$\|u\|_{L^r(B_\rho)} \leq C \, \|u\|^{1-\theta}_{L^{q+1}} \left(\|\nabla u\|_{L^p} + \|u\|_{L^{q+1}}\right)^{\theta}$$

with θ and r as in the statement of the theorem, and applying the interpolation-trace lemma of [13], we conclude that

$$E + b \leq C \left(\frac{dE}{\partial\rho}\right)^{(p-1)H/p} + \varepsilon(\rho-\rho_0)_+^{(1/(1-\sigma)}$$

for

$$H = \frac{1}{1 - \left(\frac{\theta}{p} + \frac{1-\theta}{q+1}\right)} \, .$$

The proof of Theorem 3 ends with the following lemma:

<u>LEMMA 3</u>: Let $y(\rho) \geq 0$ be such that

$$\phi(y(\rho)) \leq C_0 \frac{dy}{d\rho}(\rho) + G((\rho-\rho_0)_+), \text{ a.e. } \rho \in (0,\rho_0+\delta), \tag{25}$$

where ϕ is a nondecreasing continuous function such that $\phi(0) = 0$ and $1/\phi(\cdot) \in L^1(0,1)$. Given $\mu > 0$ we define θ_μ and η_μ as in Lemma 1. We also assume that

$$\exists\bar{\mu} > 0 \text{ such that } G(s) \leqq (1-C_0\bar{\mu})\phi(\eta_{\bar{\mu}}(s)) \text{ a.e. } s \in (0,\delta) \tag{26}$$

and

$$\delta \geqq \theta_{\bar{\mu}}(M) \quad \text{with} \quad M \geqq y(\rho_0 + \rho). \tag{27}$$

Then $y(\rho) = 0$ a.e. $\rho \in [0,\rho_0]$.

PROOF OF LEMMA 3: Function $\bar{y}(\rho) = \eta_{\bar{\mu}}((\rho-\rho_0)_+)$ satisfies

$$- C_0 \frac{d\bar{y}}{d\rho} + \phi(\bar{y}) = (-C_0\,\bar{\mu} + 1)\,\phi(\eta_{\bar{\mu}}((\rho-\rho_0)_+)),$$

and so it is a supersolution of the equation. Finally, by (26) and (27), we have that $y(\rho_0+\rho) \leqq \bar{y}(\rho_0+\delta)$ and by comparison on $[0,\rho_0+\delta]$ we conclude the result. □

REMARK 8: The conclusion of Theorem 3 can be understood as a *nondiffusion of the support of u* with respect to the support of f. This property was first obtained in [12] for a special formulation of (23) and by means of a comparison argument. The case of B depending on ∇u, extensions to nonlinear systems, etc, will be given in [7].

ACKNOWLEDGEMENTS

This paper was partially written while the first author was visiting the Universidad Complutense of Madrid. His stay was partially sponsored by Project 3308/83 of the CAICYT (Spain).

References

[1] H.W. Alt and S. Luckhaus, Quasilinear elliptic parabolic differential equations, Math. Z. 183 (1983), 311-341.

[2] S.N. Antontsev, On localization of solutions of nonlinear degenerate elliptic and parabolic equations, Soviet Math. Dokl. 24 (1981), 420-424.

[3] S.N. Antontsev, Localization of Solutions for Degenerate Equations in Continuum Mechanics. Institute of Hydrodynamic Siberian Division of the USSR Academy of Sciences, Novosibirsk, 1986 (108 pp.).

[4] S.N. Antontsev, Metastability localization of the solution of degenerate parabolic equations in general form. Dynamics of Continuum Medium, Vol. 83, Novosibirsk, 1987, pp. 138-144.

[5] S.N. Antontsev and J.I. Diaz. New results of the localization of solutions of nonlinear elliptic and parabolic equations, obtained via energy methods, Soviet Math. Dokl., Vol. 303 (1988), No.3, pp.524-528.

[6] S.N. Antontsev and J.I. Diaz. Applications of the energy method for localization of the solutions of equations of continuum mechanics, Soviet Math. Dokl., Vol. 303 (1988), No. 2, pp. 320-325.

[7] S.N. Antontsev and J.I. Diaz. Article in preparation.

[8] A. Bamberger, Etude d'une equation doublement non-lineaire, J. Funct. Analysis, 24 (1977), 148-155.

[9] Ph. Benilan, Sur un problème d'evolution non monotone dans $L^2(\Omega)$. Publ. Math. Fac. Sc. Besançon, No. 2 (1976).

[10] F. Bernis, Qualitative properties for some nonlinear higher order degenerate parabolic equations. IMA Preprint 184, University of Minnesota 1985 (to appear in Houston J. Math).

[11] F. Bernis and J.I. Diaz, Article in preparation.

[12] J.I. Diaz, Nonlinear Partial Differential Equations and Free Boundaries, Vol. 1. Elliptic Equations. Pitman Research Notes in Math. Vol. 106, Pitman, 1985.

[13] J.I. Diaz and L. Veron, Local vanishing properties of solutions of elliptic and parabolic equations, Trans. Amer. Math. Soc. 290 (1985), 787-814.

[14] A.S. Kalashnikov, Some problems of the qualitative theory of nonlinear degenerate second-order parabolic equations, Uspekhi Mat. Nauk, 42 (1987), 135-176.

[15] I.B. Polymsky, Some qualitative properties of the solutions of nonlinear heat conductivity with absorption In Chislennye metody mekhaniki sploshnoi sredy, Vol. 16, No. 1, Novosibisrsk, Institute of Theoretical and Applied Mechanics, Siberian Division of the USSR Academy of Sciences, 1985, pp. 136-145.

[16] M. Tsutsumi, On solutions of some doubly nonlinear degenerate parabolic equations with absorption, Journal of Mathematical Analysis and Applications, 132 (1988), 187-212.

S.N. Antontsev
Lavrentyev Institute of Hydrodynamics
Academy of Sciences
Novosibirsk 630090
USSR

J.I. Diaz
Departamento de Matemàtica Aplicada
Universidad Complutense de Madrid
28040 Madrid
Spain

C. BANDLE

Blow up in exterior domains

1.

Let D_i be a finite collection of bounded domains and denote by $D := \mathbb{R}^N - \bar{D}_i$ its complement. Consider the problem

$$\begin{cases} u_t - \Delta u = t^q |x|^\sigma u^p & \text{in } D \times \mathbb{R}^+, \\ u = 0 & \text{on } \partial D \times \mathbb{R}^+, \\ u(x,0) = u_0(x) \geqq 0, & \\ u(x,t) \geqq 0, & \end{cases} \tag{1}$$

where $q > 0$, $p > 1$ and are arbitrary real numbers. Since we shall be interested in the case where local classical solutions exist, we shall assume that $0 \notin D$ when $\sigma < 0$.

A solution is called "*global*" if it is defined for all $t > 0$. By the maximum principle we have either $u(x,t) > 0$ or $u(x,t) \equiv 0$ for all $t > 0$. The nonlinearity is defined for all $u > 0$, a solution thus ceases to exist if it blows up, that is, if $\lim_{t\uparrow T} \sup_{x\in D} u(x,t) = \infty$ for some $T < \infty$. If $D = \mathbb{R}^N$ and $q = \sigma = 0$ we have the well-known result [4], [5], [6]:

> If $1 < p \leqq 1 + 2/N$ every nontrivial solution blows up in finite time, whereas if $p > 1 + 2/N$ there exist global solutions for suitable initial conditions.

This result has been extended in various way [5], [11], [8], [15], [16], [19].

The blow up effect persists when we replace the nonlinearity u^p by either $t^q u^p$ or $|x|^\sigma u^p$, $\sigma > 0$ [2], [16]. In both cases there exists a critical number p^* such that problem (1) with $D = \mathbb{R}^N$ has no global solution if $p \in (1,p^*)$, whereas for $p > p^*$ there are always global solutions for sufficiently small initial data. The relation between p^*, N and q or σ is surprisingly simple, namely [2], [14]

$$p^* = \begin{cases} 1 + (2+2q)/N & \text{if } \sigma = 0 \\ 1 + (2+\sigma)/N & \text{if } q = 0. \end{cases} \tag{2}$$

It is not yet known what happens if $p = p^*$.

In bounded domains D the situation is different. In order to avoid a technical discussion concerning the existence of local solutions for singular nonlinearities we assume that when σ is negative, D does not contain the origin. Depending upon the size of the initial conditions a solution may exist for all t or may blow up at finite time. A number of papers has been devoted to this question [10], [11], [7], [15], [16]. The fact that for every $p > 1$ global solutions exist, is easily seen by means of upper solutions. Indeed, for appropriate μ and $c > 0$ the function $\bar{u}(x,t) = ce^{-\mu t}\phi(x)$, $\phi > 0$, being the first eigenfunction of $\Delta\phi + \lambda\phi = 0$ in D, $\phi = 0$ on ∂D, is an upper solution, and hence for every solution $u(x,t)$ with $u_0 \leqq c\phi(x)$ we have $u(x,t) \leqq \bar{u}(x,t)$. The question arises what happens if D is an unbounded subset of R^N.

If D is a cylindrical domain of the form $D^k \times R^{N-k}$ where $D^k \subset R^k$ is a bounded domain and if we just look at solutions in $D \times \mathbb{R}^+$ without imposing any decay at infinity, the situation is the same as for bounded domains. In fact solutions of (1) in $D^k \times \mathbb{R}^+$ are also solutions in $D \times \mathbb{R}^+$.

Several papers have been devoted to cone-like domains [2], [8], [14], exterior domains however seem to have been considered only in [2].

An interesting connexion between the critical exponent p^* and the asymptotic behaviour of the solutions of the linear heat equation

$$\begin{cases} w_t - \Delta w = 0 & \text{in } D \times \mathbb{R}^+, \\ w = 0 & \text{on } \partial D \times \mathbb{R}^+, \\ w(x,0) = w_0(x), & \end{cases} \tag{3}$$

was found by Meier [15] who showed that for $\sigma = 0$

$$p^* = 1 + \frac{q+1}{s^*},$$

$$s^* := \sup \left\{ s : \begin{array}{l} \text{there exist solutions } w \not\equiv 0 \text{ of (3) such that} \\ \limsup_{t\to\infty} t^s |w(.,t)|_\infty < \infty. \end{array} \right\}$$

Let us mention that a quite complete survey of the present situation concerning blow up is given in [13].

2.

In this section we shall prove a nonexistence result concerning global solutions of (1).

DEFINITION: A solution is called "*regular*", if for all $k > 0$, $u(x,t)e^{-k|x|} \to 0$ and $u_r(x,t)e^{-k|x|} \to 0$ as $|x| \to \infty$.

THEOREM 2.1: In the following cases every regular solution of (1) blows up in finite time:

(a) $1 < p < 1 + (2+2q+\sigma)/N$ if $< \sigma < N(p-1)$;

(b) $1 < p$ if $\sigma \geq N(p-1)$.

Before we proceed to the proof we shall collect some technical preliminaries. Let us use the notation

$$\delta := \sigma/(p-1),$$

$$\omega := \delta(N-2-\delta),$$

$$L := r^{-(N-1-2\delta)} \frac{\partial}{\partial r} \left(r^{N-1-2\delta} \frac{\partial}{\partial r}\right) - \omega r^{-2}.$$

The next result was derived in [2].

LEMMA 2.1: Let $k > 0$, $\lambda > 0$ and m be arbitrary numbers. If

$$(k^2 + \lambda)(m^2 + (N-2\delta-2)m - \omega) \geq (m + (N-2\delta-1)/2)^2 k^2, \qquad (4)$$

then the function $\phi(r) = r^m e^{-kr}$ satisfies $L\phi \geqq -\lambda\phi$.

Let $\psi > 0$ be the first eigenfunction of $L\psi + \Lambda\psi = 0$ in (R,ρ), $\psi(R) = \psi(\rho) = 0$. It is well known that the eigenvalues depend monotonically on ρ, in addition, $\Lambda(\rho) \to \infty$ as $\rho \to R$ and either $\Lambda(\rho) \to 0$ as $\rho \to \infty$ or $\Lambda(\rho)$ becomes negative for large ρ. In any case, for any $\lambda > 0$ there exists a value $\rho = \rho_\lambda$ such that, $L\psi + \lambda\psi \geqq 0$ in (R,ρ_λ).

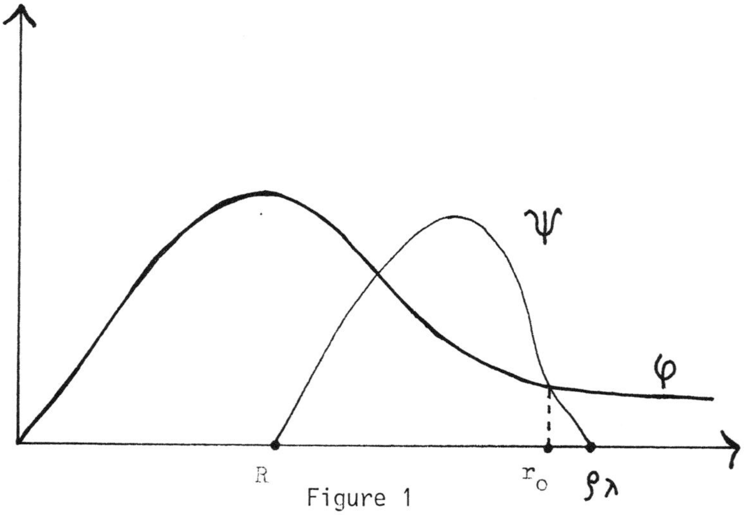

Figure 1

Suppose that $\psi > \phi$ in some subset of (R,ρ_λ) and let $r_0 = \inf\{\xi < \rho_\lambda : \psi(r) < \phi(r)$ in $(\xi,\rho_\lambda)\}$. Clearly $r_0 < \rho_\lambda$. The normalization of ψ is chosen such that

$$\int_R^{r_0} r^{N-1-2\delta}\psi dr = \int_R^{r_0} r^{N-1-2\delta}\phi(r)dr.$$

Define

$$\Phi_0(r) := \begin{cases} \psi(r) & \text{in } [R,r_0], \\ \phi(r) & \text{if } r \geqq r_0, \end{cases}$$

and

$$c := \int_R^\infty r^{N-1-2\delta}\Phi_0(r)dr.$$

Observe that $c = c(k)$ is a decreasing function of k with the following asymptotic behaviour at zero

$$c(k) \sim \begin{cases} c_1 k^{-(N-2\delta+m)} & \text{if } N - 2\delta + m > 0, \\ c_1 & \text{if } N - 2\delta + m \leqq 0, \end{cases} \qquad (5)$$

as $k \to 0$.

In summary, we have shown

LEMMA 2.2: For any $k > 0$, $\lambda > 0$ and m, the function $\Phi := \Phi_0/c$ satisfies $L\Phi + \lambda\Phi \geqq 0$ in (R,∞), $\Phi(R) = 0$ in a weak sense.

Next we consider the differential inequality

$$\frac{dG}{dt} + \lambda G \geqq h(t)G^p, \quad t > 0, \qquad (6)$$

where h is subject to the following conditions

$$h(t) \geqq 0 \quad \text{for } t \geqq 0, \qquad (7)$$

$$h(t) \geqq t^q \text{ for } t \geqq t_0,\ q \geqq 0. \qquad (8)$$

LEMMA 2.3: If

$$G(0) \geqq \lambda^{1/(p-1)} t_0^{-q/(p-1)} e^{\lambda t_0}, \qquad (9)$$

then $G(t)$ blows up in finite time T. If $t^* := q/(p-1)\lambda \geqq t_0$ the condition (9) can be improved as follows

$$G(0) \geqq \lambda^{(1+q)/(p-1)} [(p - 1)e/q]^{q/(p-1)} =: \tilde{w}_0. \qquad (10)$$

PROOF: From (7) it follows that $(d/dt)(e^{\lambda t}G(t)) \geqq 0$ and hence $G(t) \geqq G(0)e^{-\lambda t} =: w(t)$. Two possibilities may occur, namely either $G(t)$ blows up in finite time T or it is global.

Assume that G exists in $[0,t_0]$. In view of (8) G satisfies

$$\dot{G} + \lambda G \geqq t_0^q G^p \quad \text{for} \quad t \geqq t_0. \qquad (11)$$

If

$$t_0^q G^{p-1}(t_0) > \lambda, \tag{12}$$

then G blows up in finite time

$$t_0 < T < t_0 + \int_{G(t_0)} (t_0^q G^p - \lambda G)^{-1}\, dG.$$

Since $t^q G^{p-1}(t) \geq t^q w^{p-1}(t)$, (12) obviously holds when (9) is satisfied. If $t^* > t_0$ we can increase the range of initial values for which blow up occurs by observing that $t^q w^{p-1}(t) \leq (t^*)$ for all $t \geq 0$. The same arguments with t_0 replaced by t^* establish the second assertion. □

Since the regular solutions of (1) do not grow to fast at infinity the parabolic maximum principle applies and yields the following useful comparison result.

LEMMA 2.4 (Comparison Lemma): Let u(x,t) and U(x,t) be regular solutions of (1) in $D \times \mathbb{R}^+$ or $\tilde{D} \times \mathbb{R}^+$, respectively. Assume further that $\tilde{D} \supseteq D$ and $U(x,0) \geq u(x,0)$ in D. Then $U(x,t) \geq u(x,t)$.

PROOF OF THEOREM 2.1: The arguments parallel to those developed in [2]. First we note that we can restrict ourselves to domains exterior of a sphere. Indeed if $C_R := \{x: |x| > R\}$ is contained in D and if blow up takes place in C_R, then by comparison it occurs so much more in D.

Jensen's inequality yields for the mean value

$$\tilde{u}(r,t) = \frac{1}{s_{N-1} r^{N-1}} \oint_{|x|=r} u(x,t)\,ds, \quad s_N = \oint_{|x|=1} ds,$$

the differential inequality

$$\begin{cases} \tilde{u}_t - \Delta\tilde{u} \geq t^q r^\sigma \tilde{u}^p & \text{in} \quad C_R \times \mathbb{R}^+, \\ \tilde{u}(R,t) = 0, \\ \tilde{u}(r,0) = \tilde{u}_0(r). \end{cases}$$

Put $v = r^{\sigma/(p-1)}\tilde{u}$. Then a straightforward calculation yields

$$v_t - Lv \geqq t^q v^p \quad \text{in} \quad C_R \times \mathbf{R}^+. \tag{13}$$

Multiplying this inequality by $r^{N-1-2\delta}\Phi$(see Lemma 2.2) we obtain after integration by parts

$$\frac{d}{dt}\int_R^M v\Phi r^{N-1-2\delta}\,dr - \int_R^M vL\Phi r^{N-1-2\delta}\,dr +$$

$$+ v(r_0)r_0^{N-1-2\delta}[\Phi_r(r_0-0) - \Phi_r(r_0+0)] + v(M)M^{N-1-2\delta}\Phi_r(M) -$$

$$- \Phi(M)M^{N-1-2\delta}v_r(M) \geqq t^q \int_R^M v^p\Phi r^{N-1-2\delta}\,dr.$$

Observe that by construction $\Phi_r(r_0 - 0) - \Phi_r(r_0 + 0) < 0$. Moreover the normalization of Φ has been chosen such that Jensen's inequality applies. Hence if we let M tend to infinity we get for the function

$$G(t) := \int_R^\infty v\Phi r^{N-1-2\delta}\,dr$$

the inequality

$$\frac{dG}{dt} + \lambda G(t) \geqq t^q G^p(t).$$

According to Lemma 2.3, $G(t)$ blows up in finite time if $G(0)$ satisfies (10). Next we show that under the present conditions it is always possible to determine k, m and λ such that (4) and (10) hold. There is no loss of generality in assuming that $v(r,0) = v_0(r) \not\equiv 0$ is of compact support. For fixed λ and m

$$0 < \gamma_0 \leqq \int_R^\infty v_0\Phi_0 r^{N-1-2\delta}dr \text{ for all } k \leqq k_0$$

and thus the inequality

$$\gamma_0/c(k) \geqq \tilde{w}_0 \tag{14}$$

implies (10).

We now proceed as follows:

(a) Choose m such that

$$\delta < m < \begin{cases} 2(\frac{1+q}{p-1}) + 2\delta - N & \text{if } \delta < N, \\ 2\delta - N & \text{if } \delta \geqq N. \end{cases} \quad (15)$$

This is possible in view of our assumptions. The lower bound guarantees that

$$m^2 + (N - 2\delta - 2)M - \omega > 0. \quad (16)$$

(b) Fix $\beta := \lambda/k^2 \geqq \beta_0 > 0$ such that (4) holds. This is possible in view of (16).

(c) Here we distinguish between two cases.

(i) $N - 2\delta + m > 0$: Since $m > \delta$, this is always achieved for $\delta < N$. Then by (5) $c(k) < dk^{-(N-2\delta+m)}$ for $k \leqq k_0$ and some positive constant d.
By (15) and after suitable rescaling the inequality

$$k^{(N-2\delta+m)} \frac{\gamma_0}{d} \geqq \tilde{w}_0 = (k^2\beta)^{(1+q)/(p-1)} w_0 \quad (17)$$

is satisfied for all $k \leqq k'$. Determine $k \leqq \min(k_0,k')$
Thus with the choice of k and λ, (14) and consequently (10) are satisfied. This proves assertion (a) of Theorem 2.1.

(ii) $N - 2\delta + m \leqq 0$: Since $m > \delta$ this inequality holds only for $\delta > N$. In this case $c(k) \leqq d$ for $k \leqq k_0$ and therefore the inequaltiy $\gamma_0/d \geqq (k^2\beta)^{(1+q)/(p-1)} w_0$ implies (10).

The assertion now follows from Lemma 2.3. □

<u>REMARK</u>: If Φ satisfies the assumptions of Lemma 2.2 for some m, k_0 and λ_0 and if (10) holds, then the same is true for m, $k < k_0$ and $\lambda = \lambda_0 k^2/k_0^2 < \lambda_0$.

This observation leads to the following extension.

COROLLARY 2.5: Consider the problem (1) with the nonlinearity $t^q|x|^\sigma u^p$ replaced by $h(t)w(x)u^p$ where h satisfies (7) and (8) and w(x) is a non-negative function such that $w(x) \geq |x|^\sigma$ if $|x| \geq R$. Then the statement of Theorem 2.1 remains valid in this case.

PROOF: Let $R_0 \geq R$ be so large that $D \supset C_{R_0}$. Suppose that u(x,t) has not blown up at $t = t_0$. By the infinite speed propagation of the heat operator $u(x,t_0) > 0$ in D. Let $\mu(x,t)$ be the solution of (1) with C_R replaced by C_{R_0} and $v(x) \in C_0(C_{R_0})$, $v(x) \leq u(x,t_0)$ instead of $u_0(x)$. If $\mu(x,t)$ blows up at finite time then by comparison u(x,t) also blows up at finite time $< t_0 + \tau$. In order to apply Theorem 2.1, which is based on Lemma 2.2, we have to be sure that $t_* = q/(p - 1)\lambda > t_0$. Since, according to the previous observation, λ can be chosen as small as possible, this can always be achieved. Hence $\mu(x,t)$ can never be global and the assertion is thus established.

The arguments extend to problems of the type

$$\begin{cases} u_t - \Delta u = h(t)w(x)f(|x|^\delta u) & \text{in } D \times R^+, \\ \qquad u = 0 & \text{in } \partial D \times R^+, \end{cases} \tag{18}$$

$$u(x,0) = u_0(x),$$

where h(t) and w(x) are nonnegative functions such that

(a) $h(t) \geq t^q$ for $t \geq t_0$, $q \geq 0$;

(b) $w(x) \geq |x|^{-\delta}$ for $|x| \geq R$ and $\delta = \sigma/(p-1)$ and $f:R^+ \to R^+$ is subject to the conditions

$$f(0) = 0,\ f(s) > 0 \text{ for } s > 0, \tag{19}$$

$$f(s)/s^p \to 1 \text{ as } s \to 0, \tag{20}$$

$$f(s) \text{ is convex}, \tag{21}$$

$$\int_s^\infty f^{-1}(\sigma)d\sigma < \infty \text{ for all } s > 0. \quad \square \tag{22}$$

COROLLARY 2.6: Under the conditions of Theorem 2.1 concerning p and σ, (18) does not have global regular solutions.

The proof can be carried out exactly in the same way as for Theorem 2.1 (see also [2]) and will therefore be omitted.

REMARK: All nonexistence results for global solutions in exterior domains are obviously true for the corresponding Cauchy problem, i.e. if D is replaced by $\mathbb{R}^N$ and no boundary conditions are imposed.

3.

We now turn to the question of global solutions. Following Fujita's device [4], [5] we get

THEOREM 3.1: If $p > 1 + \max\{(2+2q+\sigma)/N, (2+2q)/N\}$ then problem (1) possesses global solutions.

PROOF: By the method of upper and lower solutions it suffices to construct an upper solution $\bar{u}$ such that

$$\bar{u}_t - \Delta\bar{u} \geq t^q|x|^\sigma \bar{u}^p \text{ in } \mathbb{R}^N \times \mathbb{R}^+,$$

$$\bar{u}(x,0) \geq 0,$$

$$\bar{u} \not\equiv 0,$$

and which exists for all $t > 0$. This function $\bar{u}$ is then an upper bound for all solutions of (1) with $u_0(x) \leq \bar{u}(x,0)$. As many authors observed [14], [18] a good candidate for $\bar{u}$ is

$$\bar{u}(x,t) = \frac{c}{(t+t_0)^\gamma} e^{-r^2/4(t+t_0)}.$$

A straightforward calculation yields

$$\bar{u}_t - \Delta\bar{u} - t^q|x|^\sigma \bar{u}^p = \frac{\bar{u}}{t+t_0}\left[\frac{N}{2} - \gamma - t^q(t+t_0)|x|^\sigma\left(\frac{c}{(t+t_0)^\gamma}\right)^{p-1} e^{-(r^2/4t)(p-1)}\right]. \tag{23}$$

Consider the function

$$f(\tau) := \tau^{q+1-\gamma(p-1)} e^{-\beta/\tau},$$

where $q + 1 - \gamma(p - 1) =: \alpha < 0$. It takes its maximum at $\tau = -\beta/\alpha$. Hence

$$f_{max} = e^{\alpha}(-\beta/\alpha)^{\alpha}$$

and the last term in (23) is bounded from above by $c_1 r^{\sigma+2(q+1-\gamma(p-1))}$. Thus for $\bar{u}$ to be an upper solution in $\mathbb{R}^N \times \mathbb{R}^+$ we must have

$$\max \left\{\frac{q + 1 + \sigma/2}{p - 1}, \frac{q + 1}{p - 1}\right\} < \gamma < N/2,$$

which can be achieved only under the condition stated in the theorem. □

REMARKS:

(a) For $\sigma = 0$ we recover Meier's result [15].

(b) If $\sigma = q = 0$, $p = 1 + 2/N$ belongs to the blow up case as it was shown by several authors [1], [6], [9], [19]. In general it is not known whether or not $p = 1 + (2+2q+\sigma)/N$ falls into the blow up case.

(c) If $q = 0$ the question arises whether there exist stationary states. Some partial results are known. For more details we refer to the literature [3], [17], [18].

References

[1] Aronson, D.J. and Weinberger, H.F., Multidimensional nonlinear diffusion arising in population genetics, Advances in Math. 30 (1978), 33-76.

[2] Bandle, C. and Levine, H.A., On the existence and nonexistence of global solutions of reaction-diffusion equations in sectorial domains, Trans. Am. Math. Soc.(to appear).

[3] Bandle, C. and Marcus, M., The positive radial solutions of a class of semilinear elliptic equations (to appear).

[4] Fujita, H., On the blowing up of solutions of the Cauchy problem for $u_t = \Delta u + u^{1+\alpha}$, J. Fac. Sci., Tokyo Sect. IA, Math. 16 (1966), 109-124.

[5] Fujita, H., On some nonexistence and nonuniqueness theorems for nonlinear parabolic equations, Proc. Symp. Pure Math. 18 (1969), 105-113.

[6] Hayakawa K., On nonexistence of global solutions of some semilinear parabolic equations, Proc. Japan Acad. 49 (1973), 503-525.

[7] Kaplan, S., On the growth of solutions of quasilinear parabolic equations, Comm. Pure Appl. Math. 16 (1963), 305-333.

[8] Kavian, O., Remarks on the large time behaviour of a nonlinear diffusion equation, Ann. Inst. Henri Poincaré, Analyse Nonlinéaire 4 (1987), 423-452.

[9] Kobayashi, K., Sirao, T., and Zamaka, H., On the blowing up problem for semilinear heat equations, J. Math. Soc. Japan 29 (1977), 407-424.

[10] Lacey, A., Mathematical analysis of thermal runaway for spatially inhomogeneous reactions, SIAM J. Appl. Math. 43 (1983), 1350-1366.

[11] Lacey, A., The form of blow-up for nonlinear parabolic equations, Proc. Roy. Soc. Edinburgh Sect. A 98 (1984), 183-202.

[12] Levine, H.A., Some nonexistence and instability theorems for solutions of formally parabolic equations of the form $Pu_t = -Au + F(u)$, Arch. Rat. Mech. Anal. 51 (1973), 371-386.

[13] Levine, H.A., The Long Time Behaviour of Solutions of Reaction-Diffusion Equations in Unbounded Domains: A Survey (to appear).

[14] Levine, H.A. and Meier, P., The Value of the Critical Exponent for Reaction-Diffusion Equations in Cones (to appear).

[15] Meier, P., On the critical exponent for reaction-diffusion equations (to appear).

[16] Meier, P., Blow-up of solutions of semilinear parabolic differential equations, Z. Angew. Math. Phys. 39, (1988), 135-149.

[17] Noussair, E.S. and Swanson, C.A., Global positive solutions of semilinear elliptic equations, Can. J. Math. 35 (1983), 839-861.

[18] Swanson, C.A., Positive solutions of $\Delta u = f(x,u)$, Nonlinear Analysis T. M. A. 9 (1985), 1319-1323.

[19] Weissler, F.B., Existence and nonexistence of global solutions for a semilinear heat equation, Israeli J. Math. 38 (1981), 29-40.

C. Bandle
Mathematisches Institut
University of Basel
Rheinsprung 21
CH-4051 Basel
Switzerland

Note added in proof. A modified proof of Theorem 2.1 which applies also to negative and does not require a scaling argument is given in: C. Bandle and H.A. Levine, Fujita type results for convective-like reaction-diffusion equations in exterior domains, submitted for publication in ZAMP.

B. BILER

Blow up in the porous medium equation

1. INTRODUCTION

Our aim in this article is to describe some properties of temporally non-global positive solutions of the Cauchy problem for the porous medium equation. The problem of studying noncontinuable solutions of nonlinear diffusion equations like

$$u_t = \nabla\cdot(K(u)\nabla u) + Q(u) \tag{1}$$

is not only of pure mathematical interest but is also relevant for applications. These solutions may describe, e.g. combustion and deflagration in active non-linear media.

Observe that if $K \equiv 0$ and $\int_1^\infty 1/Q(u)\,du < \infty$ then (1) becomes an ordinary differential equation whose nontrivial solutions blow up in finite time. The same phenomenon persists even if the diffusion is added in the model equation

$$u_t = \nabla(u^m) + u^p \text{ on } R^N,\ 1 \leqq m \leqq p. \tag{2}$$

For a large class of initial data the solutions of (2) do not exist globally in time and, in general, the blow up is confined to a compact set in x-space (except for spatially homogeneous solutions). In the case $m = 1$, $1 < p < (N + 2)/(N - 2)$ blowing up solutions resemble asymptotically (near the explosion time) self-similar solutions of (2), i.e. those invariant under the Boltzmann scaling $u_\lambda(x,t) = \lambda^{2/(p-1)}u(\lambda x,\lambda^2 t)$, $\lambda > 0$. More precisely, for any local solution u defined on $\{(x,t): |x| < 1, -1 < t < 0\}$ the limit of rescaled solutions $\lim_{\lambda\to 0} (-t)^{1/(p-1)}u_\lambda(x,t)$ is equal to $K = (p-1)^{1/(1-p)}$ or 0. The latter possibility corresponds in fact to a regular solution without any singularity at $T = 0$. In the first case $\lim_{t\to T} (T-t)^{1/(p-1)}u(y_0 + y(T-T)^{1/2},t)$ $= K$ uniformly for $|y| \leqq$ const, so u behaves asymptotically for $t \to T$ in parabolic regions like the homogeneous solution $K(T-t)^{1/(1-p)}$. This does not exclude of course the possibility of localized (and hence in this case single point) blow up. Results of this type are described in [6], [12], [5] and (abundant) references therein. We adopt some ideas and techniques from Giga and Kohn's papers to study blowing up solutions of the homogeneous diffusion equation

$$u_t = \Delta(u^m). \qquad (3)$$

However, here the phenomenon of noncontinuability of solutions of the Cauchy problem is completely different from that for (2): blow up is never localized, it comes from infinity.

2. LINEAR HEAT EQUATION AND ITS BLOWING UP SOLUTIONS

The simplest linear diffusion equation (3) with $m = 1$, $u_t = \Delta u$, has a large variety of blowing up solutions even if we restrict our attention to self-similar ones, i.e. satisfying the identity $u(x,t) = \lambda^{2\gamma}u(\lambda x, T-\lambda^2(T-t))$, $\lambda > 0$, therefore of the form $u(x,t) = (T-t)^{-\gamma}w(y)$, where $x = (T-t)^{1/2}y$ and $\gamma > 0$ is a parameter. The function $w = w(y)$ satisfies the following equation in R^N

$$\nabla\cdot(\nabla w - \frac{yw}{2}) = (\gamma - \frac{2}{N})w. \qquad (4)$$

Evidently $w(y) = \exp(|y|^2/4)$ is a solution of (4) with $\gamma = N/2$ which corresponds to $u(x,t)$ resembling the fundamental solution of the backward heat equation. Note that there exist positive radial solutions of (4) for arbitrary $\gamma > 0$, e.g.

$$w(y) = C(1 + \sum_{k=1}^{\infty} |y|^{2k} \frac{\gamma(\gamma+1)\dots(\gamma+k-1)}{k!4^k(N/2)(N/2+1)\dots(N/2+k-1)}).$$

All these explicit radial solutions satisfy growth estimate $w(y) \sim C\exp(|y|^2/4)$ and in fact any nontrivial positive solution of (4) has such growth in the mean: if $w(y)\exp(-|y|^2/4) \in L^1(R^N)$, then $w \equiv 0$. These radial solutions also permit us to construct nonradial solutions of (4) with given $\gamma > 0$ according to the following property of (4).

If $w_j = w_j(y_j)$, $j = 1,2$, satisfy (4) for $y_j \in R^{N_j}$ with $\gamma_j > 0$, respectively, then $w(y) = w_1(y_1)w_2(y_2)$, $y = (y_1,y_2) \in \mathbb{R}^N$, $N = N_1 + N_2$, is a solution of (4) with $\gamma = \gamma_1 + \gamma_2$ (and w has the same exponential growth $\exp(|y|^2/4)$ as before).

The solutions of such rapid growth may be treated classically using the Weierstrass kernel. The well-known Aronson-Widder representation theorem, see [1], gives (for any positive solution u of the heat equation) the formula $u(x,t) = \int \exp(-|x-y|^2/4t)u_0(y)\,dy$ valid for all sufficiently small $t > 0$ if and only if u_0 is a measure of not too rapid growth: $\int u_0(y) \exp(-c|y|^2)dy < \infty$ for some $c \geq 0$.

The self-similar solutions of (4) have critical growth preventing their continuation beyond $T = 1$. Their characteristic feature is the uniform explosion in x: $\lim_{t\to T} u(x,t) = \infty$. However, there are some other ways for solutions of (3) to blow up.

PROPOSITION 1: There exist solutions of the one-dimensional equation (3), $m = 1$, with maximal time of existence T such that:

(i) $u(x,T)$ is finite for all x; or

(ii) $u(x,T) < \infty$ for $x < x_0$, $u(y,T) = \infty$ for $y \geq x_0$; or

(iii) $u(x,T) < \infty$ for $x \leq x_0$, $u(y,T) = \infty$ for $y > x_0$; or

(iv) $u(x,T) = \infty$ for all x.

It is quite easy to construct such u as the solution with the initial data of the type $\sum_{n=0}^{\infty} c_n\delta_n$.

Using the remarks above one can easily produce further examples of solutions with a complicated structure of sets of explosion for $t = T$ in the multidimensional case.

3. POROUS MEDIUM EQUATION AND ITS SELF-SIMILAR SOLUTIONS

In this section we will consider the nonlinear porous medium equation (3) with $m \neq 1$. The fast diffusion case $0 < m < 1$ is not interesting as there are no growth restrictions on the initial data in order to have global solvability of the Cauchy problem [10]. In the case $m > 1$, and more generally for K in (1) satisfying the condition $\int_0^1 K(u)u^{-1}\, du < \infty$ and $Q \equiv 0$, the velocity of propagation of the initial disturbances is finite.

We briefly recall from [2], [3] the results of Aronson, Bénilan, Caffarelli, Crandall, Dahlberg, Kenig and Pierre on solvability and characterization of the initial data for the Cauchy problem for (3). A continuous function $u \geq 0$, defined on $\{(x,t): x \in R^N, 0 < t \leq T\}$, is said to be in P(T) if u satisfies the integral identity

$$\int_{R^N \times (t_1,t_2)} (u^m \Delta\phi + u\phi_t)\, dx\, dt = \int_{R^N} u(x,t_2)\phi(x,t_2)\, dx$$

$$- \int_{R^N} u(x,t_1)\phi(x,t_1)\, dx$$

for all $0 < t_1 < t_2 \leq T$ and all $\phi \in C^{2,1}$ with compact support in x. For $u \in P(T)$ there exists the initial trace which is a nonnegative Borel measure of moderate growth: $\lim_{t\to 0} \int_{R^N} u(x,t)\phi(x)\, dx = \int_{R^N} \phi(x)\, d\mu(x)$ for all $\phi \in C_0(R^N)$,

$$\sup_{R \geq r} R^{-(N+2/(m-1))} \mu\{|x| \leq R\} < \infty, \quad r \geq 1. \tag{5}$$

Conversely, for every such measure μ there exists a solution of (3) (unique in P(T)) so that $u(\cdot,0) = \mu$ in the sense of trace. This result is a complete nonlinear analogue of Aronson-Widder's theorem for the linear heat equation. Let us mention a new regularity result in [4] where the authors proved continuity (after a correction on a zero set) of solutions $u \in L^m_{loc}$ - hence satisfying the weakest assumption necessary to define the distribution solution of (3).

Condition (5) can be rewritten as

$$\sup_{R \geq 1} R^{-N} \int_{|x| \leq R} u_0(x)(1 + |x|^2)^{-1/(m-1)}\, dx < \infty$$

for the initial data, u_0 being a L^1_{loc} function. The length of the maximal interval of existence of the solution is estimated from below by $T = T(u_0) \geq C(N,m)/\ell(u_0)^{m-1}$ where $\ell(u_0)$ is the limit of the expression in (5) as $r \to \infty$. This estimate is sharp, as the explicit self-similar Barenblatt's solutions

$$u(x,t) = (T-t)^{-k/(m-1)}(AT^k + BT|x|^2/(T-t)^{1-k})^{1/(m-1)} \tag{6}$$

with $u_0(x) = (A + B|x|^2)^{1/(m-1)}$, show. Here $k = N(m-1)/(N(m-1) + 2)$, $c = k/(2mN)$, $T = c/B$, $A \geqq 0$.

We would like to show that these special solutions display typical ways of blow up of "many" solutions of (3). The significance of this vague statement will be revealed in the sequel. Note that in view of the results in [14] - specific for the one-dimensional case and somewhat similar to Proposition 1(ii)-(iv) - it seems impossible to characterize all blowing up solutions of (3) without some supplementary structure assumptions.

We consider self-similar solutions of (3) defined for $t < T$ of the scale invariant form $u(x,t) = \lambda^\beta u(\lambda x, T-\lambda^\alpha(T-t))$ so $u(x,t) = (T-t)^{-\gamma}\, w(y)$ with $x = (T-t)^{1/\alpha}y$, $y \in \mathbb{R}^N$. Clearly $\beta = (\alpha-2)/(m-1)$ and $\gamma = \beta/\alpha$. In this notation w is a solution of the nonlinear elliptic equation

$$\gamma w + \alpha^{-1} y\cdot\nabla w = \Delta(w^m) \tag{7}$$

and $z = w^{m-1}$ satisfies

$$(m-1)\gamma z + \alpha^{-1} y\cdot\nabla z = m/(m-1)|\nabla z|^2 + mz\Delta z. \tag{8}$$

Restricting our attention to the case $\alpha = 2/(1-k) = N(m-1) + 2$, $\beta = N$, $\gamma = k/(m-1)$ (these parameters, as in (6), correspond in fact to the maximal growth in time of blowing up solutions: $(T-t)^{-1/(m-1)}$ - note that for the linear heat equation $\gamma > 0$ was arbitrary) we get a divergence form equation

$$\nabla\cdot(z^{1/(m-1)}(\nabla z - 2cy)) = 0. \tag{9}$$

Parallel to a study of self-similar solutions of (3) satisfying (8) we will consider an equation satisfied by any solution of the form $u(x,t) = (T-t)^{-\gamma}w(y,s)$, where $x = (T-t)^{1/\alpha}y$, $y \in \mathbb{R}^N$, $(T-t) = e^{-s}$, $t < T$, $s \in \mathbb{R}$. Substituting this into (3) we obtain the nonlinear parabolic equation

$$w_s + \gamma w + \alpha^{-1} y\cdot\nabla w = \Delta(w^m) \tag{10}$$

or as a counterpart of (9)

$$(1-1/m)\,\frac{d}{ds}\,(z^{1/(m-1)}) = \nabla\cdot(z^{1/(m-1)}(\nabla z - 2cy)). \tag{11}$$

Observe an interesting property of w: rescaling u to u_λ corresponds to shifting w(y,s) in s to w(y,s-αlog λ).

4. UNIQUENESS RESULTS FOR SELF-SIMILAR BLOWING UP SOLUTIONS

In this section we describe several conditional uniqueness results for nonnegative solutions of (7) or (9) defined in $\mathbb{R}^N$. The need to introduce some supplementary hypotheses in order to get a local uniqueness or a stability result is evident since one can produce many solutions of (8) other than $z(y) = c|y|^2$, e.g. $z_J(y) = c_J|\bar{y}_J|^2$, where $J = 1,\ldots,N-1$, $c_J = c(2 + N(m-1))/(2 + J(m-1))$ and $\bar{y}_J \in \mathbb{R}^N$ has J coordinates of y and N-J zeros. These "lower dimensional" solutions do not have uniform growth in y and they present a serious obstacle in obtaining any general uniqueness result. We begin with two simple facts concerning uniqueness of the special solutions of (9).

PROPOSITION 2: If $z(y) = g(r)$, $r = |y|$, is a radial solution of (9) such that $\lim_{r\to 0} r^{N-1}g'(r) = 0$, then $g(r) = cr^2 + C$ with an arbitrary constant $C \geqq 0$.

This follows from the equation $(g^{1/(m-1)}r^{N-1}(g'-2cr))' = 0$ satisfied by radial solutions and from the regularity assumption at 0.

PROPOSITION 3: If z solves (9) and $\Delta z \geqq 2cN = k/m$ in $\mathcal{D}'$,

$$\lim_{R\to\infty} \int_{|y|=R} \exp(-M|y|^2)\nabla z(y)\cdot\bar{n}(y)\; dS(y) = 0 \text{ for some } M > 0,$$

then $z = c|y|^2 + C$ for some positive constant C.

For the proof we consider the equation

$$z\Delta Z + (m-1)^{-1}|\nabla Z|^2 + 2c(m-1)^{-1}y\cdot\nabla Z = 0 \tag{12}$$

satisfied by $Z = z - c|y|^2$. We multiply it by $\exp(-M|y|^2)$ with sufficiently large $M > 0$, integrate over the ball of radius R and after some integration by parts we get $\nabla Z = 0$ a.e.

The assumption $\nabla Z \geqq 0$ is restrictive but not unexpected. Namely, from the Aronson-Bénilan inequality for $u \in P(T)$ with the initial data imposed at $T_0 = 0$: $\Delta(u^{m-1})(\cdot,t) \geqq -k/mt$ in $\mathcal{D}'(R^N)$, $0 < t < T$. Hence for self-similar solutions, shifting T_0 to $-\infty$, we get $\Delta z \geqq 0$. The condition $\Delta Z \geqq 0$ or $\Delta z \geqq k/m$ is of course much stronger and characterizes Barenblatt's solutions.

The main result in this section is the uniqueness of sufficiently regular solutions of (9) satisfying certain estimates on minimal growth in order to separate the special solutions (6). Remark that variational methods very useful in [6] are not available for our equations and standard Liouville theorems for linear elliptic equations fail here because of too rapid growth of the coefficients.

<u>THEOREM 1</u>: If $Z \in C^3$ satisfies (12) and Z is bounded from below then Z is a constant.

The idea of the proof is based on classical arguments (Bernstein-type estimates) in [11], Theorem 4.7, Remarks, pp. 93 and 98 and [13], see also [7]. Nevertheless, our proof is somewhat different because equations (7)-(9) do not fit into the schemes in [11], [7]. An important ingredient of the proof is the following lemma.

If $Z \in C^3$ satisfies (12) and $Z(y) = o(|y|)$, then $\nabla Z(y) = o(1)$ for $|y| < \infty$.

Its proof is quite similar to that in [13] where the auxiliary function $v = \zeta|\nabla Z|^2 + Z^2$ is considered with the cutoff function $\zeta(y)\ C(R)(1-|y-y_0|^2/R^2)_+^2$. Then a more precise estimate from Theorem 4.6 in [11] $|\nabla Z(y_0)| \leq CR^{-1}$. $\sup_{R\leq|y|\leq 3R} (Z(y_0)-Z(y))$ is obtained using a more complicated auxiliary function $v = \zeta|\nabla(\phi^{-1}(Z))|^2$ with $\phi(r) = \text{const} - e^{-r}$.

Karp's method in [9] based on differential inequalities for certain integral functionals depending on the solution does not require even a C^2 regularity assumption ([9], p. 87). It is more flexible than Bernstein or Harnack inequality methods and allows us to rederive our result.

5. ASYMPTOTIC UNIQUENESS FOR THE NONLINEAR PARABOLIC EQUATION (11)

Equation (3) does not depend on x and therefore translating special solutions (6) we obtain a lot of new solutions of the form

$$v(x,t) = (T-t)^{-k/(m-1)}(AT^k + C|x-x_0|^2/(T-t)^{1-k})^{1/(m-1)} \tag{13}$$

with fixed $x_0 \in \mathbb{R}^N$. They lead to $z(y,s) = AT^k + c|y-e^{s/\alpha}\, x_0|^2$ and $Z(y,s) = z(y,s) - c|y|^2 = AT^k + c(e^{2s/\alpha}|x_0|^2 - 2e^{s/\alpha}x_0\cdot y) = \mathcal{O}(|y|)$ uniformly for all $s \leqq s_0$, $s_0 \in \mathbb{R}$ fixed, satisfying the equation

$$(1-1/m)\,\frac{d}{ds}\,(z^{1/(m-1)}) = \nabla\cdot(z^{1/(m-1)}\nabla Z). \tag{14}$$

Similarly, the translates of z_J yield new nonstationary solutions of (14). Hence the best uniqueness result (of Liouville-type for parabolic equations) we can expect is

> If Z satisfies (14) for $s \leqq 0$, $y \in \mathbb{R}^N$ and $Z(y,s) = o(|y|)$ for $|y| \to \infty$, then Z = const.

We are able to prove only the weaker property: if Z is bounded for $y \in \mathbb{R}^N$, $s \leqq 0$, then Z = const.

Finally, we would like to present an asymptotic stability result which applies to, e.g. special solutions like (6), (13). The approach is based on a fairly standard method of proving uniqueness due to Kalashnikov [8] and on some calculations in [2], Proposition 2.1, where this method was generalized to the multidimensional situation. We employ, however, quite different test functions - supersolutions of the conjugate problem.

To formulate a reasonable hypothesis guaranteeing asymptotic uniqueness, first let us observe that for v in (13) and the corresponding centred self-similar solution u in (6) we have $u^{m-1}(x,t) - v^{m-1}(x,t) = \mathcal{O}((T-t)^{-k}) + \mathcal{O}(|x|/(T-t))$, so (a weaker condition) $= \mathcal{O}((T-t)^{-k}) + \mathcal{O}(|x|/(T-t)^{(1+k)/2})$ for $x \in \mathbb{R}^N$ and $t \to -\infty$. Now we can state

<u>THEOREM 2</u>: Let $N \geqq 2$ and u,v be the solutions of (3) on $\mathbb{R}^N \times (-\infty,T)$ both blowing up at time T, such that

$$u^{m-1}(x,t) - v^{m-1}(x,t) = o((T-t)^{-k}) + o(|x|/(T-t)^{(1+k)/2}),$$

$$u(x,t) - v(x,t) = o(|x|^{2/(m-1)-2}(T-t)^{2/\alpha-1/(m-1)})$$

$$+ o(|x|^{2/(m-1)-1}(T-t)^{1/\alpha-1/(m-1)}),$$

for $x \in \mathbb{R}^N$, $t \leq T-1$ (the Landau symbols "o" are understood here for $t \to -\infty$ and $x/(T-t)^{1/\alpha} = y$ bounded). Then u is identically equal to v.

REMARKS: The first terms in the hypotheses of Theorem 2 correspond to adjusting the constant A in (6), the second ones control the behaviour for large x. The second hypothesis (on u-v) is the consequence of the first one for $m < 2$ or for regularly growing (in spatial variable) solutions. Namely $u-v = (m-1)^{-1}(\tilde{u}^{m-1})^{1/(m-1)-1}(u^{m-1}-v^{m-1})$ with $\tilde{u}$ between u and v, so for corresponding $\tilde{z}(y,s) = (T-t)^k\, \tilde{u}(x,t)^{m-1} \geq \varepsilon|y|^2$ with a positive ε. Hence $\tilde{u}^{2-m}$ is bounded from above by $C((T-t)^{-k}|x|^2/(T-t)^{2/\alpha})^{1/(m-1)-1} = C(|x|^2/(T-t))^{1/(m-1)-1}$. The case $N = 1$ requires a different argument, see the results in [14].

The great variety of solutions of type (13) makes the dynamical systems approach of Giga and Kohn hopeless (they considered, using variational methods, an analogue of (14) constructed for (2), $m = 1$, in a weighted Sobolev space). A study of attractors corresponding to these special solutions would be essential in establishing a fine description of blowing up solutions. These attractors seem to have a fairly complicated structure.

ACKNOWLEDGEMENT: The author wishes to thank Michel Pierre for helpful remarks during his visit to Wrocław in October 1986, and for his kind hospitality and stimulating discussions during the author's stay at the University of Nancy 1 in May 1987.

References

[1] D.G. Aronson, Widder's inversion theorem and the initial distribution problem, SIAM J. Math. Anal. 12 (1981), 639-651.

[2] P. Bénilan, M.G. Crandall and M. Pierre, Solutions of the porous medium equation in R^N under optimal conditions on initial values, Indiana Univ. Math. J. 33 (1984), 51-87.

[3] B.E.J. Dahlberg and C.E. Kenig, Non-negative solutions of the porous medium equation, Comm. Part. Diff. Eq. 9 (1984), 409-437.

[4] B.E.J. Dahlberg and C.E. Kenig, Weak solutions of the porous medium equation, preprint Göteborg 1987/34.

[5] V.A. Galaktionov, V.A. Dorodnicyn, G.G. Elenin, S.P. Kurdjumov and A.A. Samarskii, Quasilinear heat equation with source: strained regime, localization, symmetry, exact solutions, asymptotics, structures (in Russian), Itogi nauki i techniki, Sovremennye problemy matematiki-novejšie dostiženia 28 95-205, ed. R.V. Gamkrelidze, VINITI, Moskva, 1986.

[6] Y. Giga and R.V. Kohn, Characterizing blowup using similarity variables, Indiana Univ. Math. J. 36 (1987), 1-40.

[7] A.V. Ivanov, Quasilinear degenerate and nonuniformly elliptic and parabolic equations of second order (in Russian), Trudy Mat. Inst. Akad. Nauk V.A. Steklova 160, Leningrad, Nauka (1982), 1-285.

[8] A.S. Kalashnikov, The Cauchy problem in a class of growing functions for equations of nonsteady filtration type (in Russian), Vestnik MGU 18:6 (1963), 17-27.

[9] L. Karp, Asymptotic behavior of solutions of elliptic equations I: Liouville type theorems for linear and nonlinear equations on R^n, II: Analogues of Liouville's theorem for solutions of inequalities on R^n, $n \geqq 3$, J. Analyse Math. 39 (1981), 75-102, 103-115.

[10] M. Pierre, Nonlinear fast diffusion with measures as data, Nonlinear parabolic equations - qualitative properties of solutions, Rome, 1985; pp. 179-188, ed. L. Boccardo, A. Tesei, RNM Vol. 149, Longman, London, 1987.

[11] L.A. Peletier and J. Serrin, Gradient bounds and Liouville theorems for quasilinear elliptic equations, Ann. Sc. Norm. Sup. Pisa IV, 5 (1978), 65-104.

[12] P.E. Sacks, Global behavior for a class of nonlinear evolution equations, SIAM J. Math. Anal. 16 (1985), 233-250.

[13] I.N. Tavkhelidze, Liouville theorems for elliptic and parabolic equations of second order (in Russian), Vestnik MGU 31:4 (1976), 28-35.

[14] J.L. Vázquez, The interfaces of one-dimensional flows in porous media, Trans. AMS 285 (1984), 717-737.

Piotr Biler
Laboratoire d'Analyse Numérique
Université de Paris-Sud
Bât. 425
91405 Orsay
France

Permanent address:

Mathematical Institute
University of Wrocław
pl. Grunwaldzki 2/4
50-384 Wrocław
Poland

V. CASELLES

Duality and nonlinear equations governed by accretive operators

This article is a contribution to the study of the observed fact that many nonlinear evolution equations (which can be solved by using a nonlinear semigroup generated by an accretive operator in some Banach space) have, roughly speaking, a dual equation which can also be solved by a nonlinear semigroup generated by an accretive operator in some dual space*. Mainly, we are thinking of nonlinear diffusion problems dual to other nonlinear diffusion problems and the pairing between (nonlinear) scalar conservation laws and Hamilton-Jacobi equations. To be more precise:

Let Ω be a bounded domain in R^n with smooth boundary and let β be a continuous nondecreasing real function such that $\beta(0) = 0$. Consider the following problems:

$$(P_1) \quad \begin{cases} u_t - \Delta\beta(u) = 0 & \text{on} \quad [0, +\infty) \times \Omega, \\ \beta(u) = & \text{on} \quad [0, +\infty) \times \partial\Omega, \\ u(0,x) = u_0(x), & \end{cases}$$

$$(P_2) \quad \begin{cases} u_t - \Delta_p\beta(u) = 0 & \text{on} \quad [0, +\infty) \propto \Omega, \\ \beta(u) = 0 & \text{on} \quad [0, +\infty) \propto \partial\Omega, \\ u(0) = u_0(x), & \end{cases}$$

where $p > 1$ and Δ_p is the p-Laplacian given formally by $\Delta_p u = \operatorname{div}(|\nabla u|^{p-2} \nabla u)$

---

* Supported by a grant from the Ministerio de Educacion y Ciencia de España.

$$(P_1^*)\quad \begin{cases} u_t + \beta(-\Delta u) = 0 & \text{on } [0, +\infty) \times \Omega, \\ u = 0 & \text{on } [0, +\infty) \times \partial\Omega, \\ u(0,x) = u_0 \end{cases}$$

$$(P_2^*)\quad \begin{cases} u_t + \beta(-\Delta_p u) = 0 & \text{on } [0, +\infty) \times \Omega, \\ u = 0 & \text{on } [0, +\infty) \times \partial\Omega, \\ u(0,x) = u_0(x) \end{cases}$$

Let us give one more example. Let ϕ be a real continuous function that $\phi(0) = 0$. Consider the following problems:

$$(P_3)\quad \begin{cases} u_t + \phi(u)_x = 0 & \text{on } [0,+\infty) \times \mathbb{R}, \\ u(0,x) = u_o(x),\ x \in \mathbb{R}, \end{cases}$$

$$(P_3^*)\quad \begin{cases} u_t + \phi(-u_x) = 0 & \text{on } [0, +\infty) \times \mathbb{R}, \\ u(0,x) = u_o(x),\ x \in \mathbb{R}, \end{cases}$$

It is well known that (P_i) are well posed in $L^1(\Omega)$, $L^1(\Omega)$, $L^1(\mathbb{R})$, respectively, for i = 1,2,3 ([1], [13], [1] or [9]) and (P_i^*) are well posed in $L^\infty(\Omega)$, $L^\infty(\Omega)$ and $B\,U\,C(\mathbb{R})$, respectively, for i = 1,2,3 ([14], [14], [10], [11]). By this we mean that there exists an m-accretive operator on the corresponding L^1-space for (P_i) or L^∞-space for (P_i^*) which is related to the formal differential operator on (P_i), (P_i^*), thus generating a nonlinear semigroup in the L^1 or L^∞-spaces. This settles the question of existence, uniqueness and continuous dependence of solutions on initial data for problems (P_i) and (P_i^*). Recall that the solution of (P_3) in $L^1(\mathbb{R})$ relies on the notion of Kruzkov's solution of (P_3) ([1], [9], [15]) and the solution of (P_3^*) in $B\,U\,C(\mathbb{R})$ relies on the notion of viscosity solution for (P_3^*) introduced by Crandall and Lions in [11] (see also [10]). This "sophisticated" (but necessary) notion of solution makes the problems (P_3), (P_3^*) "different" to the problems (P_1), (P_2), (P_1^*) or (P_2^*).

To describe our purpose in what follows, let us consider this problem in a more abstract setting. Let E be a Banach space. When saying a (nonlinear) operator on E we mean a multivalued operator on E defined through its graph, i.e. a subset of $E \times E$. Its domain and range are defined in the usual way. Given two operators A, B on E, its product is defined by $A \cdot B = \{[u,v] \in E \times E: \exists w \in E \text{ with } [u,w] \in B \text{ and } [w,v] \in A\}$.

Let us clarify a preliminary question: Given an operator A on E, what can be reasonably considered as the adjoint operator A^t of A? If A is a linear operator on E, A^t should be the usual adjoint operator on E*. If H is a Hilbert space $\phi: H \to]-\infty, +\infty]$ is a convex, ℓ.s.c., proper function and $A = \partial\phi$, then its adjoint should again be $\partial\phi$. Similarly, if $\phi: H \to R$ is a C^1-function and $A = \nabla\phi$, then A^t should be $\nabla\phi$. That is, subgradients of convex, ℓ.s.c. functions or gradients of C^1-functions should be considered as self-adjoint operators. Supporting this is the fact that if $\phi: H \to R$ is C^2, the linearization of $\nabla\phi$ at $x \in H$: $T_x^{\nabla\phi}$ is a linear self-adjoint operator on H. Other results on subdifferentials in [3] are in this line. Given two operators A, B on E with reasonable adjoints A^t, B^t on E^*, we stick to the following rule: The adjoint of $A \cdot B$ should be $B^t \cdot A^t$. With these rules, if H is a Hilbert space, L is a linear densely defined operator on H and $\phi,\psi: H \to]-\infty, +\infty]$ are convex, ℓ.s.c. proper functions, then the adjoints of $L\partial\phi$, $\partial\phi\cdot\partial\psi$ are, respectively, $\partial\phi\cdot L^t$ and $\partial\psi\cdot\partial\phi$ and conversely. Since these considerations suffice for the cases considered below, we stop this discussion here and refer the reader to [8] for further remarks. Thus, there should be no confusion in what follows with our terminology. Let A, B be operators on E with respective "adjoints" A^t, B^t on E^*. Consider the following problem in E:

$$(P) \qquad \begin{cases} u_t + AB(u(t)) \ni 0, \\ \qquad\qquad u(0) = u_0. \end{cases}$$

If AB has an m-accretive extension E(AB) there is a notion of "good solution" for (P) ("bonne solution" in the sense of Benilan, [1]) and there exists a unique "good solution" of (P) for each $u_0 \in \overline{D(E(AB))}$. In this context we raise the following questions:

(Q1) Is $B^t A^t$ accretive in E^*?

(Q2) Does there exist an m-accretive extension of it? Is it unique?

Our purpose here is to give an answer to question (Q1), in a slightly different framework, under some technical conditions. To close these remarks, let us mention that the operators appearing in (P_1), (P_2), (P_1^*), (P_2^*) are formally of type $L\,\partial\phi$, $\partial\phi_1\cdot\partial\phi_2$, $\partial\phi\cdot L^t$ and $\partial\phi_2\cdot\partial\phi_1$, respectively. The operators in (P_3), (P_3^*) could be written formally as $L\,\nabla\phi$, $\nabla\phi\cdot L^t$, respectively. These considerations will be further developed in the examples below. Thus, the goal is to obtain the well-posedness of the problems (P_i^*) in L^∞ as a consequence of the well-posedness of (P_i) in the corresponding L^1-space or conversely. This can be done for (P_1)-(P_1^*) and (P_2)-(P_2^*) but the results for (P_3)-(P_3^*) are not satisfactory. Let us start with the problem.

THE MAIN RESULTS

The notions and terminology used here are standard in the literature. For the standard definitions used below we refer the reader to [4] or to the forthcoming book [2].

Let us prepare the statement of Theorem 2 with the following definition:

DEFINITION 1: Let H be a Hilbert space. Let A be an operator from $\underbrace{H \times \cdots \times H}_{n}$ into H. We say that A is cyclically monotone on each variable if for all n-ples $(u_1,\ldots,u_n) \in D(A)$, the operators $A^i_{u_1\ldots u_n}: H \to H$, $A^i_{u_1\ldots u_n}(u) = A(u_1,\ldots,u_{i-1},u,u_{i+1},\ldots,u_n)$ with domain $D(A^i_{u_1\ldots u_n}) := \{u \in H: (u_1\ldots u_{i-1},u,u_{i+1},\ldots,u_n) \in D(A)\}$ are cyclically monotone.

By Rockafellar (see [4], Theorem 2.5) we know that if an operator B on H is cyclically monotone, then there exists a convex, ℓ.s.c., proper function $\phi: H \to]-\infty, +\infty]$ such that $B \subseteq \partial\phi$. Moreover, if B is maximal monotone, then $B = \partial\phi$.

Some further remarks. First, recall that if $N: H \to [0, +\infty]$ is a convex, ℓs.c., proper function and A is an operator on H, A is called N-accretive if $N(x-\hat{x}) \leq N(x-\hat{x} + \lambda y - \lambda\hat{y})$ for all $[x,y]$, $[\hat{x},\hat{y}] \in A$ and all $\lambda \in]0,\lambda_0[$ for some $\lambda_0 > 0$. Finally, let us mention that the results below were motivated and inspired by Corollary 2.1 in [3].

Finally, let $N: H \to [0, +\infty]$ be convex, ℓ.s.c. proper and positive homogeneous of degree two. Then the convex conjugate function of N, $N^*(x) := \sup\{(x,y) - N(y): y \in H\}$ is also convex, ℓ.s.c. and positive homogeneous of degree two. Now, we can state the main result in [8]-1:

<u>THEOREM 2:</u> Let H be a Hilbert space. Let $N: H \to [0, +\infty]$ be convex, ℓ.s.c. proper and positive homogeneous of degree two. Let $L_1, \dots, L_n$, $T_1, \dots, T_n$ be linear continuous operators on H such that $L_1, \dots, L_n$ are monotone. Let $A: \underbrace{H \times \dots \times H}_{n} \to H$ with $\mathrm{Dom}\, A = H^n$ be cyclically monotone on each variable. Suppose that for each $u_1, \dots, u_n \in H$, $i \in \{1, \dots, n\}$, $A^i_{u_1 \dots u_n}$ are Lipschitz maps on H with the Lipschitz constant independent of i, $u_1, \dots, u_n$. If for each $u_1 \dots u_n \in H$ and each $i \in \{1,2,\dots,n\}$ the maps $L_i A^i_{u_1 \dots u_n} + T_i$ are N-accretive, then the map $u \to A(L_1^t u, \dots, L_n^t u) + \sum_{i=1}^n T_i^t u$ on H is N^*-accretive.

<u>REMARKS:</u>

(a) In the proof it is actually shown that

$$N^*(x-\hat{x}) \leq N^*(x-\hat{x} + \lambda A(L_1^t x, \dots, L_n^t x) - \lambda A(L_1^t \hat{x}, \dots, L_n^t \hat{x}) + \lambda \sum_{i=1}^n T_i^t (x-\hat{x})) \tag{1}$$

holds for all $x, \hat{x} \in H$ and all $\lambda \in]0, \lambda_0[$ for some $\lambda_0 > 0$ which depends on the Lipschitz constant of $A^i_{u_1 \dots u_n}$ and the norms of L_i and T_i. But if $x - \hat{x} \in \mathrm{Dom}\, N^*$, it is shown that (1) holds for all $\lambda > 0$.

(b) The statement of Theorem 2 is modelled by the observation of the following equations: Let Ω be a bounded domain in $\mathbb{R}^n$ with smooth boundary and let

$$L_i u = \sum_{j,k} \partial_j (a^k_{jk}\, \partial_k u) + \sum_j \partial_j (a_j u) + au, \quad a_{jk}, a_j \in C^1(\bar{\Omega}), \ a \in L^\infty(\Omega)_+,$$
$$i = 1,2,\dots,m,$$

be elliptic operators satisfying the usual assumptions ([6]). Consider the following problems:

$$
(P)\quad \begin{cases} \dfrac{du_i}{dt} + L_i \max(u_1,\dots,u_m) = 0 \quad \text{on } [0,+\infty) \times \Omega,\ i = 1,2,\dots,m, \\ u_i\big|_{\partial\Omega} = 0, \\ u_i(0) = u_{i_0} \in L^1(\Omega). \end{cases}
$$

$$
(P^*)\quad \begin{cases} u_t + \max(L_1^t u,\dots,L_n^t u) = 0 \qquad \text{on } [0,+\infty) \times \Omega, \\ u\big|_{\partial\Omega} = 0 \\ u(0) = u_0 \in L^\infty(\Omega). \end{cases}
$$

Here $A\colon L^2(\Omega)^m \to L^2(\Omega)$ is given by $A(u_1,\dots,u_m) = \max(u_1,\dots,u_m)$. Observe that if we fix $(m - 1)$-entries in A, with respect to the other variable, the function $u \to A^i_{u_1\dots u_m}(u) = \max(u_1,\dots,u_{i-1},u,u_{i+1},\dots,u_m)$ is a subgradient. What Theorem 2 says is that if for some regularization of L_i e.g. the Yosida regularizaton $L_{i\lambda}$), the maps $L_{i\lambda}\ A^i_{u_1\dots u_m}$ are N-accretive, with $N(f) = \frac{1}{2}\|f\|^2_{L^1(\Omega)}$ in this case, then the operator $u \to \max(L^t_{1\lambda}u,\dots,L^t_{m\lambda}u)$ is accretive in $L^\infty(\Omega)$. Letting $\lambda \to 0$ we can get the accretivity of $u \to \max(L_1^t u,\dots,L_m^t u)$ in $L^\infty(\Omega)$.

(c) As remarked in [8]-1, the N^*-accretivity in Theorem 2 coincides with the notion of $\hat{N}$-accretivity used in [3].

<u>APPLICATION</u>: Let Ω be an open bounded subset of R^n with smooth boundary Γ. Let $\beta\colon R \to R$ be strictly increasing with $\beta(0) = 0$. We want to show that

$$
(P^*)\qquad u_t + \beta(-\Delta u) = 0,\ \frac{\partial u}{\partial n}\Big|_\Gamma = 0,\ \ u(0) = u_0,
$$

is well posed in $L^\infty(\Omega)$ as a consequence of the well-posedness of (P) in $L^1(\Omega)$:

$$
(P)\qquad u_t - \Delta\beta(u) = 0,\ \frac{\partial\beta(u)}{\partial n}\Big|_\Gamma = 0,\ u(0) = u_0.
$$

Let us check the assumptions of Theorem 2. Let $H = L^2(\Omega)$,

$$\mathrm{Dom}\, L = \{u \in H^2(\Omega): \frac{\partial u}{\partial n}\Big|_\Gamma = 0\}, \quad Lu = -\Delta u, \quad u \in \mathrm{Dom}\, L.$$

Then, L is maximal monotone on H. Let L_λ be its Yosida regularization. Let $\beta_n \in W^{1,\infty}(R)$, $\beta_n(0) = 0$, $\beta_n' > 0$, $\beta_n \to \beta$, uniformly on bounded sets of R. Let $j_n: R \to R$ be such that $j_n' = \beta_n$ and let $J_n(u) = \int_\Omega j_n(u(x))\, dx$, $u \in H$. Then $\partial J_n(u) = \beta_n(u(\cdot))$, $u \in H$.

Let $P = L^2(\Omega)_+$. Let $N(f) = \frac{1}{2}(\int_\Omega f)^2$ if $f \in P$, $+\infty$ otherwise. Since for all $\lambda > 0$, $n \in \mathbb{N}$, $L_\lambda\ \partial J_n$ is N-accretive ([6], [12], Prop. 1) then $\beta_n \cdot L_\lambda$ is N^*-accretive (Theorem 2). Since $N^*(f) = \frac{1}{2}\|f^+\|_\infty^2$, then

$$\|(f-g)^+\|_\infty \leq \|(f-g + \mu\beta_n L_\lambda f - \mu\beta_n L_\lambda g)^+\|_\infty, \quad f,g \in L^\infty(\Omega),\ n = 1,2,\dots,\lambda > 0. \tag{2}$$

Let $L_0 = \{[u,v] \in L^\infty \times L^\infty(\Omega): v = Lu\}$. Then L_0 is m-accretive in $L^\infty(\Omega)$ and $L_\lambda f \to L_0 f$ in $L^\infty(\Omega)$ as $\lambda \to 0$ for all $f \in \mathrm{Dom}\, L_0$. Let $f,g \in \mathrm{Dom}\, L_0$. Letting $\lambda \to 0$ and $n \to \infty$ in this order in (2) we obtain the T-accretivity of $Bf := \beta(L_0 f)$, $f \in \mathrm{Dom}\, L_0$ in $L^\infty(\Omega)$. Moreover, $\mathrm{Ran}(1 + \mu B) \supseteq \mathrm{Dom}\, L_0$ follows as a consequence of [1], Prop. 11.2.6., i.e. if $f \in \mathrm{Dom}\, L_0$ and u is the solution of $u + \mu L\beta(u) = L_0 f$, then $w := f - \mu v \in \mathrm{Dom}\, L_0$ is the solution of $w + \mu Bw = f$.

Let us give some converse result of Theorem 2. The statement we choose is motivated by our next application below to the problems (P_2)-(P_2^*) of the introduction. A different statement is given in [8]-1.

<u>THEOREM 3:</u> Let H be a Hilbert space. Let $N:H \to [0, +\infty]$ be convex, ℓ.s.c. proper and positive homogeneous of degree two. Let L be a bounded linear operator on H which is invertible and monotone. Let $\phi:H \to R$ be convex continuous with $\partial\phi$ Lipschitz on H. If $\partial\phi \cdot L$ is N-accretive, then $L^t\ \partial\phi$ is N^*-accretive.

<u>PROOF:</u> Let $\varepsilon > 0$, $N^\varepsilon(x) = N(x) + (\varepsilon/2)\|x\|_H^2$. Since Dom L = H and L is maximal monotone, for $x, \hat{x} \in H$, $\varepsilon > 0$, $\mu > 0$, the equation:

$$Lu_{\mu,\varepsilon} + \mu\, \partial N^\varepsilon(u_{\mu,\varepsilon}) \ni L(x - \hat{x}) \tag{3}$$

is solvable in H. Fix x, $\hat{x} \in H$. Let $u_{\mu,\varepsilon}$ be its solution, let $\bar{u}_{\mu,\varepsilon} = L^{-1}u_{\mu,\varepsilon}$ and let $w_{\mu,\varepsilon} := x - \hat{x} - u_{\mu,\varepsilon} - \varepsilon\mu\bar{u}_{\mu,\varepsilon}$. For simplicity of notation, we delete the subscripts μ,ε. Let $u(t) := \hat{x} + (t/\mu)w + u$, $v(t) := \hat{x} + (t/\mu)w$. Notice that: $\dot{u}(t) = \dot{v}(t) = w/\mu \in L^{-1}\ \partial N(u(t) - v(t))$. Then:

$$\frac{d}{dt}[(\phi\cdot L)(u(t)) - (\phi\cdot L)(v(t))] = \langle\partial\phi(Lu(t)) - \partial\phi(Lv(t)), \frac{Lw}{\mu}\rangle \geqq 0. \quad (4)$$

Integrating (4) from 0 to μ and computing $u(\mu)$, $v(\mu)$, $u(0)$, $v(0)$ we get:

$$(\phi\cdot L)(x - u - \varepsilon\mu\bar{u}) + (\phi\cdot L)(\hat{x}+u) \leqq (\phi\cdot L)(x-\varepsilon\mu\bar{u}) + (\phi\cdot L)(\hat{x}). \quad (5)$$

Using the definition of subgradient

$$(\phi\cdot L)(x - u - \varepsilon\mu\bar{u}) \geqq (\phi\cdot L)(x - \varepsilon\mu\bar{u}) + \langle\partial\phi(Lx-\varepsilon\mu L\bar{u}), -Lu\rangle, \quad (6)$$

$$(\phi\cdot L)(\hat{x} + u) \geqq (\phi\cdot L)(\hat{x}) + \langle\partial\phi(Lx),Lu\rangle. \quad (7)$$

Adding (6) and (7) and using (5) we get:

$$\langle\partial\phi(Lx - \varepsilon\mu u) - \partial\phi(L\hat{x}),Lu\rangle \geqq 0. \quad (8)$$

Multiply (3) by u, use $(Lu,u) \geqq 0$ to get $\varepsilon\mu\|u\|_H^2 \leqq \|Lx - L\hat{x}\|_H$. Then, for $\varepsilon > 0$ fixed, $\{\mu u: \mu > 0\}$ is bounded in H. Thus, $Lu \to 0$ in H as $\mu \to \infty$. Thus, from $\mu u \in \partial(N^{\varepsilon})^*(Lx - L\hat{x} - Lu)$ it follows that $\mu u \to \omega_{\varepsilon} := \partial(N^{\varepsilon})^*(Lx-L\hat{x})$ in H as $\mu \to \infty$. Multiply (8) by μ and let $\mu \to \infty$. Then $\langle L^t\ \partial\phi(Lx-\varepsilon\omega_{\varepsilon}) - L^t\ \partial\phi(L\hat{x}),\omega_{\varepsilon}\rangle \geqq 0$.

Let y, $\hat{y} \in H$. Let $x := L^{-1}y$, $\hat{x} := L^{-1}\hat{y}$. Let $z(\varepsilon) := \langle L^t\ \partial\phi(y - \varepsilon\omega_{\varepsilon}) - L^t\ \partial\phi(y),\omega_{\varepsilon}\rangle$. Then (8) can be written as:

$$\langle L^t\partial\phi(y) - L^t\partial\phi(\hat{y}),\omega_{\varepsilon}\rangle + z(\varepsilon) \geqq 0. \quad (9)$$

Since $\partial\phi$ is Lipschitz, $|z(\varepsilon)| \leqq c\ \varepsilon\|\omega_{\varepsilon}\|_H^2$ for some $c > 0$. The, from (9),

$$N_{\varepsilon}^*(y-\hat{y}) \leqq N_{\varepsilon}^*(y-\hat{y} + \lambda L^t\ \partial\phi(y) - \lambda L^t\ \partial\phi(\hat{y})) + c\ \varepsilon\|\omega_{\varepsilon}\|_H^2,\ \lambda > 0.$$

If $y-\hat{y} \in \mathrm{Dom}\ N^*$, then $\varepsilon\|\omega_{\varepsilon}\|_H^2 \to 0$ as $\varepsilon \to 0$ ([8],1, Theorem 2.2) and it follows that

$$N^*(y-\hat{y}) \leqq N^*(y-\hat{y} + \lambda L^t\partial\phi(y) - \lambda L^t\partial\phi(\hat{y})) \text{ holds for all } \lambda > 0. \qquad (10)$$

If $y, \hat{y} \in H$, then (10) holds for $\lambda > 0$ sufficiently small (as in [8], 1, Theorem 2.2) and the theorem is proved. □

To avoid unnecessary technical complications in what follows, let us suppose that (Ω,Σ,μ) is a finite measure space. Let $H = L^2(\Omega)$, $\rho(f) = \|f\|_{L^1}$ and $\hat{\rho}(f) = \|f\|_{L^\infty}$. Let $A: H \to H$ be a continuous operator such that for all $u \in L^\infty(\Omega)$:

(a1) A is Fréchet differentiable at u in H (and in $L^\infty(\Omega)$) with derivative T_u^A $(T_u^A|_{L^\infty(\Omega)})$.

(a2) T_u^A is a symmetric coercive operator on H (thus invertible).

(a3) For $x,y \in L^\infty(\Omega)$, $A|_{[x,y]} : [x,y] \to L^\infty(\Omega)$ is absolutely continuous where $[x,y]$ is the segment joining x to y.

Notice that (a1) implies (a3). Then, the following theorem holds:

<u>THEOREM 4</u>: Let (Ω,Σ,μ), H, ρ, $\hat{\rho}$ be as above. Let A be as above satisfying (a1), (a2), (a3). Let $\phi: H \to \mathbf{R}$ be convex, continuous such that $\partial\phi(L^\infty(\Omega)) \subseteq L^\infty(\Omega)$ and $\partial\phi|_{L^\infty(\Omega)} : L^\infty(\Omega) \to L^\infty(\Omega)$ is Lipschitz. If for any $t > 0$ and $w \in H$, $\partial\phi(tA(\cdot) + w)$ is ρ-accretive, then $A\,\partial\phi$ is accretive as an operator in $L^\infty(\Omega)$.

The proof is a special case of the results in [8]-[11] and it is based in the three following steps: (1) Linearization of A: then $\partial\phi\cdot T_u^A$ is ρ-accretive for all $u \in L^\infty(\Omega)$; (2) Use of Theorem 3: then $T_u^A\partial\phi$ is $\hat{\rho}$-accretive; (3) Integration of $\{T_u^A\partial\phi: u \in L^\infty(\Omega)\}$ to get that $A\,\partial\phi$ is accretive in $L^\infty(\Omega)$.

<u>APPLICATION</u>: Let Ω be a bounded domain in $\mathbf{R}^n$ with smooth boundary Γ. We

want to show that (P_2^*) is well-posed in $L^\infty(\Omega)$ as a consequence of the well-posedness of (P_2) in $L^1(\Omega)$. Let $H = L^2(\Omega)$, $\phi: H \to [0,+\infty]$ be given by $\Phi(f) = (1/p)\int_\Omega |\nabla f|^p$ if $f \in W_0^{1,p}(\Omega) \cap H$, $+\infty$ otherwise, where $p > 1$. Then $\partial\Phi(u) = -\Delta_p u$, $u \in \mathrm{Dom}\,\partial\Phi$. Let Φ_λ be the Moreau-Yosida regularization of Φ. Then $\partial\Phi$, $\partial\Phi_\lambda$ are completely accretive (thus accretive in all $L^q(\Omega)$, $1 \leqq q \leqq \infty$) ([3], [13]). Thus Φ_λ satisfies the assumptions for Φ in Theorem 4. Let $\beta \in W^{1,\infty}(R) \cap C^2(R)$ with $\beta' \geqq c > 0$ for some $c > 0$ (more general β could be approached by β_n satisfying these assumptions). Then $T_\beta: H \to H$, $T_\beta(u) = \beta(u)$, $u \in H$, satisfies (a1), (a2), (a3) above. Take $A = T_\beta$ in Theorem 4. A simple adaptation of the arguments in [1], Prop. 2.5 shows that $\partial\Phi_\lambda(t\beta(\cdot) + w)$ is ρ-accretive, $\forall t > 0$, $\forall w \in H$. By Theorem 4, $\beta(\partial\Phi_\lambda(\cdot))$ is accretive in $L^\infty(\Omega)$. Letting $\lambda \to 0$ we get that if

$$B := \{[u,v] \in L^\infty \times L^\infty(\Omega): u \in \mathrm{Dom}(\partial\Phi|_{L^\infty(\Omega)}),\ v = \beta(\partial\Phi(u)\},$$

then B is accretive in $L^\infty(\Omega)$. We have to check that $\mathrm{Ran}(1 + \lambda B) \supseteq \mathrm{Dom}\,B$. Let $T: H \to H$ be given by $Tu = -\beta^{-1}(f-u)$ where $f \in H$ is fixed. T is monotone in H with domain H. Since $A_p := -\Delta_p$ + Dirichlet boundary conditions is also maximal monotone, then $T + A_p$ is maximal monotone. Since T is a subgradient: $\mathrm{Int\,Ran}(T + A_p) = \mathrm{Int}(\mathrm{Ran}\,T + \mathrm{Ran}\,A_p) = H$ ([5], Theorem 4). In particular, $0 \in \mathrm{Ran}(T + A_p)$. Let $u \in H$ be the solution of

$$-\beta^{-1}(f-u) - \Delta_p u = 0, \quad u|_{\partial\Omega} = 0.$$

Then it is easy to see that $u \in L^\infty(\Omega)$ if $f \in L^\infty(\Omega)$. Thus, $-\Delta_p u \in L^\infty(\Omega)$ and $u + \beta(-\Delta_p u) = f$, $u|_{\partial\Omega} = 0$ with u, $\Delta_p u \in L^\infty(\Omega)$.

Thus $u + Bu = f$. Therefore, $\mathrm{Ran}(1 + B) = L^\infty(\Omega) \supseteq \mathrm{Dom}\,B$. Similarly, $\mathrm{Ran}(1 + \ B) \supseteq \mathrm{Dom}\,B$ for all $\lambda > 0$. Thus B is m-accretive in $L^\infty(\Omega)$ and the Crandall-Liggett theorem guarantees that (P_2^*) is well-posed in $L^\infty(\Omega)$.

ACKNOWLEDGEMENT: I would like to thank Philippe Bénilan who encouraged me to consider this problem and openly shared many discussions with me. I would also like to thank M.G. Crandall for a stimulating discussion.

References

[1] Ph. Bénilan, Thesis, Univ. Paris XI, Orsay, 1972.

[2] Ph. Bénilan, M.G. Crandall and A. Pazy, Book in preparation.

[3] Ph. Bénilan and C. Picard, Quelques aspects non linéaires du principe du maximum. Séminaire de Théorie du Potentiel. Paris, No. 4, Springer-Verlag LNM Vol 713, 1979, pp. 1-37.

[4] H. Brézis, Opérateurs maximaux monotones. North-Holland Math. Stud., Vol. 5, 1973.

[5] H. Brézis, A. Haraux, Image d'une somme d'opérateurs monotones et applications, Israel J. Math. 23 (1976), 165-186.

[6] H. Brézis and W. Strauss, Semilinear elliptic equations in L^1, J. Math. Soc. Japan 25 (1973), 15-26.

[7] B. Calvert and C. Picard, Opérateurs accrétifs et ϕ-accrétifs dans un espace de Banach, Hiroshima Math. J. 8 (1978), 11-30.

[8] V. Caselles, Duality and nonlinear equations governed by accretive operators, I, II. Preprints.

[9] M.G. Crandall, The semigroup approach to first-order quasilinear equations in several space variables, Israel J. Math. 12 (1972), 108-132.

[10] M.G. Crandall, L.C. Evans and P.L. Lions, Some properties of viscosity solutions of Hamilton-Jacobi equations, Trans. Amer. Math. Soc. 282 (1984), 487-502.

[11] M.G. Crandall and P.L. Lions, Viscosity solutions of Hamilton-Jacobi equations, Trans. Amer. Math. Soc. 277 (1983), 1-42.

[12] M.G. Crandall and M. Pierre, Regularity effects for $u_t + A\phi(u) = 0$ in L^1, J. Funct. Anal. 45 (1982), 194-212.

[13] J.I. Diaz, Nonlinear PDE's and free boundaries, Vol. 1: Elliptic equations, Pitman Res. Notes in Math. Vol. 106, 1985.

[14] K. Ha, Sur des semigroupes non linéaires dans les espaces $L^\infty(\Omega)$, These 3eme cycle. Univ. Paris VI. 1976.

[15] S.N. Kruzkov, First order quasilinear equations in several independent variables, Math. USSR-Sb. 10 (1970), 217-243.

V. Caselles
Université de Franche Comté
Département de Mathématiques
Route de Gray
25030 Besancon Cedex
France

F. CONRAD, D. HILHORST AND T.I. SEIDMAN

On a reaction-diffusion equation with a moving boundary

1. INTRODUCTION

In this article we study the well-posedness of a moving boundary problem arising in a dissolution-growth process.

Consider a solid-liquid system consisting of a small spherical grain exchanging matter with a liquid solution by a dissolution-growth phenomenon. Let $C = C(r,t)$ denote the concentration in the liquid phase and let $R = R(t)$ denote the radius of the grain. Then C and R satisfy the following rescaled equations:

$$
(P_0)\quad \begin{cases}
C_t = \frac{1}{r^2}\frac{\partial}{\partial r} r^2 \frac{\partial C}{\partial r} + F(C), & t > 0,\ r \in (R(t),1), \quad (1.1)\\[2mm]
\frac{\partial C}{\partial r} = (1-C)H(R,C), & t > 0,\ r = R(t), \quad (1.2)\\[2mm]
C(1,t) = \gamma(t), & t > 0, \quad (1.3)\\[2mm]
\dot{R}(t) = H(R(t),C(R(t),t)), & t > 0, \quad (1.4)\\[2mm]
R(0) = R_0 \in (0,1), & \quad (1.5)\\[2mm]
C(r,0) = C_0(r), & r \in R_0,1), \quad (1.6)
\end{cases}
$$

where

$\gamma \in H^1(\mathbb{R}^+)$ and $0 \leqq \gamma \leqq 1$,

$C_0 \in L^\infty(R_0,1)$ with $0 \leqq C_0 \leqq 1$ a.e. in $(R_0,1)$,

$F \in C(\mathbb{R})$ is nonincreasing and $F(0) \geqq 0 \geqq F(1)$,

$H \in C^{0,1}([0,1] \times \mathbb{R})$ satisfies $H(R,0) \leqq 0$.

We refer to Kaleydjian and Cournil [4] for a physical derivation of

Problem P_0 and to Conrad and Yebari [3] for a thorough study of the stationary problem for H having a specific form of physical interest.

Our method of proof is based on the study of the map T defined as follows:

(i) for a function ω given on a certain time interval $(0,\tau_0)$, solve the initial value problem

$$\begin{cases} \dot{R} = H(R,\omega) \text{ on } (0,\tau_0), \\ R(0) = R_0; \end{cases}$$

(ii) solve the auxiliary problem (1.1), (1.2), (1.3), (1.6) which we denote by(P)for $t \in (0,\tau_0)$;

(iii) set $T\omega := C(R(t),t)$.

The organization of the paper is as follows:

In Section 2 we indicate our choice of τ_0, perform a coordinate transformation converting problem (P) into a problem on a fixed domain and give some estimates on differences of solutions.

In Section 3 we show that the map T is a strict contraction when defined on a suitable exponentially weighted space so that problem (P_0) has a unique solution for $t \in [0,\tau_0]$.

Finally, we give a characterization of the maximum time of existence of the solution in Section 4.

We refer to Conrad, Hilhorst and Seidman [2] for the detailed proofs of the results stated in this paper, as well as for results under slightly weaker assumptions on the data.

2. PRELIMINARIES. STUDY OF THE AUXILIARY PROBLEM P

We set for $(r,s) \in [0,1] \times \mathbb{R}$

$$\tilde{H}_0(r,s) = \begin{cases} H(r,0) & \text{if } s \leq 0, \\ H(r,1) & \text{if } s \geq 1, \\ H(r,s) & \text{otherwise,} \end{cases}$$

and

$$H_0(r,s) = \chi(s)\tilde{H}_0(r,s),$$

where $\chi \in C^\infty(\mathbb{R})$, $\chi(s) = 1$ for $s \in [0,1]$, $\chi(s) = 0$ for $s \leqq -2$ and $s \geqq 3$, $0 \leqq \chi'(s) \leqq 1$ for $s \in [-2,0]$ and $-1 \leqq \chi'(s) \leqq 0$ for $s \in [1,3]$,

and

$$I(s) = \begin{cases} (1-s)/2 & \text{if } s > 1, \\ 1-s & \text{if } s \leqq 1. \end{cases}$$

Further we replace H by H_0 and the Neumann boundary condition (1.2) by

$$\frac{\partial C}{\partial r} = I(C)H_0(R,C), \quad t > 0, \quad r = R(t),$$

in problem (P_0) and denote by $(\hat{P}_0)$ the corresponding problem. Clearly, any solution (C,R) of $(\hat{P}_0)$ such that $0 \leqq C \leqq 1$ is also a solution of problem (P_0). Similarly, we denote by $(\hat{P})$ the auxiliary problem corresponding to problem $(\hat{P}_0)$, by Λ_0 the sup norm of H_0 and by Λ_1 its Lipschitz constant and we set

$$\tau_0 = \min\{R_0/2\Lambda_0, (1-R_0)/2\Lambda_0\}.$$

Since R satisfies

$$\text{(IVP)} \quad \dot{R} = H_0(R,\omega) \text{ on } (0,\tau_0), \ R(0) = R_0,$$

we deduce that

$$R_0/2 \leqq R(t) \leqq (R_0+1)/2$$

for all $t \in [0,\tau_0]$.

Next we transform problem $(\hat{P})$ into a problem on a fixed domain. We set

$$\rho(t) = 1/(1-R(t)), \quad y = \rho(t)(1-r),$$

and

$$u(y,t) = C(r,t).$$

Problem $(\hat{P})$ becomes

$$(\tilde{P}) \quad \begin{cases} u_t = \rho^2 u_{yy} - \Psi(y,t)u_y + F(u) & \text{in } Q_0 := (0,1) \times (0,\tau_0), \\ \rho^2 u_y(1,t) = \phi(t,u(1,t)) & t \in (0,\tau_0), \\ u(0,t) = \gamma(t) & t \in (0,\tau_0), \\ u(y,0) = u_0(y) := C_0(1-y+R_0 y) & y \in (0,1), \end{cases}$$

where

$$\Psi(y,t) = \rho(t)\left(\frac{2\rho(t)}{\rho(t)-y} + \dot{R}(t)y\right) \quad \text{for } (y,t) \in \bar{Q}_0$$

and

$$\phi(t,s) = -\rho(t)I(s)H_0(R(t),s) \quad \text{for } (t,s) \in [0,\tau_0] \times \mathbb{R}.$$

We remark, in particular, that since $R_0/2 \leqq R(t) \leqq (R_0+1)/2$ there exist constants δ and α which depend only on Λ_0 and R_0 such that $0 < \delta \leqq \rho^2 \leqq \alpha$, $\|\Psi\|_{L^\infty(Q_0)} \leqq \alpha$, $\|\phi\|_{L^\infty((0,\tau_0)\times\mathbb{R})} \leqq \alpha$.

It follows from standard theory that problem $(\tilde{P})$ has a unique solution u in the following sense:

(i) $u - \gamma \in L^2(0,\tau_0;V)$, $u_t \in L^2(0,\tau_0;V')$;

(ii) $u(0) = u_0$;

(iii) u satisfies the integral equation

$$\int_0^{\tau_0} \langle u_t,\zeta\rangle = \iint_{Q_0} \{-\rho^2 u_y \zeta_y + (-\Psi u_y + F(u))\zeta\} + \int_0^{\tau_0} \phi(t,u(1,t))\zeta(1,t)dt \tag{2.1}$$

for all $\zeta \in L^2(0,\tau_0;V)$ where $V = \{v \in H^1(0,1), v(0) = 0\}$, V' denotes the dual space of V and $\langle\cdot,\cdot\rangle$ is the duality product between V and V'.

Furthermore, the norms $\|u-\gamma\|_{L^2(0,\tau_0;V)}$ and $\|u_t\|_{L^2(0,\tau_0;V')}$ depend only on the constants Λ_0 and R_0.

Next we give some estimates for differences of solutions of problem (P).

LEMMA 2.1: Let ω_1 and $\omega_2 \in L^2(0,\tau_0)$ and let R_1 and R_2 be the corresponding solutions of (IVP). We set $\tilde{\omega} = \omega_1-\omega_2$ and $\tilde{R} = R_1 - R_2$. Then there exists $\lambda_0 > 0$ such that for all $\lambda \geqq \lambda_0$

$$\int_0^t e^{-\lambda s}(\dot{\tilde{R}}(s))^2 ds \leqq \tilde{C} \int_0^t e^{-\lambda s}(\tilde{\omega}(s))^2\, ds \tag{2.2}$$

for all $t \in (0,\tau_0]$ where the constant $\tilde{C}$ does not depend on λ.

THEOREM 2.2: Let u_1 and u_2 be the two solutions of problem $(\tilde{P})$ corresponding to R_1 and R_2. Let $\tilde{u} = u_1-u_2$ and $\tilde{R} = R_1-R_2$. For each $\theta > 0$ there exists $\lambda_0 > 0$ such that for each $\lambda \geqq \lambda_0$

$$e^{-\lambda t}\,\|\tilde{u}(t)\|^2 + \frac{\lambda}{2}\int_0^t e^{-\lambda s}\,\|\tilde{u}(s)\|^2 ds + \delta\int_0^t e^{-\lambda s}\,\|\tilde{u}_y(s)\|^2 ds$$

$$\leqq \left(\frac{\tilde{C}}{\lambda} + \theta\right)\int_0^t e^{-\lambda s}(\dot{\tilde{R}}(s))^2 ds + \frac{\tilde{C}}{\theta}\int_0^t e^{-\lambda s}\,\|u_2y(s)\|^2\,\|\tilde{u}(s)\|^2 ds \tag{2.3}$$

for all $t \in (0,\tau_0]$ where $\|\cdot\|$ denotes the norm in $L^2(0,1)$ and the constant $\tilde{C}$ does not depend on λ and θ.

In order to prove Theorem 2.2 one multiplies the difference of the equations for u_1 and u_2 by $e^{-\lambda t}(u_1-u_2)$ and integrates by parts. Next we remark that using Gronwall's inequality one can deduce from (2.3) and (2.2) the simpler result

COROLLARY 2.3: Let ω_1 and $\omega_2 \in L^2(0,\tau_0)$, let R_1 and R_2 be the corresponding solutions of (IVP) and let u_1 and u_2 be the two solutions of problem $(\tilde{P})$ corresponding to R_1 and R_2. We set $\tilde{\omega} = \omega_1-\omega_2$ and $\tilde{u} = u_1-u_2$. There exists $\lambda_0 > 0$ such that for all $\lambda \geqq \lambda_0$ there exists a constant $\tilde{C}_1 > 0$ such that

$$e^{-\lambda t} \|\tilde{u}(t)\|^2 + \frac{\lambda}{2}\int_0^t e^{-\lambda s} \|\tilde{u}(s)\|^2 ds + \delta\int_0^t e^{-\lambda s} \|\tilde{u}_y(s)\|^2 ds$$

$$\leq \tilde{C}_1 \int_0^t e^{-\lambda s}(\tilde{\omega}(s))^2 ds. \qquad (2.4)$$

However, the constant $\tilde{C}_1$ in (2.4) is not a priori very small so that inequality (2.4) will not be as useful as inequality (2.3) in proving that T defines a strict contraction. Note that in (2.3) the constant $(\tilde{C}/\lambda + \theta)$ can be made arbitrarily small. However, we still have to control the second term on the right-hand side of (2.3). To that purpose we multiply (2.3) by $e^{-\mu t}$ where μ is a positive constant and obtain the following result:

THEOREM 2.4: For all $\theta > 0$ there exists $\lambda_0 > 0$ such that for all $\lambda \geq \lambda_0$ and all $\mu > 0$

$$e^{-(\lambda+\mu)t} \|\tilde{u}(t)\|^2 + e^{-\mu t}\int_0^t e^{-\lambda s}(\delta \|\tilde{u}_y(s)\|^2 + \frac{\lambda}{2}\|\tilde{u}(s)\|^2)ds$$

$$\leq (\frac{\tilde{C}}{\lambda} + \theta)e^{-\mu t}\int_0^t e^{-\lambda s}(\dot{\tilde{R}}(s))^2 ds$$

$$+ \frac{\tilde{C}}{\theta} \sup_{s\in[0,\tau_0]} (e^{-(\lambda+\mu)s} \|\tilde{u}(s)\|^2) \int_0^t e^{-\mu(t-s)} \|u_{2y}(s)\|^2 ds,$$

where $\tilde{C}$ does not depend on λ, μ and θ.

Theorem 2.4 together with Lemma 2.5 below are the main ingredients for the proof of existence and uniqueness of the solution of problem $(\hat{P}_0)$ in Section 3.

LEMMA 2.5: Let Σ be a compact subset of $L^2(0,\tau_0)$. Then for all $\sigma \in \Sigma$ and all $t \in (0,\tau_0]$

$$\lim_{\mu\to 0} \int_0^t e^{-\mu(t-s)} \sigma^2(s)\, ds = 0$$

uniformly with respect to $(t,\sigma) \in (0,\tau_0] \times \Sigma$.

3. EXISTENCE AND UNIQUENESS OF THE SOLUTION OF PROBLEM (P_0) FOR $t \in [0,\tau_0]$

We start with some compactness results.

LEMMA 3.1: Let $\omega \in L^2(0,\tau_0)$. The function $T\omega$ belongs to a compact subset C of $L^2(0,\tau_0)$.

PROOF: Set $v = u-\gamma$. Then $T\omega = v(1,\cdot) + \gamma$. Since v is bounded in $L^2(0,\tau_0;V)$ and v_t is bounded in $L^2(0,\tau_0;V')$ and since $V \subset C([0,1]) \subset V'$ where both injections are continuous and the first injection is compact, it follows from Aubin's compactness theorem [1] that v and thus also u belong to a compact subset of $L^2(0,\tau_0;C([0,1]))$ which in turn implies that $u(1,\cdot)$ belongs to a compact subset C of $L^2(0,\tau_0)$. Note that C is invariant under the map T. □

From now on we restrict the map T to C. The following result is then a straightforward consequence of (2.4).

LEMMA 3.2: Let $\omega \in C$. Then $\|u_y(\cdot)\|_{L^2(0,1)}$ belongs to a compact subset of $L^2(0,\tau_0)$.

Next we denote by $\|\cdot\|_*$ the following norm on $L^2(0,\tau_0)$

$$\|v\|_* = \sup_{t\in[0,\tau_0]} (e^{-\mu t}\int_0^t e^{-\lambda s}\, v^2(s)ds)^{\frac{1}{2}}.$$

Using the results of the Lemmas 3.2, 2.5 and 2.4 one can deduce the main result of this paper.

THEOREM 3.3: The map T is a strict contraction from C to C with the exponentially weighted norm $\|\cdot\|_*$.

Next we denote by $(\tilde{P}_0)$ the problem consisting of problem $(\tilde{P})$ together with the initial value problem (IVP) with $\omega = u(1,\cdot)$. By Theorem 3.3, T has a unique fixed point $\omega \in C$ to which corresponds a unique solution (u,R) of problem $(\tilde{P}_0)$ such that $u-\gamma \in L^2(0,\tau_0;V)$, $u_t \in L^2(0,\tau_0;V')$ and $R \in C^{0,1}([0,\tau_0])$. Furthermore, one can show that $u - \gamma \in H^1((0,1)\times(\eta,\tau_0)) \cap L^\infty(\eta,\tau_0;V)$ and that $R \in C^1([\eta,\tau_0])$ for each $\eta \in (0,\tau_0)$.

Finally, we return to the original variables. We set $r = 1-y/\rho$ and $C(r,t) = u(y,t)$. Then (C,R) is the unique solution of problem $(\hat{P}_0)$.

THEOREM 3.4: The solution (C,R) of problem $(\hat{P}_0)$ satisfies $0 \leqq C \leqq 1$ so that (C,R) is a solution of problem (P_0).

In order to prove Theorem 3.4 one uses arguments based on the maximum principle.

4. THE MAXIMUM TIME OF EXISTENCE

In this section we use arguments similar to those used to prove the global existence of solutions of ordinary differential equations to characterize the maximum time of existence of the solution of problem (P_0). Let A be the set of triples (C,R,T) such that $T > 0$ and

(i) $R \in C^{0,1}([0,T])$ and $0 < R(t) < 1$, $t \in [0,T]$;

(ii) (C,R) is a solution of problem (P_0) for $t \in [0,T]$ such that $0 \leqq C \leqq 1$.

One can easily prove the following result:

LEMMA 4.1: Let (C_1,R_1,T_1) and $(C_2,R_2,T_2) \in A$ and set $T = \min(T_1,T_2)$. Then $R_1 = R_2$ and $C_1 = C_2$ for $t \in [0,T]$.

Consequently A is ordered by T. We define

$$T^* = \sup\,(T > 0 \text{ such that there exists } (C,R,T) \in A).$$

Next we give the main result of this section.

THEOREM 4.2: Suppose that T^* is finite. Then either $R(t) \to 0$ or $R(t) \to 1$ as $t \uparrow T^*$.

PROOF: For each $t \in [0,T^*)$ we have $0 < R(t) < 1$. Suppose that neither $R(t) \to 0$ nor $R(t) \to 1$ as $t \uparrow T^*$. Then there exists $T_k \uparrow T^*$ such that $\varepsilon \leqq R(T_k) \leqq L-\varepsilon$ for some $\varepsilon \in (0,1/4)$. Set $\tau_1 = \varepsilon/2\Lambda_0$ and let k be such that $T_k \geq T^*-\tau_1/2$. Then by the local theory developed in the previous sections there exists a solution of problem (P_0) on $[T_k, T_k + \tau_1]$ and since $T_k + \tau_1 > T^*$ this contradicts the definition of T^*. □

Finally, we give some further properties of T^*.

LEMMA 4.3: Let H be nondecreasing in its second variable. Then

(i) if $R(t) \to 0$ as $t \uparrow T^*$, $T^* \geq \int_0^{R_0} - \frac{dr}{H(r,0)}$;

(ii) if $R(t) \to 1$ as $t \uparrow T^*$, $T^* \geq \int_{R_0}^{1} \frac{dr}{(H(r,1))^+}$.

In particular if both the integrals in Lemma 4.3 diverge, then $T^* = \infty$.

LEMMA 4.4: Suppose that H is nondecreasing in both its variables and that $H(1,1) \leq 0$. Then R is nonincreasing and if $T^* < \infty$, $R(t) \downarrow 0$ as $t \uparrow T^*$. If $H(1,1) < 0$, then $T^* < \infty$.

References

[1] J.P. Aubin, Un théorème de compacité, C.R. Acad. Sci. Paris, 265 (1963), 5042-5045.

[2] F. Conrad, D. Hilhorst and T. Seidman, On the well-posedness of a moving boundary problem arising in a dissolution-growth process, in preparation.

[3] F. Conrad and N. Yebari, Multiple solutions in reaction-diffusion systems with a free boundary, SIAM J. Appl. Math. (to appear).

[4] F. Kaleydjian and M. Cournil, Stability of steady states in some solid-liquid systems, Reactivity of Solids, 2 (1986), 1-21.

F. Conrad
Departement de Mathématiques
Université de Nancy II
42, Av.de la Libération
54000 Nancy Cedex
France

D. Hilhorst
Laboratoire d'Analyse numérique
Université de Paris Sud
Bat. 425, 91405 Orsay
France

T.I. Seidman
Department of Mathematics
University of Maryland Baltimore County
Baltimore, MD., 21228
USA

J. GONCERZEWICZ

Porous medium-type equations with irregular boundary data

1. INTRODUCTION

In this article we consider the initial boundary value problem

$$u_t = \phi(u)_{xx} + b(x)\phi(u)_x \quad \text{in } S_T = (0,\infty) \times (0,T],\ T > 0, \tag{1}$$

$$u(x,0) = u_0(x) \quad \text{for } x \in (0,\infty), \tag{2}$$

$$u(0,t) = u_1(t) \quad \text{for } t \in (0,T], \tag{3}$$

under the assumptions

$$\phi \in C^1([0,\infty)) \cap C^{2+\alpha}((0,\infty)) \text{ for some } \alpha \in (0,1] \text{ and}$$
$$\phi(0) = \phi'(0) = 0,\ \phi''(s) > 0 \text{ for } s > 0. \tag{A1(i)}$$

$$\int_0^1 \frac{\phi'(s)}{s}\, ds < \infty, \tag{ii}$$

$$0 < \beta = \inf_{0<s<1} \frac{1}{\phi'(s)} \int_0^s \frac{\phi'(\sigma)}{\sigma}\, d\sigma, \tag{iii}$$

$$b \in C^{1+\alpha}([0,\infty)),\ b \geq 0,\ B_0 = \sup_{0\leq x<\infty} b(x) < \infty \text{ and} \tag{A2}$$

$$B_1 = \sup_{0\leq x<\infty} |b'(x)| < \infty,$$

$$u_0 \in L^\infty(0,\infty),\ \text{ess inf } u_0 \geq 0, \tag{A3}$$

$$u_1 \in L^\infty(0,T),\ \text{ess inf } u_1 \geq 0.$$

Our purpose is to study the existence, uniqueness and regularity properties of (weak) solutions of problem (1)-(3).

Equation (1) includes, as particular cases, the one-dimensioanl porous medium equation

$$u_t = (u^m)_{xx}, \quad m > 1, \tag{4}$$

and

$$u_t = (u^m)_{rr} + \frac{N-1}{r}(u^m)_r, \quad m > 1,$$

if $r > \xi > 0$, which is satisfied by classical radial solutions of $u_t = \Delta(u^m)$, the N-dimensional analogue of (4).

There exists an extensive literature devoted to porous medium-type equations. (We refer to [14] for a comprehensive bibliography or to [5] and [16].) However, most of the existence results concerning initial boundary value problems have been obtained under one of the following sets of assumptions on the data: Either both initial and boundary functions are sufficiently smooth (or at last continuous) and satisfy suitable compatibility conditions [6], [8], [9], [10], [12], [15], or the initial function is in L^1 and the boundary function vanishes [2], [3], [4]. (We mention only results where equations "close" to ours are considered). We get rid of the above restrictions.

It is well known that only solutions of (1) in some generalized sense can be expected. We define the solution of (1)-(3) as follows.

<u>DEFINITION</u>: A function u defined on S_T is said to be a weak solution of problem (1)-(3) with the data u_0, u_1 if:

(i) u is nonnegative, bounded and continuous on S_T;

(ii) u satisfies the identity

$$\iint_{S_T} \{u\zeta_t + \phi(u)[\zeta_{xx} - b(x)\zeta_x - b'(x)\zeta]\}\,dx\,dt$$
$$+ \int_0^\infty u_0(x)\zeta(x,0)\,dx + \int_0^T \phi(u_1(t))\zeta_x(0,t)\,dt = 0 \tag{5}$$

for all nonnegative $\zeta \in C^{2,1}(\bar{S}_T)$ which vanish for $t = T$, $x = 0$ and for large x.

If in (5) the equality is replaced by $\leqq$ ($\geqq$) then we call u a weak supersolution (subsolution) of problem (1)-(3) with the data u_0 and u_1.

2. COMPARISON PRINCIPLE AND UNIQUENESS

In this section, instead of (A1), we only assume that

$\phi : [0,\infty) \to \mathbb{R}$, $\phi(0) = 0$ is locally Lipschitz continuous and increasing.

Without changes we assume (A2) and (A3).

The crucial result of this section is the following theorem.

THEOREM 2.1: Let u be a weak solution of (1)-(3) with the data u_0 and u_1 and let $\bar{u}$ (resp. $\underline{u}$) be a weak supersolution (resp. subsolution) of (1)-(3) with the data $\bar{u}_0$, $\bar{u}_1$ (resp. $\underline{u}_0$, $\underline{u}_1$). For any nonnegative $g \in C_0^\infty(S_T)$ there exist positive constants C_1 and C_2 such that

$$\iint_{S_T} (u - \bar{u})g \, dx \, dt \leqq C_1 \int_0^\infty e^{-(1+B_0)x} (u_0 - \bar{u}_0)^+ \, dx + C_2 \int_0^T (\phi(u_1) - \phi(\bar{u}_1))^+ \, dt \tag{6}$$

(resp.

$$\iint_{S_T} (\underline{u} - u)g \, dx \, dt \leqq C_1 \int_0^\infty e^{-(1+B_0)x} (\underline{u}_0 - u_0)^+ \, dx + C_2 \int_0^T (\phi(\underline{u}_1) - \phi(u_1))^+ \, dt),$$

where B_0 is the constant from (A2) and $(p)^+ = \max(p,0)$.

The proof of Theorem 2.1 follows the duality argument of [13]. Its details can be found in [11].

Here we state some obvious consequences of Theorem 2.1.

COROLLARY 2.1: Let u and $\bar{u}$ (resp. $\underline{u}$) be as in Theorem 2.1. If $u_0 \leqq \bar{u}_0$ (resp. $\underline{u}_0 \leqq u_0$) a.e. in $(0,\infty)$ and $u_1 \leqq \bar{u}_1$ (resp. $\underline{u}_1 \leqq u_1$) a.e. in (0,T) then $u \leqq \bar{u}$ (resp. $\underline{u} \leqq u$) in S_T.

COROLLARY 2.2 (Monotonical Dependence on the Data):

Let u and $\hat{u}$ be weak solutions of (1)-(3) with the data u_0, u_1 and $\hat{u}_0$, $\hat{u}_1$, respectively. If $u_0 \leqq \hat{u}_0$ a.e. in $(0,\infty)$ and $u_1 \leqq \hat{u}_1$ a.e. in (0,T) then $u \leqq \hat{u}$ in S_T.

COROLLARY 2.3: The weak solution of (1)-(3) is unique.

REMARK 2.1: The results of this section remain true for the nonhomogenous equation

$$u_t - \phi(u)_{xx} - b(x)\phi(u)_x = f(x,t)$$

with, e.g. $f \in L^1_{loc}(S_T)$. (An obvious modification of (5) must be made.)

3. EXISTENCE AND REGULARITY

We have

THEOREM 3.1: Let (A1)(i)-(ii), (A2) and (A3) hold. There exists a weak solution u of problem (1)-(3). Moreover:

(i) u is a classical solution of (1) in a neighbourhood of any point $(x_0,t_0) \in S_T$ where $u(x_0,t_0) > 0$;

(ii) the derivative $\phi(u)_x$ exists and is continuous in S_T and, if $\delta > 0$ and $0 < \tau < T$, then

$$|\phi(u)_x(x,t)| \leqq Cu(x,t)$$

for $(x,t) \in S_{\delta\tau} = [\delta,\infty) \times [\tau,T]$, where the constant C depends on δ, τ, ϕ, B_0, B_1, $\|u_0\|_\infty$, $\|u_1\|_\infty$ and does not depend on T.

PROOF: Let $Q_n = (0,n) \times (0,T]$ and $\psi(s) = \int_0^s [\phi'(\sigma)/\sigma]d\sigma$ for $s \geqq 0$. Following the already classical strategy we construct the sequence of functions u_n, $n = 2,3,\ldots$, such that

$$\frac{1}{n} \leqq u_n \leqq \max(\|u_0\|_\infty, \|u_1\|_\infty) + 1,$$

$$u_n \in C^{2,1}(\bar{Q}_n),\ \psi(u_n)_x \in C^{2,1}(Q_n),$$

$$(u_n)_t = \phi(u_n)_{xx} + b(x)\phi(u_n)_x \quad \text{in } \bar{Q}_n,$$

$$u_n(x,0) = u_{0n}(x) \quad \text{for } x \in [0,n],$$

$$u_n(0,t) = u_{1n}(t) \quad \text{for } t \in [0,T],$$

where u_{0n}, u_{1n} are (appropriately chosen) smooth functions such that $u_{0n} \to u_0$ a.e. in $(0,\infty)$ and $u_{1n} \to u_1$ a.e. in $(0,T)$ as $n \to \infty$.

Let $\delta > 0$, $0 < \tau < T$. By the argument similar to [1] we obtain

$$|\psi(u_n)_x| \leqq C_1 \tag{7}$$

on $Q_{n-1} \cap S_{\delta\tau}$, where $C_1 = C_1(\delta,\tau,\phi, B_0,B_1, \|u_0\|_\infty, \|u_1\|_\infty)$. Now it follows from [7] that $\psi(u_n)$ are equicontinuous on any compact subset of S_T. Hence, by a diagonalization argument and the Arzela-Ascoli theorem we may find a subsequence $n' \to \infty$ and a function $w \in C(S_T)$ such that $\psi(u_{n'}) \to w$ uniformly on compact subsets of S_T. Clearly $u = \psi^{-1}(w)$ (here ψ^{-1} denotes the inverse function) is a weak solution of (1)-(3).

The proof of (i) is standard.

To prove (ii) let us observe that $F(x,t) = \lim_{n'\to\infty} \phi(u_{n'})_x(x,t)$ exists and is a continuous function on S_T (for $(x,t) \in S_T$ where $u(x,t) = 0$, this follows from (7) since we have $\phi(u_{n'})_x = u_{n'}\psi(u_{n'})_x$). Now we can derive

$$\phi(u(x,t)) - \phi(u(x',t)) = \int_{x'}^{x} F(\sigma,t)d\sigma$$

for all (x,t), $(x',t) \in S_T$ and the result follows. □

Now we give a result concerning regularity up to the boundary.

THEOREM 3.2: Let (A1)-(A3) hold and let u be the weak solution of (1)-(3). If $x_0 \in (0,\infty)$ and u_0 is continuous at x_0, then

$$\lim_{\substack{(x,t)\to(x_0,0)\\(x,t)\in S_T}} u(x,t) = u_0(x_0), \tag{8}$$

if $t_0 \in (0,T]$ and u_1 is continuous at t_0 then

$$\lim_{\substack{(x,t)\to(0,t_0)\\(x,t)\in S_T}} u(x,t) = u_1(t_0). \tag{9}$$

SKETCH OF THE PROOF:

First step. We assume that both u_0 and u_1 are upper semicontinuous. In this case we are able to construct a decreasing sequence of classical solutions u_n of (1) which converges to u, and adapt the standard barrier method to show (8) and (9). The critical step in verifying (9) when $u_1(t_0) > 0$ is to find a neighbourhood N of $(0,t_0)$, $N \subset \bar{S}_T$ and $\mu > 0$ such that

$$u_n \geqq \mu$$

in N, uniformly with respect to n. We do this by comparing u_n with the classical subsolution u_* of (1) of the form

$$u_*(x,t) = \Psi(\omega(x,t;\tau))$$

in $D_\tau = \{(x,t): 0 < x < \xi + \log^{1/2}(t+\tau),\ t_1 < t \leq t_2\}$ for suitably chosen $t_1, t_2 \in [0,T]$ and for τ large enough, where

$$\xi = -\log^{1/2}(t_1+\tau),\ \omega(x,t;\tau) = [4(t+\tau)]^{-1}[1 - (x-\xi)^2 \log^{-1}(t+\tau)]$$

and Ψ is the inverse function to ψ. In the proof we need our assumption (A1)(iii).

Second step. For general u_0 and u_1 the proof consists in finding, for any fixed $\varepsilon > 0$, the functions $\underline{u}$ and $\bar{u}$, $\underline{u} \leq u \leq \bar{u}$ in S_T, both continuous at the examined boundary point P and such that

$$-\varepsilon + u_0(P) < \underline{u}(P), \quad \bar{u}(P) < u_0(P) + \varepsilon,$$

or, respectively,

$$-\varepsilon + u_1(P) < \underline{u}(P), \quad \bar{u}(P) < u_1(P) + \varepsilon.$$

As $\underline{u}$ and $\bar{u}$ we pick the solutions of (1)-(3) with appropriately chosen, upper semicontinuous initial and boundary functions. For the details of the proof see [11].

REMARK 3.1: If u_0 and u_1 are continuous at zero and $u_0(0) = u_1(0) = 1$, then one can show that $\lim_{\substack{(x,t)\to(0,0) \\ (x,t)\in S_T}} u(x,t) = 1$.

Our final result in this section is the following theorem.

THEOREM 3.3: Let (A1)-(A3) hold and let u be the weak solution of (1)-(3). If $\psi(u_0)$ is locally Lipschitz continuous on $(0,\infty)$, and $\phi(u_1)$ is absolutely continuous on $[0,T]$, then:

(i) for any $\delta > 0$

$$\int_0^\delta \int_0^T [\phi(u)_x]^2 \, dx \, dt < \infty \; ;$$

(ii) u satisfies the identity

$$\iint_{S_T} [\phi(u)_x(\zeta_x - b(x)\zeta) - u\zeta_t] \, dx \, dt = \int_0^\infty u_0(x)\zeta(x,0) \, dx$$

for all $\zeta \in C^1(\bar{S}_T)$ which vanish for $x = 0$, for large x and for $t = T$.

The proof follows the ideas of [15] (see also [8]), and we refer to [11] for the details.

REMARK 3.2: Statement (ii) of Theorem 3.3 is usually taken as a part of the definition of the solution of the problem (see, e.g. [8], [15]). Let us point out however that we do not require the compatibility condition $u_0(0) = u_1(0)$.

References

[1] Aronson, D.G., Regularity properties of flows through porous media, SIAM J. Appl. Math. 17 (1969), 461-467.

[2] Aronson, D.G., Crandall, M.G. and Peletier, L.A., Stabilization of solutions of a degenerate nonlinear diffusion problem, Nonlinear Anal. TMA, 6 (1982), 1001-1022.

[3] Bénilan, Ph., A strong L^p regularity for solutions of the porous media equation, in: Contributions to Nonlinear Partial Differential Equations, eds. C. Bardos et al., Pitman, 1983, pp. 39-58.

[4] Bénilan, Ph. and Toure, H., Sur l'equation générale $u_t = \phi(u)_{xx} - \psi(u)_x + v$, C. R. Acad. Sci. Paris 299 (1984), 919-922.

[5] Bertsch, M., and Peletier, L.A. Porous media type equations: An overview, Mathematical Institute University of Leiden, Preprint, No. 7, 1983.

[6] Diaz, J.I. and Kersner, R. On a nonlinear parabolic equation in infiltration or evaporation through a porous medium, J. Diff. Eq. 69 (1987), 368-403.

[7] Gilding, B.H., Hölder continuity of solutions of parabolic equations, J. London Math. Soc. 13 (1976), 103-106.

[8] Gilding, B.H., A nonlinear degenerate parabolic equation, Ann. Scuola Norm. Sup. Pisa, 4 (1977), 393-432.

[9] Gilding, B.H., Improved theory for a nonlinear degenerate parabolic equation, Twente University of Technology Department of Applied Mathematics, Memorandum 567, 1986.

[10] Goncerzewicz, J., On the boundary value problem arising in the radially symmetrical filtration of fluid, Zastosow. Matem. 17 (1983), 631-643.

[11] Goncerzewicz, J., On an initial boundary value problem for a certain class of degenerate parabolic equations. Preprint.

[12] Kersner, R., Some properties of generalized solutions of quasi-linear degenerate parabolic equations, Acta Math. Acad. Sci. Hungaricae 32 (1978), 301-330.

[13] Kalashnikov, A.S., The Cauchy problem in a class of growing functions for equations of unsteady filtration type, Vestnik Mosk. Univ. Ser. VI Mat. Mech. 6 (1963), 17-27.

[14] Kalashnikov, A.S., Some questions of the qualitative theory of second-order nonlinear degenerate parabolic equations, Uspekhi Mat. Nauk 42 (1987), 135-176.

[15] Oleinik, O.A., Kalashnikov, A.S., and Chzhou Yui-Lin, The Cauchy problem and boundary problems for equations of the type of unsteady filtration, Izv. Akad. Nauk SSSR Ser. Mat. 22 (1958), 667-704.

[16] Peletier, L.A., The porous media equation, in: Application of Nonlinear Analysis in the Physical Sciences, eds. H. Amann, N. Bazley and K. Kirchgassner, Pitman (1981), pp. 229-241.

J. Goncerzewicz
Institute of Mathematics
University of Wroclaw
Pl Grunwaldzki 2/4
50-384 Wroclaw
Poland

N. KENMOCHI AND M. KUBO

Periodic behaviour of solutions to a parabolic-elliptic problem

1. INTRODUCTION AND STATEMENTS OF RESULTS

This paper is concerned with the periodic behaviour of solutions to parabolic-elliptic problems with boundary conditions of mixed type (Dirichlet-Neumann-Signorini type).

Let $\Omega \subset R^N$ ($N \geqq 1$) be an open bounded set with smooth boundary Γ. Assume that Γ admits the decomposition $\Gamma = \Gamma_D \cup \Gamma_N \cup \Gamma_U$, where Γ_i, $i = D, N, U$, are mutually disjoint measurable subsets of Γ. Let ρ: $R \to R$ be a non-decreasing Lipschitz continuous function. The following system is then considered on the time interval $J = R$ or R_+:

$$
(P)\quad \begin{cases} \rho(v)_t - \Delta v = f & \text{in } J \times \Omega, \\ v = g_D & \text{on } J \times \Gamma_D, \\ \partial_\nu v = q_N & \text{on } J \times \Gamma_N, \\ v \leqq g_U,\ \partial_\nu v \leqq q_U,\ (v-g_U)(\partial_\nu v - q_U) = 0 & \text{on } J \times \Gamma_U. \end{cases}
$$

Here ∂_ν stands for the outward normal derivative on Γ, g_D, g_U, q_N are given boundary data and f is a given function on $R \times \Omega$. This kind of problem arises, for instance, from modelling the processes of flows with saturation-unsaturation in porous media. We refer to [1]-[10], [13]-[17] and their references for related topics.

In order to give a notion of a solution to problem (P) in the variational sense, let us introduce the following convex set $K(t)$ for given functions g_i: $R \to H^{1/2}(\Gamma)$, $i = D, U$:

$$K(t) = \{z \in H^1(\Omega);\ z = g_D(t,\cdot) \text{ a.e. on } \Gamma_D,\ z \leqq g_U(t,\cdot) \text{ a.e. on } \Gamma_U\}$$

for each $t \in R$.

DEFINITION: Let $J = R$ or R_+. Let $f \in L^2_{loc}(J;L^2(\Omega))$. Then a function $v \in L^2_{loc}(J;H^1(\Omega))$ is called a solution to (P) on J, if $v(t) \in K(t)$ for a.e. $t \in J$, $\rho(v) \in W^{1,2}_{loc}(J;L^2(\Omega))$ and v satisfies the following variational inequality for a.e. $t \in J$:

$$\int_\Omega (\rho(v)_t(t,x)-f(t,x))(v(t,x)-z(x))\,dx + \int_\Omega \nabla v(t,x)\cdot\nabla(v(t,x)-z(x))\,dx$$

$$\leq \int_{\Gamma_N} q_N(t,x)(v(t,x)-z(x))\,d\Gamma + \int_{\Gamma_U} q_U(t,x)(v(t,x)-z(x))\,d\Gamma$$

for all $z \in K(t)$.

The initial value problem for (P) is treated here under the following assumptions (A1) $\sim$ (A3):

(A1) meas $(\Gamma_D) > 0$ (meas denotes the surface measure on Γ).

(A2) $f \in W^{1,1}_{loc}(R;L^2(\Omega))$.

(A3) There exist functions $\bar{g}$, $\bar{q}_N$, $\bar{q}_U \in W^{1,2}_{loc}(R;H^1(\Omega))$ such that for all $t \in R$,

(i) $\bar{g}(t,x) = \begin{cases} g_D(t,x) & \text{for a.e. } x \in \Gamma_D, \\ g_U(t,x) & \text{for a.e. } x \in \Gamma_U, \end{cases}$

(ii) $\bar{q}_N(t,x) = q_N(t,x)$ for a.e. $x \in \Gamma_N$,

(iii) $\bar{q}_U(t,x) = q_U(t,x)$ for a.e. $x \in \Gamma_U$.

According to the results in [13], [15], under assumptions (A1) $\sim$ (A3), there exists a unique solution v of the initial value problem for (P) on R_+ associated with initial condition $\rho(v(0,\cdot)) = u_0$, where u_0 is a given initial datum satisfying $u_0 = \rho(v_0)$ for some v_0 in K(0).

Our main results concern the periodicity of solutions to (P) and are stated in the following theorems.

THEOREM 1: In addition to assumptions (A1) $\sim$ (A3) suppose that for a number $T > 0$, the data f, $\bar{g}$, $\bar{q}_N$, $\bar{q}_U$ are T-periodic in time t, i.e.

(A4) $f(t+T,\cdot) = f(t,\cdot)$, $\bar{g}(t+T,\cdot) = \bar{g}(t,\cdot)$, $\bar{q}_N(t+T,\cdot) = \bar{q}_N(t,\cdot)$ and

$\bar{q}_U(t+T,\cdot) = \bar{q}_U(t,\cdot) = \bar{q}_U(t,\cdot)$ a.e. in Ω and for all $t \in R$.

Then we have:

(1) There is one and only one T-periodic solution to (P) on R.

(2) For any solution v of (P) on R, it is T-periodic on R if and only if the trajectory $\{v(t,\cdot); t \in R\}$ is bounded in $L^2(\Omega)$.

The next theorem concerns the stability of the T-periodic solution of (P) on R.

THEOREM 2: Assume that (A1) $\sim$ (A4) hold. Then the T-periodic solution ω of (P) on R is asymptotically stable at infinity in the sense that for any solution v of (P) on R_+

$$\rho(v(t,\cdot)) - \rho(\omega(t,\cdot)) \to 0 \text{ in } L^2(\Omega) \text{ and weakly in } H^1(\Omega) \text{ as } t \to \infty.$$

We shall give the outline of the proofs of these theorems in the next section. For their detailed proofs, see [12] in which the same problem is studied in the case when the decomposition $\Gamma_D \cup \Gamma_N \cup \Gamma_U$ of Γ depends smoothly on time $t \in R$, say $\Gamma = \Gamma_D(t) \cup \Gamma_N(t) \cup \Gamma_U(t)$.

For simplicity, given a measurable function $v = v(t,x)$, we sometimes denote the function $v(t,\cdot)$ by $v(t)$, when $v(t,x)$ is regarded as a function of the variable x for each t.

2. OUTLINE OF PROOFS

In the proofs of the theorems the key steps are given in the following lemmas.

LEMMA 1: Assume (A1) $\sim$ (A4) hold. Let v and $\hat{v}$ be two solutions of (P) on R_+ such that $v \leq \hat{v}$ a.e. in $R_+ \times \Omega$. Then

$$\partial_\nu v(t) \geq \partial_\nu \hat{v}(t) \text{ in the sense of } H^{-1/2}(\Gamma) \text{ for a.e. } t \in R_+, \qquad (2.1)$$

that is, $\langle\partial_\nu v(t),z\rangle \geqq \langle\partial_\nu \hat{v}(t),z\rangle$ for all $z \in H^{1/2}(\Gamma)$ with $z \geqq 0$. Here $\langle\cdot,\cdot\rangle$ stands for the duality between $H^{-1/2}(\Gamma)$ and $H^{1/2}(\Gamma)$.

PROOF: Fix $t \in R_+$ and write v and $\hat{v}$ for $v(t)$ and $\hat{v}(t)$, respectively. For each $\lambda > 0$ and $\mu > 0$, let $v_{\lambda\mu} \in H^1(\Omega)$ be the solution to the problem:

$$v_{\lambda\mu} - \lambda\Delta v_{\lambda\mu} = v \text{ in } \Omega, \tag{2.2}$$

$$-\partial_\nu v_{\lambda\mu} = -\chi_N q_N - \chi_U q_U + \frac{1}{\mu}\chi_D(v_{\lambda\mu}-g_D) + \frac{1}{\mu}\chi_U(v_{\lambda\mu}-g_U)^+ \text{ on } \Gamma,$$

where χ_i ($i = D, N, U$) are the characteristic functions of the sets Γ_i. Also, let $\hat{v}_{\lambda\mu}$ be the solution of problem (2.2) with v replaced by $\hat{v}$. The boundary conditions imply that $\partial_\nu v_{\lambda\mu}, \partial_\nu v_{\lambda\mu} \in L^2(\Gamma)$. Since $v \leqq \hat{v}$, it follows that $v_{\lambda\mu} \leqq \hat{v}_{\lambda\mu}$ in Ω and hence $-\partial_\nu v_{\lambda\mu} \leqq -\partial_\nu \hat{v}_{\lambda\mu}$ on Γ. Consequently, we have (2.1), because $\partial_\nu v_{\lambda\mu} \to \partial_\nu v$ and $\partial_\nu \hat{v}_{\lambda\mu} \to \partial_\nu \hat{v}$ in $H^{-1/2}(\Gamma)$ as $\mu \downarrow 0$ and $\lambda \downarrow 0$. See [12; Proposition 4.1] for the detail. Q.E.D.

Next, using Lemma 1, we prove:

LEMMA 2: Assume (A1) $\sim$ (A4) hold. Let ω be a T-periodic solution of (P) on R and let v be any solution of (P) on R_+ such that $\omega \leqq v$ (or $\omega \geqq v$) a.e. on $R_+ \times \Omega$. Then we have

$$\rho(v)(t+nT) \to \rho(\omega)(t) \text{ weakly in } H^1(\Omega) \text{ as } n \to \infty \text{ for all } t \in R_+. \tag{2.3}$$

PROOF: We prove the lemma only in the case of $\omega \leqq v$. The other case is similarly proved.

First note (see [15]) that $t \to |[\rho(v)(t)-\rho(\omega)(t)]^+|_{L^1(\Omega)}$ is nonincreasing on R_+. Hence it follows from $v \geqq \omega$ that

$$t \to \int_\Omega \{\rho(v)(t)-\rho(\omega)(t)\}dx \text{ is nonincreasing on } R_+. \tag{2.4}$$

In particular, since ω is T-periodic, we see from (2.4) that

$$\int_\Omega \rho(v)(mT)dx \leqq \int_\Omega \rho(v)(nT)dx \text{ for all } n, m \in N, n \leqq m,$$

so that

$$\lim_{n\to\infty} \int_\Omega \rho(v)(nT)dx \text{ exists,} \tag{2.5}$$

Next, by virtue of [11; Theorem 1], $\{\rho(v)(t);\ t \in R_+\}$ is bounded in $H^1(\Omega)$. Therefore, on account of a convergence result [13; Theorem 1.4], there is a subsequence $\{n_k\}$ of $\{n\}$ with a solution v^* of (P) on R_+ such that

$$\rho(v)(t+n_kT) \to \rho(v^*)(t) \text{ weakly in } H^1(\Omega) \text{ as } k \to \infty \text{ for all } t \in R_+. \tag{2.6}$$

We are going to show that $v^* \equiv \omega$. To this purpose, from (2.5) and (2.6) we note that

$$\begin{aligned}\int_\Omega \rho(v^*)(nT)dx &= \lim_{k\to\infty} \int_\Omega \rho(v)(nT+n_kT)\ dx = \lim_{m\to\infty} \int_\Omega \rho(v)(mT)\ dx \\ &= \lim_{k\to\infty} \int_\Omega \rho(v)(n_kT)\ dx = \int_\Omega \rho(v^*)(0)dx \text{ for all } n \in N.\end{aligned} \tag{2.7}$$

Therefore, taking into account the fact that $\rho(v^*)_t - \Delta v^* = f$ and $\rho(\omega)_t - \Delta\omega = f$ a.e. in $R_+ \times \Omega$, we infer that

$$\begin{aligned}0 &= \int_0^{nT} dt \frac{d}{dt} \int_\Omega \{\rho(v^*)(t)-\rho(\omega)(t)\}dx = \int_0^{nT} dt \int_\Omega \Delta(v^*(t)-\omega(t))dx \\ &= \int_0^{nT} \langle\partial_\nu(v^*(t)-\omega(t)),\ 1\rangle dt \text{ for all } n \in N.\end{aligned} \tag{2.8}$$

Moreover, it is evident that $\omega \leqq v^*$, and hence by Lemma 2, $\partial_\nu\omega(t) \geqq \partial_\nu v^*(t)$ in the sense of $H^{-1/2}(\Gamma)$ for a.e. $t \in R_+$. This inequality for fluxes together with (2.8) implies that

$$\langle\partial_\nu(v^*(t)-\omega(t)),1\rangle = 0 \text{ for a.e. } t \in R_+,$$

so that we see

$$\partial_\nu v^*(t) = \partial_\nu\omega(t) \text{ in } H^{-1/2}(\Gamma) \text{ for a.e. } t \in R_+. \tag{2.9}$$

Now, put

$$V = \{z \in H^1(\Omega);\ z = 0 \text{ a.e. on } \Gamma_D\}.$$

By (A1), for each $t \in R_+$ there is a unique solution $u(t) \in V$ of the variational problem:

$$\int_\Omega \nabla u(t)\cdot\nabla z\, dx = \int_\Omega \{\rho(v^*)(t)-\rho(\omega)(t)\}z\, dx \text{ for all } z \in V. \qquad (2.10)$$

From (2.9) and (2.10) we observe that

$$\begin{aligned}
\frac{1}{2}\frac{d}{dt}|\nabla u(t)|^2_{L^2(\Omega)} &= \int_\Omega \nabla u'(t)\cdot\nabla u(t)\, dx \\
&= \int_\Omega \{\rho(v^*)'(t)-\rho(\omega)'(t)\}u(t)\, dx \\
&= \int_\Omega \Delta(v^*(t)-\omega(t))u(t)\, dx \\
&= -\int_\Omega \nabla(v^*(t)-\omega(t))\cdot\nabla u(t)\, dx \\
&= -\int_\Omega (v^*(t)-\omega(t))\{\rho(v^*)(t)-\rho(\omega)(t)\}\, dx.
\end{aligned}$$

Hence, using Poincaré's inequality and noting the Lipschitz continuity of ρ, we get for suitable positive constants C_1, C_2

$$\begin{aligned}
0 &= \frac{1}{2}\frac{d}{dt}|\nabla u(t)|^2_{L^2(\Omega)} + \int_\Omega (v^*(t)-\omega(t))\{\rho(v^*)(t)-\rho(\omega)(t)\}\, dx \\
&\geqq \frac{1}{2}\frac{d}{dt}|\nabla u(t)|^2_{L^2(\Omega)} + C_1|\rho(v^*)(t)-\rho(\omega)(t)|^2_{L^2(\Omega)} \\
&\geqq \frac{1}{2}\frac{d}{dt}|\nabla u(t)|^2_{L^2(\Omega)} + C_2|\nabla u(t)|^2_{L^2(\Omega)}.
\end{aligned}$$

From these inequalities it follows that

$$\frac{d}{dt}|\nabla u(t)|^2_{L^2(\Omega)} \leqq 0 \quad\text{and}\quad \int_0^\infty |\nabla u(t)|^2_{L^2(\Omega)}\, dt < \infty,$$

so that

$$|\nabla u(t)|_{L^2(\Omega)} \to 0 \text{ as } t \to \infty.$$

Combining this with (2.10), we obtain

$$\int_\Omega \{\rho(v^*)(t)-\rho(\omega)(t)\}z\, dx \to 0 \text{ as } t \to \infty \quad \text{for all } z \in V.$$

Therefore

$$\rho(v^*)(t) - \rho(\omega)(t) \to 0 \text{ weakly in } H^1(\Omega) \text{ as } t \to \infty, \qquad (2.11)$$

since $\rho(v^*)$ and $\rho(\omega)$ are in $L^\infty(R_+;H^1(\Omega))$ (see [11; Theorem 1]) and V is dense in $L^2(\Omega)$. Besides, from (2.7) and (2.11) we see that

$$\int_\Omega (\rho(v^*)(0)-\rho(\omega)(0))\, dx = \int_\Omega (\rho(v^*)(nT)-\rho(\omega)(nT))\, dx \to 0 \text{ as } n \to \infty.$$

This shows

$$\int_\Omega \rho(v^*)(0)\, dx = \int_\Omega \rho(\omega)(0)\, dx$$

from which we infer $\rho(v^*)(0) = \rho(\omega)(0)$, because $\rho(v^*) \geqq \rho(\omega)$. Consequently, the uniqueness of solution to the initial value problem for (P) results that $v^* = \omega$. The assertion (2.3) follows immediately from this fact with (2.4) and (2.6). Q.E.D.

<u>PROOF OF THE THEOREMS</u>: The existence of at least one T-periodic solution to (P) on R is due to [11; Theorem 2]. We fix a T-periodic solution ω for the moment, and first prove the asymptotic stability of ω as $t \to \infty$.

Let v be any solution of (P) on R_+. Then from the order property for solutions to (P) we can find two solutions $\underline{v}$, $\bar{v}$ on R_+ such that

$$\underline{v} \leqq v \wedge \omega \leqq v \vee \omega \leqq \bar{v} \text{ a.e. on } R_+ \times \Omega.$$

In fact, we can take as $\bar{v}$ (resp. $\underline{v}$) the solution of (P) associated with initial value $\rho(v)(0)\wedge\rho(\omega)(0)$ (resp. $\rho(v)(0)\vee\rho(\omega)(0)$). Now applying Lemma 2, we see that $\rho(\bar{v})(t+nT)-\rho(\omega)(t+nT) \to 0$ weakly in $H^1(\Omega)$ as $n \to \infty$. On the other hand, since $|\rho(\bar{v})(t)-\rho(\omega)(t)|_{L^1(\Omega)} \leqq |\rho(\bar{v})(s)-\rho(\omega)(s)|_{L^1(\Omega)}$

for $0 \leqq s \leqq t$, it follows that $\rho(\bar{v})(t)-\rho(\omega)(t) \to 0$ weakly in $H^1(\Omega)$ as $t \to \infty$. Similarly, $\rho(\underline{v})(t)-\rho(\omega)(t) \to 0$ weakly in $H^1(\Omega)$ as $t \to \infty$. Therefore,

$$\rho(v)(t) - \rho(\omega)(t) \to 0 \text{ weakly in } H^1(\Omega) \text{ as } t \to \infty,$$

because $\rho(\underline{v})-\rho(\omega) \leqq \rho(v)-\rho(\omega) \leqq \rho(\bar{v})-\rho(\omega)$. This fact implies at the same time that the T-periodic solution to (P) is unique. Thus we have proved (1) of Theorem 1, and Theorem 2. Also, the uniqueness of $L^2(\Omega)$-bounded solution on R can be similarly proved by applying Lemma 2. Thus (2) of Theorem 1 is also valid. Q.E.D.

REMARK: The results mentioned in Theorems 1 and 2 are able to be extended to the almost-periodic case; more precisely, if $\bar{g}$, $\bar{q}_N$, $\bar{q}_U$ are almost-periodic as functions from R into $H^1(\Omega)$ as well as f from R into $L^2(\Omega)$, then we obtain the statements of Theorems 1 and 2 with "T-periodic" replaced by "almost-periodic".

References

[1] H.W. Alt and S. Luckhaus, Quasilinear elliptic-parabolic differential equations, Math. Z. 183 (1983), 311-341.

[2] H.W. Alt, S. Luckhaus and A. Visintin, On nonstationary flow through porous media, Ann. Mat. Pura Appl. 136 (1984), 303-316.

[3] M. Bertsch and J. Hulshof, Regularity results for an elliptic-parabolic free boundary problem, Trans. Amer. Math. Soc. 297 (1986), 337-350.

[4] E. DiBenedetto and A. Friedman, Periodic behaviour for the evolutionary dam problem and related free boundary problems, Comm. Partial Diff. Equations 11 (1986), 1297-1377.

[5] C.J. van Duyn and L.A. Peletier, Nonstationary filtration in partially saturated porous media, Arch. Rat. Mech. Anal. 78 (1982), 173-198.

[6] A. Fasano and M. Primicerio, Rigid flow in partially saturated porous media, J. Inst. Math. Appl. 24 (1979), 503-517.

[7] U. Hornung, A parabolic-elliptic variational inequality, Manuscripta Math. 39 (1982), 155-172.

[8] J. Hulshof, A parabolic-elliptic free boundary problem: continuity of the interface, Proc. Royal Soc. Edinburgh 106 (1987), 327-339.

[9] J. Hulshof, Bounded weak solutions of an elliptic-parabolic Neumann problem, Trans. Amer. Math. Soc. 303 (1987), 211-227.

[10] J. Hulshof and L.A. Peletier, An elliptic-parabolic free boundary problems, Nonlinear Anal. 10 (1986), 1327-1346.

[11] N. Kenmochi and M. Kubo, Periodic solutions to a class of nonlinear variational inequalities with time-dependent constraints, Funk. Ekvac. 30 (1987), 333-349.

[12] N. Kenmochi and M. Kubo, Periodic behaviour of solutions to parabolic-elliptic free boundary problems, Tech. Rep. Math. Sci., Chiba Univ., No. 13, 1987.

[13] N. Kenmochi and I. Pawlow, A class of nonlinear elliptic-parabolic equations with time-dependent constraints, Nonlinear Anal. 10 (1986), 1181-1202.

[14] N. Kenmochi and I. Pawlow, Asymptotic behavior of solutions to parabolic-elliptic variational inequalities, Tech. Rep. Math. Sci., Chiba Univ., No. 3, 1987.

[15] N. Kenmochi and I. Pawlow, Parabolic-elliptic free boundary problems with time-dependent obstacles, Japan J. Appl. Math. 5 (1988), 87-121.

[16] D. Kröner, Parabolic regularization and behavior of the free boundary for unsaturated flow in a porous media, J. Reine Angew. Math. 348 (1984), 180-186.

[17] D. Kröner and J.F. Rodrigues, Global behaviour for bounded solutions of a porous media equation of elliptic-parabolic type, J. Math. Pure Appl. 64 (1985), 105-120.

N. Kenmochi
Department of Mathematics
Faculty of Education
Chiba University
Chiba-Shi
Japan

M. Kubo
Department of Mathematics
Faculty of Sciences and Engineering
Saga University
Saga-shi
Japan

M. LANGLAIS

Asymptotic behaviour in some evolution equations arising in population dynamics

We are interested in the largetime behaviour (exhibited as t goes to $+\infty$ by the solution $u = u(x,t,a)$) of the evolution equations

$$\partial_t u + \partial_a u - \Delta u + [\mu_n(a) + \mu_e(P(x,t))]u = 0, \quad x \in \Omega,\ t > 0,\ a > 0; \quad (1)$$

$$P(x,t) = \int_0^\infty u(x,t,a)da,\ x \in \Omega,\ t > 0; \quad (2)$$

$$u(x,t,0) = \beta_e(P(x,t)) \int_0^\infty \beta_n(a)u(x,t,a)da,\ x \in \Omega,\ t > 0; \quad (3)$$

$$u(x,t,a) = 0,\ x \in \partial\Omega,\ t > 0,\ a > 0; \quad (4)$$

$$u(x,0,a) = u_0(x,a),\ x \in \Omega,\ a > 0. \quad (5)$$

This problem arises in some mathematical models of population dynamics with age-structure and diffusion; in this setting $u(x,t,a)$ stands for the distribution of individuals in a single species population having age $a > 0$ at time $t > 0$ and locus x in Ω, open and bounded domain in $\mathbb{R}^N$. For more details see [1].

Our goal is to supply a direct proof of the stabilization results announced (for the one-dimensional case) in [2], avoiding unpleasant technicalities arising when μ_n and β_n depend on the variable a, thereby shedding some light on the heart of the proof.

Hence throughout this article we assume

$$\mu_n(a) = \mu \geqq 0,\ \beta_n(a) = \beta > 0 \text{ for } a > 0.$$

Under this assumption the initial condition (3) now reads

$$u(x,t,0) = \beta\beta_e(P(x,t))P(x,t), \quad x \in \Omega,\ t > 0;$$

But even in this simple situation the trick used in [3], to handle this problem when no diffusion takes place, does not work and a new idea is needed to obtain a stabilization result. This is what is done in this simple setting in Section IV below.

I. BASIC MATERIAL

We assume that μ_e and β_e are C^1-functions such that:

$$\mu_e(0) = 0,\ \beta_e(0) = 1 \text{ and } \beta_e(p) \geqq 0 \text{ for } p > 0;$$

$$\mu_e \text{ nondecreasing and } \beta_e \text{ nonincreasing.}$$

The initial condition u_0 satisfies

$$0 \leqq u_0(x,a) \leqq m_0 < +\infty,\ x \in \Omega,\ a > 0;\ u_0 \in L^2(\Omega x(0,\infty)) \cap L^1(\Omega x(0,\infty));$$

$$p_0(x) = \int_0^\infty u_0(x,a)da \leqq M_0 < +\infty,\ x \in \Omega;$$

$$\nabla p_0 \in L^2(\Omega),\ \nabla u_0 \in L^2(\Omega x(0,\infty)),\ u_0 = 0 \text{ on } \partial\Omega x(0,\infty).$$

By a solution of (1)-(5) we mean a nonnegative function $u(x,t,a) \geqq 0$ such that for each $T > 0$

$$u \in L^1(\Omega x(0,T)x(0,\infty)),\ 0 \leqq P(x,t) \leqq M(T) < +\infty,\ x \in \Omega \text{ and } 0 < t < T;$$

$$u,\ \Delta u,\ \partial_t u + \partial_a u \text{ lie in } L^2(\Omega x(0,T)x(0,\infty)),$$

u obeying the partial differential equation (1) and taking the initial and boundary conditions (3)-(5).

II. EXISTENCE, UNIQUENESS

Integrating (1) over a from $a = 0$ to $a = +\infty$ we get that for any solution u, $P = P(x,t)$ is a solution of

$$\partial_t P - \Delta P + [\mu + \mu_e(P) - \beta\beta_e(P) = 0, \qquad x \text{ in } \Omega,\ t > 0,$$

$$P(x,0) = p_0(x),\ x \text{ in } \Omega, \qquad (6)$$

$$P(x,t) = 0, \qquad x \text{ on } \partial\Omega,\ t > 0.$$

This semilinear parabolic equation has a unique, local in time, suitable solution; furthermore, it is nonnegative, globally defined and

$$0 \leqq P(x,t) \leqq \exp(\beta-\mu)t,\ x \in \Omega,\ t > 0; \qquad (7)$$

this follows from the maximum principle.

Hence P being uniquely defined, our problem (1)-(5) turns out to be a linear problem with prescribed data on $t = 0$ and $a = 0$. Using the Galerkin method, for example, straightforward computations yield a unique solution having the required properties.

III. LARGETIME BEHAVIOUR

From now on we assume

$$\mu_e(p) \to +\infty \text{ and } \beta_e(p) \to 0 \text{ as } p \to +\infty. \qquad (8)$$

Let c be the following nondecreasing function:

$$c(p) = \beta\beta_e(p) - \mu - \mu_e(p),\ p \geqq 0;$$

we have $c(0) = \beta - \mu$ and $c(p) \to -\infty$ as $p \to +\infty$.

THEOREM 1: Assume that $\beta - \mu < 0$. Then $u(\cdot,t,\cdot) \to 0$ in $L^1(\Omega x(0,\infty))$ as t goes to $+\infty$.

PROOF: By estimate (7) $P(x,t)$ is exponentially decaying towards 0 as t tends to $+\infty$. □

REMARK: It follows that 0 is a stable equilibrium.

Now we may assume $\beta - \mu \geqq 0$; this implies $c(0) \geqq 0$. Applying the maximum principle to (6) we conclude that P is bounded from above by the solution of the ordinary differential equation

$$y' = c(y)y, \ t > 0; \ y(0) = M_0.$$

By (8) it follows that y is uniformly bounded in time and so does P

$$0 \leqq P(x,t) \leqq M_1 < +\infty, \ x \in \Omega, \ t > 0. \qquad (9)$$

Using dynamical systems techniques we show that any R lying in the ω-limit set

$$\{R \in C(\bar{\Omega}), \ \exists t_n \to +\infty, \ P(\cdot,t_n) \to R(\cdot) \text{ in } C(\bar{\Omega}) \text{ as } n \to +\infty\}$$

is an equilibrium, that is, a solution of

$$\begin{aligned} -\Delta R &= c(R)R \quad \text{in } \Omega, \\ R &= 0 \quad \text{on } \partial\Omega. \end{aligned} \qquad (10)$$

It is known that (see [4])

<u>THEOREM</u>: Equation (10) has a unique nontrivial solution if and only if $c(0) > \lambda_1$, where λ_1 is the first eigenvalue of the Dirichlet problem for $-\Delta$ in Ω.

As a consequence we have

<u>THEOREM 2</u>: Assume that $\beta - \mu \leqq \lambda_1$. Then $u(\cdot,t,\cdot) \to 0$ as $t \to +\infty$ in $L^1(\Omega x(0,\infty))$ as t goes to $+\infty$.

<u>PROOF</u>: Under this assumption, (10) has no nontrivial solution and therefore $P(\cdot,t)$ goes to 0 as t goes to $+\infty$. □

<u>REMARK</u>: When $\beta - \mu < \lambda_1$ it is possible to build a supersolution for (1)-(5) exhibiting an exponential decay from which the stability of the trivial

equilibrium follows at once. This is still possible if we remove the monotonicity assumption on μ_e and β_e and merely assume $\mu_e(p) \geq 0$ and $0 \leq \beta_e(p) \leq 1$ for $p \geq 0$; thus, in this situation no nontrivial periodic solution may exist.

Now we are left with the case $\beta - \mu > \lambda_1$; first we know that $P(\cdot,t) \to R(\cdot)$ as $t \to +\infty$ where R is the unique nonnegative nontrivial solution of (10): see [5]. In Section IV we shall prove

THEOREM 3: Assume $\beta - \mu > \lambda_1$. Then $u(\cdot,t,\cdot) \to w(\cdot,\cdot)$ as $t \to +\infty$ in $L^2(\Omega x(0,A))$ for any $A > 0$, w being the unique nontrivial steadystate of (1)-(5).

This is achieved upon proving a general stabilization result independent of the size of β and μ. But due to the lack of compactness on the orbits $\{u(\cdot,t,\cdot),\ t > 0\}$ dynamical systems technique fail and instead we adapt a method devised in [6] for the porous medium equation.

By a steadystate for (1)-(5) we mean a solution w of

$$\begin{aligned}
&\partial_a w - \Delta w + [\mu + \mu_e(R(x))]w = 0, \quad x \in \Omega,\ a > 0; \\
&R(x) = \int_0^\infty w(x,a)da,\ x \in \Omega; \\
&w(x,0) = \beta\beta_e(R(x))R(x),\ x \in \Omega; \\
&w(x,a) = 0, \quad x \in \partial\Omega,\ a > 0.
\end{aligned} \tag{11}$$

Note that by integrating (11) from $a = 0$ to $a = \infty$ we check that R is a solution of (10); hence for $\beta - \mu \leq \lambda_1$, 0 is the only steadystate while for $\beta - \mu > \lambda_1$ there are two steadystate and the nontrivial one is stable by Theorem 3.

REMARK: If we replace the condition on β_e in (8) by $\beta_e(p) = 1$ for $p \geq 0$, then the nontrivial steadystate is found explicitly as

$$w(x,a) = \beta R(x) \, . \, \exp(-\beta a), \quad x \in \Omega, \; a > 0,$$

R being the nontrivial solution of (10).

IV. A STABILIZATION RESULT

By (9) P is uniformly bounded over $\Omega x(0,\infty)$; hence taking the product of (6) with $\partial_t P$ and integrating over $\Omega x(0,\infty)$ we have

$$\partial_t P \in L^2(\Omega x(0,\infty)). \tag{12}$$

Next, using again (9) and the boundedness of u_0, a weak maximum principle applied to (1)-(5) yields

$$0 \leqq u(x,t,a) \leqq \mathrm{Max}\{m_0,\beta M_1\}, \; x \in \Omega, \; t > 0, \; a > 0, \tag{13}$$

so that u is also uniformly bounded in time.

Now due to the hyperbolic type of $\partial_t + \partial_a$, u is bound to be discontinuous along the characteristic line t = a because the initial condition u_0 need not satisfy condition (3) at t = 0. Anyway, using differential quotient techniques, we show

$$\partial_t u \in L^2(\Omega x\{(t,a), \; 0 < a < t < T\}) \quad \text{for any } T > 0. \tag{14}$$

This being granted we have

THEOREM 4:

$$\partial_t u \in L^2(\Omega x(A,\infty)x(0,A)) \text{ for any } A > 0.$$

PROOF: Differentiating (1) with respect to the variable t in the domain t > a we get that $z = \partial_t u$ is a solution of

$$\partial_t z + \partial_a z - \Delta z + [\mu + \mu_e(P(x,t)]z = -u\mu_e'(P(x,t))P_t, \; x \in \Omega, \; 0 < a < t,$$

$$z(x,t,0) = \beta[\beta_e(P) + \beta_e'(P)P]\partial_t P, \; x \in \Omega, \; t > 0, \tag{15}$$

$$z(x,t,a) = 0, \quad x \in \partial\Omega, \; 0 < a < t.$$

Note that no condition is needed on the characteristic line $t = a$ for this problem to be well-posed.

Fix $A_0 > 0$ and choose A and T such as $0 < A < A_0 < T$. Taking the product of (15) with z and integrating over

$$U = \Omega x\{(t,a),\ 0 < a < t,\ 0 < t < T,\ 0 < a < A\},$$

after some calculations using (12)-(14), we arrive at

$$\int_{\Omega x(A,T)} z^2(x,t),A)\ dx\ dt + \int_U |\nabla z|^2(x,t,a)\ dx\ dt\ da \leqq \leqq \int_U z^2(x,t,a)\ dx\ dt\ da + C_1(A_0), \tag{16}$$

where $C_1(A_0)$ is a constant independent of T and A for $A < A_0 < T$. From (14) and (16) we may conclude

$$\int_{\Omega x(A_0,T)} z^2(x,t,A)\ dx\ dt \leqq \int_{\Omega x(A_0,T)x0,A)} z^2(x,t,a)\ dx\ dt\ da + C_2(A_0).$$

Hence by the Gronwall inequality there exists a constant $C_3(A_0)$ independent of T provided $T > A_0$ such that

$$\int_{\Omega x(A_0,T)x(0,A_0)} z^2(x,t,a)\ dx\ dt\ da < C_3(A_0).$$

This completes the proof. □

Let now introduce the ω-limit set for (1)-(5), that is,

$$\omega(u_0) = \{w(\cdot,\cdot) \in L^\infty(\Omega x(0,\infty)),\ \exists t_n \to \infty,\ u(\cdot,t_n,\cdot) \to w(\cdot,\cdot)$$
$$\text{weakly in } L^2(\Omega x(0,A)) \text{ for each } A > 0, \text{ as } n \to \infty\};$$

it is not empty and we are to prove

<u>THEOREM 5</u>: Any w in $\omega(u_0)$ is a solution of (11).

PROOF: Take ρ a smooth function satisfying the conditions

$$\rho(s) \geqq 0 \text{ for } s \geqq 0; \text{ supp } \rho \subset [-1,1]; \int_{-1}^{1} \rho(s)ds = 1.$$

Let $\theta = \theta(x,a)$ be a smooth function vanishing on $\partial\Omega$ and $a = A$. Following an idea of [6], we take the product of (1) with $\rho(t-t_n)$ (x,a) and integrate by parts over $\Omega x(0,\infty)x(0,A)$. Setting $s = t - t_n$ and for $t_n > A + 2$ we find

$$\int_{-1}^{1} \int_{\Omega x(0,A)} [-\theta\rho'(s)-(\partial_a\theta+\Delta\theta)\rho(s)]\, u(x,s+t_n,a)\, dx\, ds\, da +$$

$$+ \int_{-1}^{1} \int_{\Omega x(0,A)} \theta\rho(s)u(x,s+t_n,a)[\mu+\mu_e(P(x,s+t_n)]\, dx\, ds\, da = \tag{17}$$

$$= \int_{\Omega x(-1,1)} \theta(x,0)\beta\beta_e(P(x,s+t_n))P(x,s+t_n)\, dx\, ds.$$

From estimate (12) and the result of Theorem 5 we obtain

$$u(x,s+t_n,a) \to w(x,a) \text{ weakly in } L^2(\Omega x(-1,1)x(0,A)),$$

$$P(x,s+t_n) \to R(x) \text{ strongly in } L^2(\Omega x(-1,1)) \text{ as } n \to \infty.$$

Passing to the limit in (17) as $n \to \infty$ and noting that $\int_{-1}^{1} \rho'(s)ds = 0$ we see that w is a weak solution of

$$\partial_a w - \Delta w + [\mu+\mu_e(R(x))]w = 0, \quad x \in \Omega,\ a > 0;$$

$$w(x,0) = \beta\beta_e(R(x)),\ x \in \Omega; \tag{18}$$

$$w(x,a) = 0,\ x \in \partial\Omega,\ a > 0.$$

A bootstrapping argument shows w is strong. We must now prove

$$\int_0^{\infty} w(x,a)da = R(x) \text{ for } x \text{ in } \Omega; \tag{19}$$

to this end, integrating (18) from a = 0 to a = ∞, we have that

$\int_0^\infty w(x,a)da$ is a solution of the elliptic problem

$$
\begin{aligned}
&-\Delta Q + [\mu+\mu_e(R(x))]Q = \beta\beta_e(R(x))R(x), \ x \in \Omega; \\
&\qquad\qquad Q(x) = 0, \ x \in \partial\Omega.
\end{aligned} \tag{20}
$$

But R is also a solution of (20): see (10) so that by uniqueness in (20) we conclude that (18) holds.

V. FURTHER COMMENTS

When μ_n and β_n depend on the variable a, it is no more possible to handle separately the behaviour of P and that of u; the proof given above does not work as it stands, even the existence part has to be modified.

Nevertheless, the final result is roughly the same, the role of β-μ being played by r^* the root of the so-called characteristic equation

$$\int_0^\infty \beta_n(a)\cdot\exp(-\int_0^a \mu_n(\sigma)d\sigma)\cdot e^{-ra}\,da = 1;$$

of course, when $\mu_n(a) = \mu$ and $\beta_n(a) = \beta$ we have $r^* = \beta - \mu$. For a precise formulation we refer to [2] in the one-dimensional case and to a forthcoming article in a more involved setting.

VI. REFERENCES

[1] M.E. Gurtin and R.C. MacCamy, Product solutions and asymptotic behavior in age-dependent population diffusion, Math. Biosci. 62 (1982), 157-167,

[2] M. Langlais, Stabilization in a nonlinear age-dependent population model, in Biomathematics and Related Computational Problems. L. Ricciardi, editor, Kluwer Acad. Pub. (1988), 337-344.

[3] S. Busenberg and M. Iannelli, Separable models in age-dependent population dynamics, J. Math. Biol. 22 (1985), 145-173.

[4] H. Berestycki, Le nombre de solutions de certains problèmes semi-linéaires élliptiques, J. Functional Analysis 40 (1981), 1-29.

[5] D.H. Sattinger, Monotone operators in nonlinear elliptic and parabolic boundary value problems, Indiana Univ. J. Math. 21 (1972), 979-1000.

[6] M. Langlais and D. Phillips, Stabilization of solutions of nonlinear and degenerate evolution equations. Nonlinear Analysis T.M.A. 9, 4 (1985), 321-333.

M. Langlais
Institut des Sciences Humaines Appliquées
U.A. CNRS 004226
Université de Bordeaux II
3, place de la Victoire
33 000 Bordeaux
France

M.S. MOULAY AND M. PIERRE

About regularity of the solutions of some nonlinear degenerate parabolic equations

1. INTRODUCTION

It is well known (see [1]) that in dimension N = 1 the nonnegative solutions of the porous media equation

$$\frac{\partial u}{\partial t} - \Delta(u^m) = 0 \text{ on } Q = (0,\infty) \times \mathbb{R}^N,\ m > 1, \tag{1.1}$$

satisfy the property

$$\nabla(u^{m-1}) \in L^\infty_{loc}(Q). \tag{1.2}$$

Moreover, this result is sharp in the sense that this derivative is in general discontinuous. Estimate (1.2) does not hold in higher dimension $N \geqq 2$ (see [2] and also [4] where the estimate is proven to be valid in part of Q).

However the solution of (1.1) is Hölder continuous as proved in [5] or [10]. Therefore, it is natural to ask whether

$$\nabla(u^p) \in L^\infty_{loc}(Q) \tag{1.3}$$

for some p large enough. This is the case if

$$m \leqq 1 + (N-1)^{-\frac{1}{2}} \tag{1.4}$$

as noticed in [3]. This can be easily deduced from the properties of the nonlinear equation satisfied by $w = |\nabla(u^p)|^2$ where, for instance, p = m.

It is the purpose of this article to prove that property (1.3) holds for the nonnegative solutions of (1.1) assuming that, for some r > 0,

$$\nabla(u^r) \in L^{N+2+\varepsilon}_{loc}(Q),\ \varepsilon > 0. \tag{1.5}$$

It turns out, that (1.5) is satisfied for the solution of (1.1) under suitable conditions on (m,N) and, in particular, for any $m > 1$ if $N = 2$.

Here we will limit ourselves to the proof of (1.5) ⇒ (1.3). The analysis of (1.5) can be found in [7] or [9] with related results.

Our proof relies on classical iterative arguments of the Moser type applied to the equation satisfied by $w = |\nabla v|^2$ where $v = u^r$. Therefore it applies to a large class of equations containing, in particular, the case of nonregular source-terms, generalizing at the same time the one-dimensional results obtained by the same authors in [8]. However, to keep this article as simple as possible we will restrict ourselves to the homogeneous case. Extensions can be found in [7].

2. STATEMENT OF THE RESULT

Notations: Let $\mu \in \mathbb{R}$ and let ω, $\hat{\omega}$ be open subsets of $Q = (0,\infty) \times \mathbb{R}^N$ such that

$$\text{closure}(\hat{\omega}) \subset \omega \subset Q. \tag{2.1}$$

We will consider solutions of

$$\begin{cases} (2.2(a)) & v \in C^3(\omega),\ v > 0, \\ (2.2(b)) & \dfrac{\partial v}{\partial t} = v\Delta v + \mu|\nabla v|^2 \text{ on } \omega. \end{cases} \tag{2.2}$$

THEOREM: Let v be a solution of (2.2) and

$$\beta > N + 2. \tag{2.3}$$

Set $\sigma = \beta/(\beta - N - 2)$. Then for all $r > 0$ satisfying

$$\sigma r > 1, \tag{2.4}$$

there exists $C = C(\omega, \hat{\omega}, \beta, N, q, \mu, \|v\|_{L^\infty(\omega)})$ such that

$$\|\nabla(v^{\sigma r})\|^{\beta/\sigma}_{L^\infty(\hat{\omega})} \leq C[1 + \|\nabla(v^r)\|^{\beta}_{L^\beta(\omega)}]. \tag{2.5}$$

REMARK: If u is a positive solution of (1.1) on ω, then $v = mu^{m-1}$ is a solution of (2.2) with $\mu = m/(m-1)-1$. Therefore estimate (2.5) holds with $v = u^{m-1}$.

Obviously estimate (2.4) remains valid for bounded weak solutions of (2.2(b)) that can be obtained as the limit of regular solutions v_n of (2.2) for which $\nabla(v_n^r)$ remains bounded in L^β. This is the case for weak non-negative solutions of (1.1) in $W^{1,\beta}_{loc}(Q)$.

3. PROOF OF THE RESULT

Let v be a solution of (2.2) and

$$w := |\nabla v|^2, \ |\cdot| = \text{Euclidian norm in } \mathbb{R}^N. \tag{3.1}$$

LEMMA 1: The function w satisfies

$$w_t = v\Delta w - 2v \sum_{i,j} (v_{x_i x_j})^2 + 2w\Delta v + 2\mu\nabla v\nabla w. \tag{3.2}$$

PROOF: We use the identities

$$w_{x_j} = 2 \sum_i v_{x_i} v_{x_i x_j}, \ \Delta w = 2 \sum_{i,j} (v_{x_i x_j})^2 + 2\nabla v\nabla(\Delta v). \tag{3.3}$$

We take the gradient of both sides of (2.2(b)), we multiply by ∇v and we use (3.3) to obtain (3.2). □

LEMMA 2: Let $\eta \in C_0^\infty(\omega)$, $0 \leq \eta \leq 1$ and $\alpha > 1$, $p \in \mathbb{R}$ Then

$$\int_{\omega_t} v^p w^\alpha \eta^2(t) + \lambda_1\alpha \int_\omega v^{p+1} w^{\alpha-1} \sum_{i,j} (v_{x_i x_j})^2\eta^2 + \alpha(\alpha-1) \int_\omega v^{p+1} w^{\alpha-2} |\nabla w|^2\eta^2$$
$$\leq \lambda_2(\alpha+p)^2 \int_\omega v^{p-1} w^{\alpha+1}\eta^2 + 2\int_\omega v^p w^\alpha |\eta_t| + \lambda_3\alpha \int_\omega v^{p+1} w^\alpha |\nabla\eta|^2, \tag{3.4}$$

where $\omega_t = \omega \cap \{t\} \times \mathbb{R}^N$ and $\lambda_i = \lambda_i(\mu,N)$, $i = 1,2,3$.

PROOF: We multiply (3.2) by $\alpha v^p w^{\alpha-1}\eta^2$ and integrate over ω to obtain

$$\int_{\omega_t} v^p w^\alpha \eta^2(t) + 2\alpha \int_\omega v^{p+1} w^{\alpha-1} \sum_{i,j} (v_{x_i x_j})^2 \eta^2 - \int_\omega w^\alpha \frac{\partial}{\partial t}(v^p \eta^2)$$

$$+ \alpha \int_\omega \nabla w \nabla (v^{p+1} w^{\alpha-1} \eta^2) = 2\alpha \int_\omega \Delta v \, v^p w^\alpha \eta^2 + 2\mu\alpha \int_\omega v^p w^{\alpha-1} \nabla v \nabla w.$$

Then to obtain (3.4), we use

$$p \int_\omega w^\alpha v^{p-1} v_t \eta^2 = p \int_\omega w^\alpha v^{p-1} \eta^2 (v\Delta v + \mu w)$$

and the following Schwarz's inequalities with ε small enough

$$v^p w^\alpha \Delta v \leqq \varepsilon v^{p+1} w^{\alpha-1} N \sum_i (v_{x_i x_i})^2 + \frac{1}{4\varepsilon} v^{p-1} w^{\alpha+1},$$

$$v^p w^{\alpha-1} \nabla w \nabla v \leqq \varepsilon v^{p+1} w^{\alpha-2} |\nabla w|^2 + \frac{1}{4\varepsilon} v^{p-1} w^{\alpha+1},$$

$$w^{\alpha-1} \nabla w \nabla n \leqq \varepsilon w^{\alpha-2} |\nabla w|^2 + \frac{1}{4\varepsilon} w^\alpha |\nabla \eta|^2,$$

$$|\nabla w|^2 \leqq 4w \sum_{i,j} (v_{x_i x_j})^2. \quad \square$$

LEMMA 3: With the notations of Lemma 2, if $q = 2(N+2)/N$

$$\{\int_\omega (v^{p+1} w^\alpha \eta^2)^{q/2}\}^{2/q} \leqq \lambda(\alpha+p)^2 [\int_\omega v^{p-1} w^{\alpha+1} \eta^2 + v^p w^\alpha |\eta_t| + v^{p+1} w^\alpha |\nabla \eta|^2] \tag{3.5}$$

where $\lambda = \lambda(\mu, N, \|v\|_{L^\infty(\omega)})$.

PROOF: This is a consequence of Sobolev's imbedding applied to the function

$$z = (v^{p+1} w^\alpha \eta^2)^{\frac{1}{2}},$$

namely (see, for instance, [6])

$$\|z\|^2_{L^q(\omega)} \leqq C(N)\{\max_t \|z(t)\|^2_{L^2(\omega_t)} + \int_\omega |\nabla z|^2\}. \tag{3.6}$$

Here, up to a constant depending only on N

$$|\nabla z|^2 \leqq p^2 v^{p-1} w^{\alpha+1} \eta^2 + \alpha^2 v^{p+1} w^{\alpha-2} |\nabla w|^2 \eta^2 + v^{p+1} w^{\alpha} \eta^2 |\nabla \eta|^2$$

$$\|z(t)\|^2_{L^2(\omega_t)} \leq \|v\|_{L^\infty(\omega)} \int_{\omega_t} v^p w^\alpha \eta^2.$$

We plug these inequalities into (3.6) and (3.4) to obtain (3.5).

We are now ready to iterate inequality (3.5). We set

$$Q := q/(q-2) = (N+2)/2, \; \nu = q/2, \tag{3.7}$$

$$\alpha_n := (\alpha_0 - Q)\nu^n + Q, \; p_n = (p_0 + 2Q)\nu^n - 2Q, \tag{3.8}$$

where

$$\alpha_0 > Q, \; p_0 > -2Q. \tag{3.9}$$

Let $(\omega_n)_{n\geqq 0}$ be a decreasing sequence of open subsets of ω and $\eta_n \in C_0^\infty(\omega)$ with $0 \leqq \eta_n \leqq 1$ and such that

$$\omega = \omega_0 \supset \bar{\omega}_1 \supset \ldots \supset \omega_n \supset \bar{\omega}_{n+1} \supset \ldots \supset \bar{\omega}_\infty \supset \hat{\omega} \tag{3.10}$$

$$\eta_n = 1 \text{ on } \omega_n, \; \eta_n = 0 \text{ on } \bar{\omega} \setminus \omega_{n-1}, \tag{3.11}$$

$$\|\eta_{nt}\|_{L^\infty(\omega)} + \|\nabla \eta_n\|^2_{L^\infty(\omega)} \leqq C\, 2^n, \; C = C(\omega, \hat{\omega}). \quad \square \tag{3.12}$$

<u>LEMMA 4</u>: Let $a_n := \{\int_{\omega_n} v^{p_n} w^{\alpha_n}\}^{1/\nu}$. Then

$$\forall n \geqq 1, \; a_n \leqq c\, d^n [1 + a_{n-1}^\nu], \tag{3.13}$$

where $c = c(N, \mu, \alpha_0, p_0, \omega, \hat{\omega}, \|v\|_{L^\infty(\omega)})$, $d = 2\nu^2$.

<u>PROOF</u>: We apply (3.5) with $p = p_{n-1} + 1$, $\alpha = \alpha_{n-1} - 1$, $\eta = \eta_n$, so that $(p+1)\nu = p_n$, $\alpha\nu = \alpha_n$ by (3.8) and, thanks to (3.11), (3.12)

$$a_n \leqq \lambda(\alpha_0+p_0)^2\nu^{2n}[a_{n-1}^{\nu}+c\ 2^n \int_{\omega_{n-1}} v^{p_{n-1}+1}w^{\alpha_{n-1}-1}(1+v)]. \tag{3.14}$$

We use that

$$\int_{\omega_{n-1}} v^p w^\alpha \leqq \int_{(v^p w^\alpha<1)\cap\omega_{n-1}} v^p w^\alpha + \int_{[v^p w^\alpha>1]\cap\omega_{n-1}} (v^p w^\alpha)^{(\alpha+1)/\alpha}$$

$$\leqq \text{mes}\ \omega + \|v\|^{p/\alpha+1}_{L^\infty(\omega)} \int_{\omega_{n-1}} v^{p-1}w^{\alpha+1}, \tag{3.15}$$

where $p/\alpha = (p_{n-1}+1)/(\alpha_{n-1}-1) \leqq (p_0 + Q)/(\alpha_0 - Q)$. Using this remark in (3.14) yields (3.13). □

PROOF OF THE THEOREM: We use the notations of Lemmas 1-4. From Lemma 4 and the Technical Lemma 5.6, page 95 in [6], we obtain

$$\forall n \geqq 1,\quad a_n \leqq c_1^{\nu^n}\{\max(a_0,1)\}^{\nu^n},\quad c_1 = c_1(c,d).$$

Therefore, for α_0, p_0 satisfying (3.9), we have

$$\|v^{p_0+2Q}w^{\alpha_0-Q}\|^{1/\nu}_{L^\infty(\omega)} \leqq \limsup_{n\to\infty} a_n^{-\nu^n} \leqq c_1[1 + \{\int_\omega v^{p_0}w^{\alpha_0}\}^{1/\nu}]. \tag{3.16}$$

Now let $\beta > N + 2$, $\sigma = \beta/(\beta - N - 2)$ and $r > 0$ satisfying (2.4). We set $\alpha_0 = \beta/2 > Q$ and $p_0 = 2\alpha_0(r-1)$. By (2.4)

$$p_0 > 2\alpha_0\ (\frac{2\alpha_0-N-2}{2\alpha_0} - 1) = -(N + 2) = -2Q.$$

Thus (3.16) holds and taking its ν-th power yields (2.5). □

References

[1] Aronson, D.G., Regularity properties of flows through porous media, SIAM J. Applied Math. 17 (1969), 461-468.

[2] Aronson, D.G., The porous medium equation in nonlinear diffusion problem, Lecture Notes in Math. 1224, CIME Series, Springer-Verlag 1986.

[3] Benilan, Ph., Pointwise estimates for the porous media type equations, Cours University of Kentucky, 1981.

[4] Caffarelli, L.A., Vazquez, J.L. and Wolanski, N.I., Lipschitz continuity of solutions and interfaces of the N-dimensional porous medium equation, Indiana Univ. Math. J. 36, No. 2 (1987).

[5] Caffarelli, L.A. and Friedman, A., Regularity of the free boundary of a gas flow in an n-dimensional porous medium, Indiana Univ. Math. J. 29 (1980), 361-391.

[6] Ladyzhenskaya, O.A., Solonnikov, V.A. and Ural'ceva, N.N., Linear and quasi-linear equations of parabolic type, Transl. Math. Monographs, Vol. 23, A.M.S. Providence, 1968.

[7] Moulay, M.S., Thèse d'Etat, Alger, 1988.

[8] Moulay, M.S. and Pierre, M., Regularizing effect for a nonlinear diffusion with irregular source-term, J. Nonlinear Analysis, TMA (to appear).

[9] Moulay, M.S. and Pierre, M., in preparation.

[10] Di Benedetto, E. and Friedman, A., Hölder estimates for nonlinear degenerate parabolic systems, J. für die Reine und Angew. Math. 357 (1985), 1-22.

M. Moulay and M. Pierre
Département de Mathématiques
BP 239 54506 Vandoeuvre-lès-Nancy Cedex
France

2. TIME-DEPENDENT SYSTEMS

P. ACQUISTAPACE AND B. TERRENI

Fully nonlinear parabolic systems

PRELIMINARIES

This article is concerned with the existence of continuously differentiable solutions of fully nonlinear parabolic systems. For the sake of simplicity we just consider here second-order systems of this kind:

$$\begin{aligned} &D_t u^h = f^h(t,x,u,Du,D^2u), && (t,x) \in [0,T] \times \bar{\Omega}, \\ &g^h(t,x,u,Du) = 0, && (t,x) \in [0,T] \times \partial\Omega, \\ &u^h(0,x) = \phi^h(x), && x \in \bar{\Omega} \quad (h = 1,\ldots,N), \end{aligned} \tag{0.1}$$

where $T > 0$, Ω is a bounded, smooth domain of R^n and f, g are smooth $\mathbb{C}^N$-valued functions.

The first result on local existence for fully nonlinear parabolic problems is due to Hudjaev [7] (see also Eidel'man [5, Section III.4]), by quasi-linearization. The technique of linearization was first used by Sopolov [12] and Kruzhkov, Castro and Lopez [8], [9], [10] in the case of boundary conditions of Dirichlet or quasi-linear oblique derivative type. Linearization is also used, in an abstract framework based on semigroup theory, by Da Prato and Grisvard [4] and Lunardiand Sinestrari [11]; these papers, however, concern a smaller class of nonlinear parabolic problems.

In this article we also use linearization, but our technique seems simpler than the previous ones and allows us to consider more general (nonlinear) boundary conditions. It has been previously used (Acquistapace and Terreni [1], [2]) for quasi-linear parabolic systems.

Here is the plan of this paper. After a survey of the main properties of the linear autonomous version of (0.1) (Section 1), we will show in Section 2 that problem (0.1) possesses a unique local solution, whereas in Section 3 we will prove the regularity properties of the maximal solution of (0.1) with respect to the initial datum.

We assume the following hypotheses:

HYPOTHESIS 0.1 (Regularity): The boundary of Ω and the functions $f(t,x,u,p,q)$, $g(t,x,u,p)$, $\phi(x)$ satisfy

$$\partial\Omega \in C^{2+2\alpha}, \qquad \phi \in C^{2+2\alpha}(\bar{\Omega},\mathbb{C}^N),$$
$$f \in C^2(\Lambda,\mathbb{C}^N),\ g \in C^3(\Lambda',\mathbb{C}^N),\quad (\alpha \in]0,1/2[), \tag{0.2}$$

where

$$\Lambda := [0,\infty[\times \bar{\Omega} \times \mathbb{C}^N \times \mathbb{C}^{nN} \times \mathbb{C}^{n^2N},\quad \Lambda' := [0,\infty[\times \bar{\Omega} \times \mathbb{C}^N \times \mathbb{C}^{nN}. \tag{0.3}$$

HYPOTHESIS 0.2 (Parabolicity): The complex-valued functions $(s,\ j = 1,\ldots,n;\ h,\ k = 1,\ldots,N)$

$$A_{sj}^{hk}(x) := \frac{\partial f^h}{\partial q_{sj}^k}\ (0,x,\phi(x),D\phi(x),D^2\phi(x)), \tag{0.4}$$

$$B_j^{hk}(x) := \frac{\partial g^h}{\partial p_j^k}\ (0,x,\phi(x),D\phi(x))$$

satisfy the following ellipticity assumptions: there exists $\theta_0 \in]\pi/2,\pi[$ such that:

(θ_0-root condition) for each $\theta \in]-\theta_0,\theta_0[$ the operator $L_\theta(x,D_t,D_x)$ defined by (0.5)

$$L_\theta(x,D_t,D_x)v := \{A_{sj}^{hk}(x)D_sD_jv^k(t,x) + e^{i\theta}\delta^{hk}D_t^2v(t,x)\}_{h=1,\ldots,N}$$

is properly elliptic in $\mathbb{R} \times \bar{\Omega}$ (see Geymonat and Grisvard [6, Hypothesis $\widetilde{II}$, page 163 and $(\widetilde{AN},\theta)$, page 167]);

(θ_0-complementing condition) for each $\theta \in]-\theta_0,\theta_0[$ the boundary (0.6) operator $\Gamma(x,D_t,D_x)$ defined by

$$\Gamma(x,D_t,D_x)v := \{B_j^{hk}(x)D_jv^k(t,x)\}_{h=1,\ldots,N}$$

fulfills the complementing condition with respect to $L_\theta(x,D_t,D_x)$ in $\partial(R \times \Omega)$ see Geymonat and Grisvard [6, Hypotheses $\widetilde{III}$, page 164 and $(\widetilde{AN},\theta)$, page 167]);

<u>HYPOTHESIS 0.3</u> (Compatibility): The initial datum ϕ satisfies

$$g^h(0,x,\phi(x),D\phi(x)) = 0 \quad \text{on} \quad \partial\Omega, \quad h = 1,\ldots,N. \tag{0.7}$$

<u>REMARKS 0.4</u>:

(a) The order of the parabolic system (0.1) might be 2m rather than 2; this would only require longer calculations. Moreover, the regularity exponent α in (0.2) might be any number in $]0,1[\setminus\{1/2\}$, but when $\alpha > 1/2$ a further compatibility condition along $\partial\Omega$ appears. A more general study, covering these cases, will be published elsewhere.

(b) In (0.4) and (0.6) it is tacitly supposed that the function g(t,x,u,p) really depends on p, i.e. the boundary operator is of first order; however, it would be possible to consider boundary operators of different orders, such as:

$$g^h = \begin{cases} g^h(t,x,u,p), & h = 1,\ldots,r_0, \\ g^h(t,x,u), & h = r_0 + 1,\ldots,N, \end{cases}$$

(see Terreni [13] for the linear case).

(c) We shall need in Section 3 a stronger version of Hypotheses 0.1, 0.2.

We will look for solutions u of (0.1) in the Banach space

$$\begin{aligned} E_T &:= C^{1+\alpha,2+2\alpha}([0,T] \times \bar{\Omega}, \mathbb{C}^N) \qquad (0.8) \\ &\equiv C^{1+\alpha}([0,T], C^0(\bar{\Omega},\mathbb{C}^N)) \cap L^\infty(0,T;C^{2+2\alpha}(\bar{\Omega},\mathbb{C}^N)) \end{aligned}$$

for a suitable $T > 0$, endowed with the norm (obviously equivalent to the usual one)

$$\|u\|_{E_T} := \sup_{t\in[0,T]} \|u(t,\cdot)\|_C + [D_t u]_{C^\alpha(C)} + \sup_{t\in[0,T]} [D_x^2 u(t,\cdot)]_{C^{2\alpha}}. \quad (0.9)$$

Here C, $C^\alpha(C)$, $C^{2\alpha}$ stand for $C(\bar{\Omega},\mathbb{C}^N)$, $C^\alpha([0,T],C(\bar{\Omega},\mathbb{C}^N))$, $C^{2\alpha}(\bar{\Omega},\mathbb{C}^N)$.

We start by introducing an auxiliary function z which will be useful later on. First of all consider the differential operator

$$L(x,D_x)v := \{A_{sj}^{hk}(x)D_sD_jv(t,x)\}_{h=1,\dots,N}, \quad (0.10)$$

where the functions A_{sj}^{hk} were defined in (0.4).

<u>LEMMA 0.5</u>: There exists a function $z : [0,T] \times \bar{\Omega} \to \mathbb{C}^N$ such that:

$$z(0,\cdot) = \phi, \quad D_t z(0,\cdot) = h := f(0,\cdot,\phi,D\phi,D^2\phi) \text{ in } \bar{\Omega}, \quad (0.11)$$

$$\|z\|_{C^{1+\alpha,2+2\alpha}} + \|D_t z\|_{C^{\alpha,2\alpha}} + \|D_x z\|_{C^{1/2+\alpha,1+2\alpha}} + \|D_x^2 z\|_{C^{\alpha,2\alpha}} \leq \quad (0.12)$$

$$\leq C_0(\alpha,\Omega,f,\phi,T).$$

<u>PROOF:</u> As, by (0.2), (0.3) and (0.11), $h \in C^{2\alpha}(\bar{\Omega},\mathbb{C}^N)$, we can find a function $H \in C^{2\alpha}(R^n,\mathbb{C}^N)$ which extends h and is such that

$$\|H\|_{C^{2\alpha}(R^n,\mathbb{C}^N)} \leq C_1(\alpha,\Omega) \|h\|_{C^{2\alpha}(\bar{\Omega},\mathbb{C}^N)}. \quad (0.13)$$

Define now

$$z(t,x) := \phi(x) + t^{(2-n)/2} \int_{R^n} H(y)\eta(t^{-1/2}(x-y))dy \quad (0.14)$$

$$= \phi(x) + t\cdot(\eta_{\sqrt{t}} * H)(x),$$

where η is a nonnegative, even function from $C_0^\infty(R^n)$ such that $\int_{R^n} \eta(x)dx = 1$; the desired properties then follow in a straightforward way. □

1. <u>THE LINEAR AUTONOMOUS PROBLEM</u>

Consider the problem

$$
\begin{cases}
D_t u^h - A_{sj}^{hk}(x) D_s u^k = f^h(t,x), & (t,x) \in [0,T] \times \bar{\Omega}, \\
B_j^{hk}(x) D_j u^k = g^h(t,x) & (t,x) \in [0,T] \times \partial\Omega, \\
u^h(0,x) = \phi^h(x), & x \in \bar{\Omega} \quad (h = 1,\dots,N).
\end{cases} \tag{1.1}
$$

Suppose that

$$A_{sj}^{hk} \in C^{2\alpha}(\bar{\Omega}), \quad B_j^{hk} \in C^{1+2\alpha}(\bar{\Omega}) \text{ and } (0.5),\ (0.6) \text{ hold}, \tag{1.2}$$

$$f \in C^{\alpha,2\alpha}([0,T] \times \bar{\Omega}, \mathbb{C}^N), \quad g \in C^{\frac{1}{2}+\alpha,1+2\alpha}([0,T] \times \bar{\Omega}, \mathbb{C}^N), \tag{1.3}$$

$$B_j^{hk}(x) D_j \phi^k(x) = g^h(0,x), \ \forall x \in \partial\Omega. \tag{1.4}$$

Due to Lemma 0.5, it is not restrictive for our purposes to suppose moreover that

$$f(0,x) \equiv \phi(x) \equiv 0 \text{ on } \bar{\Omega}, \quad g(0,x) \equiv 0 \text{ on } \partial\Omega. \tag{1.5}$$

<u>THEOREM 1.1</u>: Assume (1.2),...,(1.5). Then for any $T > 0$ problem (1.1) has a unique solution $u \in E_T$, and the following estimate holds:

$$
\begin{aligned}
\|u\|_{E_T} \leq c_2(\alpha,\Omega,A,B)\{[f]_{C^{\alpha}(C)} &+ \sup_{t\in[0,T]} [f(t,\cdot)]_{C^{2\alpha}} + \\
&+ [g]_{C^{\alpha+1/2}(C)} + \sup_{t\in[0,T]} [D_x g(t,\cdot)]_{C^{2\alpha}}\},
\end{aligned} \tag{1.6}
$$

where the constant c_2 does not depend on T.

<u>PROOF</u>: Existence and uniqueness of the solution in E_T were proved in Terreni, [13, Theorem 4.1]. The estimate (1.6) is a consequence of that result and of (1.5). Finally, in order to verify that c_2 is independent of T, we need to revisit the proof of Terreni [13, Lemma 2.4-2.5 and Theorem 4.1], taking into account that we may assume, by Lemma 0.5, $u(0,x) \equiv D_t u(0,x) \equiv 0$ in $\bar{\Omega}$. □

2. THE NONLINEAR PROBLEM, LOCAL EXISTENCE

Consider now problem (0.1). The main result of this section is:

THEOREM 2.1: Suppose that Hypotheses 0.1, 0.2 and 0.3 hold. Then there exists $\tau_* > 0$ such that problem (0.1) has a unique solution $u \in E_{\tau_*}$.

PROOF: It consists of two steps: linearization and use of the contraction principle.

Step 1. Let h be the function defined in (0.11) and let z be the function constructed in Lemma 0.5. For each $M > 0$ consider the (nonempty) set

$$B_{M,T} := \{w \in E_T : \|w-z\|_{E_T} \leq M,\ w(0,\cdot) = \phi,\ D_t w(0,\cdot) = h\} \tag{2.1}$$

which is closed in E_T. This is the set where we will linearize problem (0.1). For fixed $w \in B_{M,T}$ consider the linear autonomous problem

$$\begin{cases} D_t v^h - \dfrac{\partial f^h}{\partial q^k_{sj}}(0,x,\phi,D\phi,D^2\phi)D_s D_j v^k = f^h(t,x,w,Dw,D^2w) \\ \qquad - \dfrac{\partial f^h}{\partial q^k_{sj}}(0,x,\phi,D\phi,D^2\phi)D_s D_j w^k, \quad (t,x) \in [0,T] \times \bar{\Omega}, \\ \dfrac{\partial g^h}{\partial p^k_j}(0,x,\phi,D\phi)D_j v^k = - g^h(t,x,w,Dw) + \dfrac{\partial g^h}{\partial p^k_j}(0,x,\phi,D\phi)D_j v^k, \\ (t,x) \in [0,T] \times \partial\Omega, \\ v^h(0,x) = \phi^h(x), \qquad x \in \bar{\Omega} \qquad (h = 1,\ldots,N). \end{cases} \tag{2.2}$$

The data of problem (2.2) satisfy, for every $w \in B_{M,T}$, the hypotheses of Terreni [13, Theorem 4.1]. Therefore we can find a unique solution $v \in E_T$ of (2.1), so that we have a well-defined map

$$S : B_{M,T} \to E_T, \qquad S(w) := v. \tag{2.3}$$

Step 2. The main properties of the map S are expressed in the following lemma (which is an easy consequence of Lemma 3.6 below). □

LEMMA 2.2: For each $M > 0$ we have:

$$\|S(w) - z\|_{E_T} \leqq c_3(\alpha,\Omega,f,g,\phi) + c_4(\alpha,\Omega,f,g,\phi,M)T^\alpha \quad \forall w \in B_{M,T}; \qquad (2.4)$$

$$\|S(w_1)-S(w_2)\|_{E_T} \leqq c_5(\alpha,\Omega,f,g,\phi,M)T^\alpha \|w_1-w_2\|_{E_T} \quad \forall w_1,w_2 \in B_{M,T}. \qquad (2.5)$$

The inequalities (2.4) and (2.5) show that the map S satisfies

$$S(w) \in B_{M,\tau_*} \qquad \forall w \in B_{M,\tau^*},$$

$$\|S(w_1)-S(w_2)\|_{E_{\tau*}} \leqq \frac{1}{2} \|w_1-w_2\|_{E_{\tau*}} \qquad \forall w_1,w_2 \in B_{M,\tau_*},$$

provided we fix in advance M and τ_* such that

$$M > C_3(\alpha,\Omega,f,g,\phi),$$

$$T \leqq [\min \{[2C_5(\alpha,\Omega,f,g,\phi,M)]^{-1}, \frac{M - c_3(\alpha,f,g,\phi)}{c_4(\alpha,\Omega,f,g,\phi,M)}\}]^{1/\alpha}.$$

Hence we get a unique fixed point $u \in B_{M,\tau_*}$ of the map S, i.e. a unique solution in E_{τ_*} of problem (0.1). □

3. CONTINUOUS DEPENDENCE ON DATA

Consider again problem (0.1), replacing Hypotheses 0.1 and 0.2 by the following stronger ones:

HYPOTHESIS 3:1: $\partial\Omega \in C^{2+2\alpha}$, $\phi \in C^{2+2\alpha}$, $f \in C^3(\Lambda,\mathbb{C}^N)$, $g \in C^4(\Lambda',\mathbb{C}^N)$ ($\alpha \in]0,1/2[$), where Λ, Λ' are defined by (0.3).

HYPOTHESIS 3.2: The complex-valued functions

$$A_{sj}^{hk}(x) := \frac{\partial f^h}{\partial q_{sj}^k}(t,x,u,p,q), \qquad B_j^{hk}(x) := \frac{\partial g^h}{\partial p_j^k}(t,x,u,p),$$

satisfy Hypothesis 0.2 uniformly in (t,u,p,q) on compact subsets of $[0,\infty[\times \mathbb{C}^N \times \mathbb{C}^{nN} \times \mathbb{C}^{n^2N}$.

Moreover, we reformulate Hypothesis 0.3 as follows:

HYPOTHESIS 3.3: Consider for $\tau > 0$ the closed submanifold of $C^{2+2\alpha}(\bar{\Omega},\mathbb{C}^N)$ defined by

$$V_g(\tau) := \{\phi \in C^{2+2\alpha}(\bar{\Omega},\mathbb{C}^N) : g(\tau,x,\phi(x),D\phi(x)) = 0, \forall x \in \partial\Omega\}; \qquad (3.2)$$

we assume that $V_g(0)$ is not empty.

If we now choose any $\phi \in V_g(0)$, Theorem 2.1 yields a unique solution $u \in E_{\tau_*}$ of (0.1); hence we have

$$u(\tau_*,\cdot) \in V_g(\tau_*).$$

This condition and Hypotheses 3.1 and 3.2 allow us to extend the solution of (0.1) to an interval $[\tau_*,\tau_* + \tau_1]$. At the point τ_1 we can start again, and so on. In short, we can consider the maximal interval of existence $[0,\tau_\phi[$, by defining:

$$\tau_\phi := \sup \{\tau > 0: \text{problem (0.1) has a unique solution in } E_\tau\}, \qquad (3.3)$$

$$u_\phi := \text{the unique solution of (0.1) in } [0,\tau_\phi[. \qquad (3.4)$$

Consider now the set $D \subset V_g(0) \times [0,\infty[$ defined by

$$D := \{(\phi,t) \in V_g(0) \times [0,\infty[: t < \tau_\phi\}. \qquad (3.5)$$

THEOREM 3.4: Under Hypotheses 3.1, 3.2 and 3.3 we have:

(i) D is open in $V_g(0) \times [0,\infty[$ in the topology of $C^{2+2\alpha}(\bar{\Omega},\mathbb{C}^N) \times [0,\infty[$;

(ii) the map

$$U : D \to C^{2+2\alpha}(\bar{\Omega},\mathbb{C}^N), \quad u(\phi,t) = u_\phi(t,\cdot) \qquad (3.6)$$

is continuous as a function from D into $C^{2+2\delta}(\bar{\Omega},\mathbb{C}^N)$ for each $\delta \in [0,\alpha[$.

PROOF: We need a more general version of Theorem 2.1. First of all, we remark that

$$\|U_\phi\|_{E_{\tau_\phi-\varepsilon}} \leq C_6(\alpha,\Omega,f,g,\phi,\varepsilon) \qquad \forall\varepsilon \in]0,\tau_\phi[. \tag{3.7}$$

Next, for $t_0 \geq 0$, $r_0 > 0$, $\phi_0 \in C^{2+2\alpha}(\bar{\Omega},\mathbb{C}^N)$ we set

$$Q(\phi_0,t_0,r_0) := \{\psi \in V_g(t_0) : \|\psi-\phi_0\|_{C^{2+2\alpha}} \leq r_0\}. \tag{3.8}$$

Of course, if $\phi \in V_g(0)$ and $t_0 \in [0,\tau_\phi[$ then

$$u_\phi(t_0,\cdot) \in Q(u_\phi(t_0,\cdot),t_0,r_0) \qquad \forall r_0 > 0.$$

Now for $0 \leq t_0 < t_0 + \sigma =: \tau$ define (analogously to (0.8), (0.9)) the Banach space

$$E_{t_0,\tau} := C^{1+\alpha,2+2\alpha}([t_0,\tau] \times \bar{\Omega},\mathbb{C}^N).$$

PROPOSITION 3.5: Let $\phi \in V_g(0)$. For each $\varepsilon \in]0,\tau_\phi[$ and $r_0 > 0$ there exists $\sigma \in]0,\varepsilon]$ depending on $\alpha,\Omega,f,g,\phi,r_0,\varepsilon$, such that for every $t_0 \in [0,\tau_\phi - \varepsilon]$ and $\psi \in Q(u_\phi(t_0,\cdot),t_0,r_0)$ the problem

$$\begin{cases} D_t v^h = f^h(t,x,v,Dv,D^2v), & (t,x) \in [t_0,\tau] \times \bar{\Omega}, \\ g^h(t,x,v,Dv) = 0, & (t,x) \in [t_0,\tau] \times \partial\Omega, \\ v^h(t_0,x) = \psi^h(x), & x \in \bar{\Omega} \quad (h = 1,\dots,N), \end{cases} \tag{3.9}$$

where $\tau := t_0 + \sigma$, has a unique solution $v_\psi \in E_{t_0,\tau}$: moreover, there exists $c_7(\alpha,\Omega,f,g,\phi,r_0,\varepsilon) \geq 1$ such that

$$\|v_\psi-v_\chi\|_{E_{t_0,\tau}} \leq c_7(\alpha,\Omega,f,g,\phi,r_0,\varepsilon)\|\psi-\chi\|_{C^{2+2\alpha}} \quad \forall\psi,\chi \in Q(u_\phi(t,\cdot),t_0,r_0). \tag{3.10}$$

PROOF: We repeat, "mutatis mutandis", the proof of Theorem 2.1.

Fix $\varepsilon \in]0,\tau_\phi[$, $r_0 > 0$, $t_0 \in [0,\tau_\phi-\varepsilon]$, $\psi \in Q(u_\phi(t_0,\cdot),t_0,r_0)$ and let $\sigma \in]0,\varepsilon]$. Choose a function $z_\psi \in E_{t_0,\tau}$ such that

$$z_\psi(t_0,\cdot) = \psi, \quad D_t z_\psi(t_0,\cdot) = h := f(t_0,\cdot,\psi,D\psi,D^2\psi) \text{ in } \bar{\Omega} \tag{3.11}$$

(this can be done as in Lemma 0.5). Finally set

$$B_{M,T_0,\tau}(\psi) := \{v \in E_{t_0,\tau}: \|v-z_\psi\|_{E_{t_0,\tau}} \leq M,\ v(t_0,\cdot) = \psi,\ D_t v(t_0,\cdot) = h\}. \tag{3.12}$$

For fixed $w \in B_{M,T_0,\tau}(\psi)$ we linearize problem (3.9) in the following way:

$$\begin{aligned} & D_t v^h - \frac{\partial f^h}{\partial q^k_{sj}}(t_0,x,\psi,D\psi,D^2\psi) D_s D_j v^k \\ & \quad = F_w(t,x), && (t,x) \in [t_0,\tau] \times \bar{\Omega}, \\ & \frac{\partial g^h}{\partial p^k_j}(t_0,x,\psi,D\psi) D_j v^k = G_w(t,x), && (t,x) \in [t_0,\tau] \times \partial\Omega, \\ & v^h(t_0,x) = \psi^h(x), && z \in \bar{\Omega} \quad (h = 1,\dots,N), \end{aligned} \tag{3.13}$$

where

$$\begin{aligned} F_w(t,x) &:= \{f^h(t,x,w,Dw,D^2w) - \frac{\partial f^h}{\partial q^k_{sj}}(t_0,x,w,Dw,D^2w) D_s D_j w\}_{h=1,\dots,N}, \\ G_w(t,x) &:= \{-g^h(t,x,w,Dw) + \frac{\partial g^h}{\partial p^k_j}(t_0,x,\psi,D\psi\ ) D_j w\}_{h=1,\dots,N}. \end{aligned} \tag{3.14}$$

We remark that by (3.8), (3.7) we have

$$\begin{aligned} \|\psi\|_{C^{2+2\alpha}} &\leq \|\psi-u_\phi(t_0,\cdot)\|_{C^{2+2\alpha}} + \|u_\phi(t_0,\cdot)\|_{C^{2+2\alpha}} \leq \\ &\leq r_0 + c_6(\alpha,\Omega,f,g,\phi,\varepsilon) \quad \forall\psi \in Q(u_\phi(t_0,\cdot),t_0,r_0). \end{aligned} \tag{3.15}$$

Next, by Hypothesis 3.2 we see that the coefficients of the linear problem

(3.13), as well as their derivatives, are bounded by constants depending on f,g,ϕ,r_0,ε but not on t_0; the same holds for the constants involved by ellipticity and by the complementing condition. Moreover, it is easy to see that

$$F_w \in C^{\alpha,2\alpha}([t_0,\tau] \times \bar{\Omega},\mathbb{C}^N), \; G_w \in C^{\frac{1}{2}+\alpha,1+2\alpha}([t_0,\tau] \times \bar{\Omega},\mathbb{C}^N), \; \psi \in C^{2+2\alpha}(\bar{\Omega},\mathbb{C}^N)$$

and that the compatibility condition

$$\frac{\partial g}{\partial p_j^k}(t_0,x,\psi(x),D\psi(x))D_j\psi^k(x) = G_w(0,x) \qquad \forall x \in \partial\Omega,$$

holds. Hence by revisiting the proof of Terreni, [13, Theorem 4.1] we deduce the existence of a unique solution $v \in E_{t_0,\tau}$ of problem (3.13). Thus the map

$$S_\psi : B_{M,t_0,\tau}(\psi) \to E_{t_0,\tau}, \quad S_\psi(w) = v, \tag{3.16}$$

is well defined; in addition, the analogue of Lemma 2.2 holds (whose proof is straightforward although very tedious):

<u>LEMMA 3.6</u>: For each $M > 0$ we have:

$$\|S_\psi(w) - z_\psi\|_{E_{t_0,\tau}} \leq c_8(\alpha,\Omega,f,g,\phi,r_0,\varepsilon) + c_9(\alpha,\Omega,f,g,\phi,r_0,\varepsilon,M)(\tau-t_0)^\alpha \tag{3.17}$$

$$\forall\psi \in Q(u_\phi(t_0,\cdot),t_0,r_0), \qquad \forall w \in B_{M,t_0,\tau}(\psi);$$

$$\|S_\psi(v) - S_\chi(w)\|_{E_{t_0,\tau}} \leq c_{10}(\alpha,\Omega,f,g,\phi,r_0,\varepsilon,M)\{\|\psi-\chi\|_{C^{2+2\alpha}} + (\tau-t_0)^\alpha \|v-w\|_{E_{t_0,\tau}}\} \tag{3.18}$$

$$\forall\psi,\chi \in Q(u_\phi(t_0,\cdot),t_0,r_0), \quad \forall v \in B_{M,t_0,\tau}(\psi), \quad \forall w \in B_{M,t_0,\tau}(\chi).$$

As in Section 2, the above lemma implies that there exists $\sigma \in]0,\varepsilon]$ depending on $(\alpha,\Omega,f,g,\phi,r_0,\varepsilon)$ such that problem (3.9) has a unique solution $v_\psi \equiv S_\psi(v\psi) \in E_{t_0,\tau}$.

Finally (3.10) follows by (3.18). This completes the proof of Proposition 3.5. □

Let us prove now the first part of Theorem 3.4. Fix $(\phi,t_*) \in D$, which means $\phi \in V_g(0)$ and $t_* \in [0,\tau_\phi[$. Fix $r_0 > 0$ and $\varepsilon \in]0,\frac{1}{2}(\tau_\phi - t_*)[$, and let $\sigma \in]0,\varepsilon]$ be the number determined by Proposition 3.5. Define

$$s := [(t_*+\varepsilon)/\sigma], \quad \tau_k := k\sigma, \quad k = 0,1,\ldots,s+1;$$

then, in particular,

$$t_* + \varepsilon < \tau_{s+1} = (s + 1)\sigma \leq (\frac{t_* + \varepsilon}{\sigma} + 1)\sigma = t_* + \varepsilon + \sigma \leq t_* + 2\varepsilon < \tau_\phi .$$

Now let $\psi \in V_g(0)$ be such that

$$||\psi - \phi||_{C^{2+2\alpha}} < r_0[c_7(\alpha,\Omega,f,g,\phi,r_0,\varepsilon)]^{-1-s} \leq r_0;$$

according to (3.4), denote the maximal solutions of (0.1) with initial data ϕ, ψ by u_ϕ, u_ψ. By applying Proposition 3.5 with $t_0 = \tau_0 = 0$, we get, since $\psi \in Q(\phi,0,r_0)$,

$$\|u_\phi - u_\psi\|_{E_{0,\tau_1}} \leq c_7 \|\phi-\psi\|_{C^{2+2\alpha}} \leq r_0 c_7^{-s} \leq r_0,$$

which implies, in particular, $u_\psi(\tau_1,\cdot) \in Q(u_\phi(\tau_1,\cdot), \tau_1,r_0)$. An iteration of this argument shows, since the solution of (3.9) is unique, that

$$\|u_\phi - u_\psi\|_{E_{\tau_k,\tau_{k+1}}} \leq r_0 c_7^{-s+k} \leq r_0, \quad k = 0,1,\ldots,s. \tag{3.19}$$

The last step proves that u_ψ is defined in the whole interval $[0,\tau_{s+1}]$ and hence in $[0,t_* + \varepsilon]$; consequently, by definition, $t_* + \varepsilon < \tau_\phi$. We have shown that $(\psi,t) \in D$ provided $(\phi,t_*) \in D$ and

$$\psi \in V_g(0), \ \|\psi-\phi\|_{C^{2+2\alpha}} < r_0[c_7(\alpha,\Omega,f,g,\phi,r_0,\varepsilon)]^{-[(t_*+\varepsilon)/\sigma]-1},$$

$$|t-t_*| < \varepsilon, \quad \varepsilon \in]0,\frac{1}{2}(\tau_\phi-t_*)[.$$

This show that D is open in $V_g(0) \times [0,\infty[$.

Let us prove now the second part of Theorem 3.4. Fix $r_0 > 0$, $(\phi,t) \in D$ and let

$$(\psi,s) \in I := \{\psi \in V_g(0): \|\psi-\phi\|_{C^{2+2\alpha}} < \varepsilon\} \times]t-\varepsilon, \ t+\varepsilon[,$$

where $\varepsilon \in]0,\min\{t,\tau_\phi-t\}[$. Then (3.10) yields for each $s \in]0,t+\varepsilon[$:

$$\|u_\phi(s,\cdot)-u_\psi(s,\cdot)\|_{C^{2+2\alpha}} \leq c_7(\alpha,\Omega,f,g,\phi,r_0,\varepsilon)\|\phi-\psi\|_{C^{2+2\alpha}}. \tag{3.20}$$

On the other hand, we recall that, by interpolation (see Terreni, [13, Lemma 2.5])

$$E_{0,\tau_\phi-\varepsilon} \hookrightarrow C^{\alpha-\delta}([0,\tau_\phi-\varepsilon],C^{2+2\delta}(\bar{\Omega},\mathbb{C}^N)) \quad \forall\delta \in]0,\alpha[. \tag{3.21}$$

By (3.21) and (3.20) we conclude that

$$\begin{aligned}\|u_\phi(t,\cdot) - u_\psi(s,\cdot)\|_{C^{2+2\delta}} \\ &\leq \|u_\phi(t,\cdot)-u_\phi(s,\cdot)\|_{C^{2+2\delta}} + \|u_\phi(s,\cdot) - u_\psi(s,\cdot)\|_{C^{2+2\alpha}} \\ &\leq c_{11}(\alpha,\Omega,f,g,\phi,r_0,\varepsilon) \{|t-s|^{\alpha-\delta} + \|\phi-\psi\|_{C^{2+2\alpha}}\}.\end{aligned}$$

The proof of Theorem 3.4 is complete. □

REMARKS 3.7

(a) Hypothesis 3.1 is required just in the proof of (3.18); in order to get (3.17) Hypothesis 0.1 (together with Hypotheses 3.2, 3.3) would be sufficient.

(b) In order to have continuity of the map U (see (3.6)) in the topology of $C^{2+2\alpha}(\bar{\Omega},\mathbb{C}^N)$ we have to take $\phi \in h^{2+2\alpha}(\bar{\Omega},\mathbb{C}^N)$ and to repeat the proof in the Banach space

$$\begin{aligned}E_{t_0,\tau} &:= h^{1+\alpha,2+2\alpha}([t_0,\tau] \times \bar{\Omega},\mathbb{C}^N) \\ &\equiv h^{1+\alpha}([t_0,\tau],C(\bar{\Omega},\mathbb{C}^N) \cap C([t_0,\tau],h^{2+2\alpha}(\bar{\Omega},\mathbb{C}^N));\end{aligned}$$

here h^{α} means "little α-Hölder continuous", i.e.

$$F \in h^{\alpha} \Longleftrightarrow |F(x) - F(y)| = o(|x-y|^{\alpha}) \quad \text{as } |x-y|^{\alpha} \downarrow 0.$$

(c) If the boundary condition is autonomous, i.e.

$$\frac{\partial g}{\partial t}(t,x,u,p) = 0 \quad \forall (t,x,u,p) \in [0,\infty[\times \partial\Omega \times \mathbb{C}^N \times \mathbb{C}^{nN},$$

then the map U generates a semiflow (see Amann, [3], which is bounded in $C^{2+2\alpha}(\bar{\Omega},\mathbb{C}^N)$ and continuous in $C^{2+2\delta}(\bar{\Omega},\mathbb{C}^N)$ $\forall \delta \in]0,\alpha[$ (or in $h^{2+2\alpha}(\bar{\Omega},\mathbb{C}^N)$).

References

[1] P. Acquistapace and B. Terreni, On quasilinear parabolic systems Math. Ann. 282 (1988), 315-335.

[2] P. Acquistapace and B. Terreni, Quasilinear parabolic integrodifferential systems with fully nonlinear boundary conditions. Proceedings of the Meeting "Volterra integrodifferential equations in Banach spaces and applications", Trento 1987, Pitman (in press).

[3] H. Amann, Quasilinear parabolic systems under nonlinear boundary conditions, Arch. Rat. Mech. An. 92 (1986), 153-192.

[4] G. Da Prato and P. Grisvard, Équations d'evolution abstraites nonlinéaires de type parabolyque, Ann. Mat. Pures Appl. (4) 120, (1979), 329-396.

[5] S.D. Eidel'man, Parabolic Systems, North-Holland, Amsterdam/London; Noordhoff, Gröningen, 1969.

[6] G. Geymonat and P. Grisvard, Alcuni risultati di teoria spettrale per i problemi ai limiti lineari ellittici, Rend. Sem. Mat. Univ. Padova 38 (1967), 121-173.

[7] S.I. Hudjaev, The first boundary-value problem for nonlinear parabolic equations, Dokl. Akad. Nauk 149 (1963), 535-538 (Russian); English transl.: Soviet Math. Dokl. 4 (1963), 441-445.

[8] S.N. Kruzhkov, A. Castro and M. Lopez, Schauder-type estimates and existence theorems for the solution of basic problems for linear and nonlinear parabolic equations, Dokl. Akad. Nauk 220 (1975), 277-280 (Russian); English transl.: Soviet Math. Dokl. 16 (1975), 60-64.

[9] S.N. Kruzhkov, A. Castro and M. Lopez, Mayoraciones de Schauder y teoremas de existencia de las soluciones del problema de Cauchy para ecuaciones parabolicas lineales y no lineales, I, Cienc. Mat. (Havana) 1 (1980), 55-76.

[10] S.N. Kruzhkov, A. Castro and M. Lopez, Mayoraciones de Schauder y teoremas de existencia de las soluciones del problema de Cauchy para ecuaciones parabolicas lineales y no lineales, II, Cienc. Mat. (Havana) 3 (1982), 37-56.

[11] A. Lunardi and E. Sinestrari, Fully nonlinear integrodifferential equations in general Banach space, Math. Z. 190 (1985), 225-248.

[12] N.N. Sopolov, The first boundary value problem for nonlinear parabolic equations of arbitrary order, C.R. Acad. Bulgare Sci. 23 (1970), 899-902 (Russian).

[13] B. Terreni, Hölder regularity results for non homogeneous parabolic initial-boundary value problems, Proceedings of the Meeting "Trends in semigroup theory and applications", Trieste 1987, Lecture Notes in Pure Appl. Math., M. Dekker (in press).

P. Acquistapace
Università di Roma "La Sapienza"
Dipartimento M.M.M.S.A.
Via A. Scarpa 10
00161 Roma
Italy

B. Terreni
Dipartimento di Matematica
Università di Milano
via C. Saldini 50
20133 Milano
Italy

S. AMRAOUI

On the bifurcation at double critical values for the Brusselator

We are interested in the bifurcation for the following stationary reaction-diffusion system:

$$\begin{cases} d_1 \, \Delta u = (b+1)u - u^2 v - a, \\ d_2 \, \Delta v = -bu + u^2 v \quad \text{on } \Omega, \end{cases} \tag{0.1}$$

where Ω is a bounded open set in R^N, and a, b, d_1, d_2 are positive constants. We consider different boundary conditions and will make it precise later; however, the boundary conditions will be satisfied by the constant "trivial" solution of (0.1): u = a, v = b/a.

Many articles are related to the system (0.1); we mention here only those in which the bifurcation with respect to the trivial solution is studied. In the case N = 1, existence of bifurcation at single critical values has been proved in [3] and [9] for Dirichlet boundary conditions. In a former paper [1], we have proved the same kind of results for some cases at double critical values. In this paper we are interested by the cases when the techniques of [1] do not apply. We mention also that bifurcation at a double critical value has been studied, in particular, in [5], [6], [1] and [8], but the results cannot apply to the situations we consider.

In order to make precise the results we obtain, we will focus on the particular example where Ω =]-L,L[periodic boundary conditions

$$\begin{cases} u(-L) = u(L), \quad v(-L) = v(L), \\ u'(-L) = u'(L), \quad v'(-L) = v'(L). \end{cases} \tag{0.2}$$

For this problem we may state:

<u>THEOREM 0</u>: Consider the problem (0.1)-(0.2) where Ω =]-L, +L[and take L as parameter. Then there is bifurcation with respect to the branch of

trivial solution $\{(L,a,\ b/a);\ L > 0\}$ if and only if the following condition holds:

$$b > (1 + a\sqrt{\frac{d_1}{d_2}})^2. \tag{0.3}$$

In Section 1 we study the critical values for the bifurcation with respect to the trivial solutions, taking the "size" of Ω as a parameter: we will see that for almost any choice of boundary conditions, the relation

$$b \geqq (1 + a\sqrt{\frac{d_1}{d_2}})^2$$

is a necessary condition for bifurcation. In the case $\Omega =]-L,L[$ with periodic boundary conditions, we list all the critical values with their order which may be 2 or 4.

In Section 2 we prove Theorem 0 and describe the bifurcation from the double critical values. Other cases can be studied with the same techniques: we give in Section 3 some examples of such results without proofs (see [2]). In Sections 1 and 2, we take the "size" L as parameter of bifurcation; in Section 3, we also consider b as a parameter.

1. CRITICAL POINTS

By taking the size of Ω as parameter, we mean that after dilatation, the problem is reduced to the system

$$\begin{cases} \dfrac{d_1}{L^2}\Delta u = (b+1)u - u^2 v - a \\[2ex] \dfrac{d_2}{L^2}\Delta v = -bu + u^2 v \quad \text{on } \Omega, \end{cases} \tag{1.1}$$

where Ω is now a fixed bounded open set in R^N and L is a parameter.

Let U be the Banach algebra of the continuous functions on $\bar{\Omega}$ and let D(A) be a linear subspace of $\{w \in U;\ \Delta w \in U\}$ which defines the boundary conditions:

$$u - a \in D(A), \quad v - \frac{b}{a} \in D(A). \tag{1.2}$$

Set $w = d_1(u-a) + d_2(v - b/a)$. Then (u,v) is the solution of (1.1)-(1.2) if and only if

$$\Delta w = \lambda(u-a), \quad w = d_1(u-a) + d_2(v - \frac{b}{a}), \tag{1.3}$$

where w is the solution of

$$T(\lambda)w + H(\lambda,w) = 0 \tag{1.4}$$

with

$$\begin{cases} T(\lambda) = A^2 - 2\lambda \underline{a}A + \lambda^2 a^2 I, \\ H(\lambda,w) = h_1 \lambda w A w + h_2 (Aw)^2 + h_3 w (Aw)^2 + \frac{h_4}{\lambda}(Aw)^3, \end{cases} \tag{1.5}$$

$$\begin{cases} A = -\sqrt{d_1 d_2}\Delta \text{ with domain } D(A), \\ \lambda = L^2, \ \underline{a} = \frac{1}{2}\left[\sqrt{\frac{d_2}{d_1}}(b-1) - \sqrt{\frac{d_1}{d_2}}\, a^2\right], \\ h_1 = -\frac{2a}{\sqrt{d_1 d_2}}, \ h_2 = \frac{b}{a d_1} - \frac{2a}{d_2}, \ h_3 = \frac{1}{d_1 d_2}, \ h_4 = \frac{1}{d_2\sqrt{d_1 d_2}} . \end{cases} \tag{1.6}$$

The map $(L,u,v) \to (\lambda,w)$ is a one-to-one correspondence between the solutions (L,u,v) of (1.1)-(1.2) and the solutions (λ,w) of (1.4) which applies the trivial branch $\{(L,a,b/a)\}$ onto the trivial branch $\{(\lambda,0)\}$. Then, by the implicit function theorem, the following result is immediate:

PROPOSITION 1.1: Let $L_0 > 0$. Then a necessary condition for $(L_0,a,b/a)$ to be a bifurcation point with respect to the trivial branch $\{(L,a,b/a)\}$ of solutions of (1.1)-(1.2) is the existence of $\mu \in \sigma(A)$ and δ root of the equation $\delta^2 - 2\underline{a}\delta + a^2 = 0$ such that $\lambda_0 = L_0^2 = \mu/\delta$.

Assume now the boundary conditions, i.e. $D(A)$, are such that A is symmetric nonnegative with compact resolvent. Then the spectrum $\sigma(A)$ is contained in $[0,\infty[$ and a necessary condition for existence of a bifurcation point $(L_0,a,b/a)$ is the existence of a positive root of $\delta^2 - 2\underline{a}\delta + a^2 = 0$, that is, $\underline{a} \geqq a$. Then we may state

COROLLARY 1.2: Assume A is symmetric nonnegative with compact resolvent. Then a necessary condition for existence of bifurcation with respect to the trivial branch of solution of (1.1)-(1.2) is

$$b \geqq \left(1 + a\sqrt{\frac{d_1}{d_2}}\right)^2. \tag{1.7}$$

Now let $\Omega =]-1,+1[$ and $D(A) = \{w \in C^2([-1,+1]);\ w(-1) = w(1), w'(-1) = w'(1)\}$ which corresponds to the problem (0.1)-(0.2). It is clear that A is symmetric nonnegative with compact resolvent and has the sequence of double eigenvalues

$$\mu_n = \sqrt{d_1 d_2}\, n^2\pi^2 \tag{1.8}$$

with the associated eigenfunctions

$$\phi_n^1(x) = \cos n\pi x,\ \phi_n^2(x) = \sin n\pi x \tag{1.9}$$

Then it follows:

COROLLARY 1.3: The condition (1.7) is necessary for the existence of bifurcation with respect to the trivial branch $\{(L,a,b/a)\}$ of solutions of (0.1)-(0.2). Assume (1.7) holds and set

$$\delta_i = \underline{a} + (-1)^i\,(\underline{a}^2 - a^2)^{\frac{1}{2}} \text{ for } i = 1,2, \tag{1.10}$$

$$L_i^n = \frac{(d_1d_2)^{1/4}}{\delta_i^{1/2}}\, n\pi \text{ for } i = 1,2,\ n = 1,2,\ldots\ . \tag{1.11}$$

Then the values L_i^n are critical of order 2 except in the case

$$b > \left(1 + a\sqrt{\frac{d_1}{d_2}}\right)^2,\quad n\left(\frac{\delta_i}{\delta_j}\right)^{\frac{1}{2}} = m \in \mathbb{N} \text{ with } j = 1,2,\ j \neq i, \tag{1.12}$$

where $L_i^n = L_j^m$ is a critical value of order 4.

We state now the main result of this article.

<u>THEOREM 1.4</u>: With the notations above, assume (1.7) holds. Let $i \in \{1,2\}$, $n \in \{1,2,...\}$ and assume (1.12) does not hold. Then $(L_i^n, a, b/a)$ is a bifurcation point with respect to the trivial branch of solutions of (0.1)-(0.2) if and only if (0.3) holds.

Theorem 0 is an immediate consequence of Corollary 1.3 and Theorem 1.4. We prove Theorem 1.4 and describe the bifurcation in the next section.

2. BIFURCATION AT DOUBLE CRITICAL VALUES

We use the notation of Section 1, assuming A symmetric nonnegative with compact resolvent. Let $\mu \in \sigma(A)$, $\mu \neq 0$ and $N_0 = \ker(A-\mu)$; on the other hand, assume (1.7) holds and let δ be a solution of $\delta^2 - 2\underline{a}\delta + a^2 = 0$. Set $\lambda_0 = \mu/\delta$ and assume

$$\underline{a} = a \quad \text{or} \quad \frac{\lambda_0 a}{\delta} \notin \sigma(A) \tag{2.1}$$

such that $N_0 = \ker T(\lambda_0)$, $R_0 = N_0^\perp = \operatorname{Im} T(\lambda_0)$ and for $\lambda = \lambda_0 + \gamma/a$ with γ small enough, $(I-P)T(\lambda)$ is an isomorphism of R_0 onto R_0, where P is the orthogonal projection onto N_0; set $R(\gamma) = [(I-P)T(\lambda_0 + \lambda/a)_{|R_0}]^{-1}$.

Let $w = p + q$ with $p \in N_0$, $q \in R_0$; by projection on R_0 and N_0, with the notation above, the equation (1.4) is equivalent to the Lyapunov-Schmidt system

$$q + R(\gamma)(I-P)H(\lambda_0 + \frac{\gamma}{a}, p + q) = 0, \tag{2.2}$$

$$\gamma(\gamma-\underline{\gamma})p + PH(\lambda_0 + \frac{\gamma}{a}, \quad p + q) = 0, \tag{2.3}$$

with

$$\underline{\gamma} = 2\mu(\frac{\underline{a}}{a} - \frac{a}{\delta}) = \mu(\frac{\delta}{a} - \frac{a}{\delta}). \tag{2.4}$$

Now (2.2) may be solved by the implicit function theorem, and there exists Z a neighbourhood of 0 in $\mathbb{R} \times N_0$, V a neighbourhood of 0 in R_0, f a C^∞ mapping from Z into V such that $(\gamma,p,q) \in Z \times V$ is a solution of (2.2) if and only if $(\gamma,p) \in Z$ and $q = f(\gamma,p)$. Then the problem is reduced to the "bifurcation equation"

$$\gamma(\gamma-\underline{\gamma})p + PH(\lambda_0 + \frac{\gamma}{a},\ p + f(\gamma,p)) = 0. \tag{2.5}$$

All this was standard bifurcation theory. Now by (2.2), using the particular form of $H(\lambda,w)$ (see (1.5)), one gets

$$f(\gamma,p) = M(\gamma)p^2 + |p|^2\ g(\gamma,p), \tag{2.6}$$

where $|p|$ is a norm of p in N_0, g is a C^∞ mapping from Z into R_0 and

$$M(\gamma) = -R(\gamma)(I-P)(h_1\lambda\mu + h_2\mu^2) \tag{2.7}$$

Then plugging (2.6) into the particular form of H,

$$\begin{aligned} H(\lambda,p + f(\gamma,p)) &= (h_1\lambda + h_2\mu)\mu p^2 + \\ &+ p[(h_1\lambda\mu + (h_1\lambda + 2h_2\mu)A)M(\gamma)p^2] + \\ &+ (h_3 + \frac{h_4\mu}{\lambda})\mu^2 p^3 + |p|^4\ G(\gamma,p) \end{aligned} \tag{2.8}$$

with G a continuous mapping from Z into U.

In order to solve the bifurcation equation (2.5), we now enter into the particular situation of our problem, where $\mu = \mu_n$ given by (1.8) and N_0 is the space spanned by $\{\phi_n^1,\phi_n^2\}$ given by (1.9), with some positive integer n. We then have the following properties for $p = |p| \cos n\pi(x-\underline{x}) \in N_0$:

$$\begin{cases} p^2 \in R_0 \cap D(A) \text{ and } Ap^2 = 4\mu p^2 - 2\mu|p|^2, \\ Pp^3 = \frac{3}{4}|p|^2 p. \end{cases} \tag{2.9}$$

It follows that for $p \in N_0$, $c \in \mathbb{R}$,

$$T(\lambda)(p^2 + c|p|^2 = 16\mu^2) - 8\lambda\underline{a}\mu + \lambda^2 a^2)p^2$$

$$+ (\lambda^2 a^2 c + 4\lambda\underline{a}\mu - 8\mu^2)|p|^2,$$

and then

$$R(\gamma)p^2 = (16\mu^2 - 8\lambda\underline{a}\mu + \lambda^2 a^2)^{-1}[p^2 + \frac{4\mu}{\lambda^2 a^2}(2\mu - \lambda\underline{a})|p|^2]. \tag{2.10}$$

According to (2.7), (2.8), (2.9) and (2.10) the bifurcation equation (2.5) may be written now as

$$[\gamma(\gamma-\underline{\gamma}) + \mu^2 c(\gamma)|p|^2]p + |p|^4 PG(\gamma,p) = 0, \tag{2.11}$$

where $c(\gamma)$ is a C^∞ function in a neighbourhood of 0 with

$$c(0) = \frac{h_1 + h_2\delta}{3(4\delta^2 - a^2)}[2h_2 + (\frac{7}{4\delta} + \frac{4(2\delta - \underline{a})}{\bar{a}^2}h_1] + \frac{3}{4}(h_3 + h_4\delta). \tag{2.12}$$

Assume first $\underline{\gamma} = 0$, that is, $b = (1 + a\sqrt{d_1/d_2})^2$ and let us show that there is no bifurcation at $(\lambda_0,0)$. By contradiction assume there exists $(\gamma_k,p_k) \in Z$ solutions of (2.11) with $p_k \neq 0$ and $r_k^2 = \gamma_k^2 + |p_k|^2 \to 0$. Then by (2.11) we have

$$\frac{p_k^2}{|p_k|^2} \to 1 - \mu^2 c(0) \tag{2.13}$$

such that $c(0) \leqq 0$. But in the case $\underline{\gamma} = 0$, we have $\delta = \underline{a} = a$ and then by (2.12),

$$c(0) = \frac{h_1 + h_2 a}{9a^2}(2h_2 + \frac{23}{4}\frac{h_1}{a}) + \frac{3}{4}(h_3 + h_4 a)$$

which is strictly positive according to the values of h_1, h_2, h_3, h_4 and $b = (1 + a\sqrt{d_1/d_2})^2$. So in this case there is no bifurcation at $(\lambda_0,0)$.

Assume now $\underline{\gamma} \neq 0$ and let us show there is bifurcation at $(\lambda_0,0)$. Set

$$\sigma_0 = \mu^2 c(0)\underline{\gamma}^{-1}, \quad t(r,\sigma) = (1-r^2\sigma^2)^{\frac{1}{2}}$$

and for $\xi \in N_0$ with $|\xi| = 1$, let F be the function

$$F(r,\sigma) = [\sigma(r^2\sigma - \underline{\gamma}) + \mu^2 c(r^2\sigma)t(r,\sigma)^2]$$

$$+ rt(r,\sigma)^4 G(r^2\sigma, rt(r,\sigma)\xi)\cdot\xi$$

defined on

$$W = \{(r,\sigma) \in \mathbb{R}^2; \ r^2\sigma^2 < 1, \ (r^2\sigma, rt(r,\sigma)\xi) \in Z\}.$$

It is clear that W is an open set in $\mathbb{R}^2$, F is C^∞, $F(0,\sigma_0) = 0$ and

$$\frac{\partial F}{\partial\sigma}(0,\sigma_0) - \underline{\gamma} \neq 0.$$

By the implicit function theorem there exists $r_0 > 0$, $\eta > 0$ and $\sigma(r)$ a C^∞ function on $]-r_0,r_0[$ such that $W_0 =]-r_0,r_0[\times]\sigma_0 - \eta, \sigma_0 + \eta[\subset W$ and $(r,\sigma(r))$ is the unique solution of $F(r,\sigma) = 0$ in W_0. Returning to the original problem, this shows that $(r^2\sigma(r), r.t.(r,\sigma(r))\xi)$ is the unique solution of (2.11) in the set $\{(r^2\sigma, rt(r,\sigma)\xi); (r,\sigma) \in W_0\}$. It is clear that $\sigma(r)$ depends smoothly on ξ. Then we have proved the following result:

THEOREM 2: Assume (0.3) holds; let $i \in \{1,2\}$, $n \in \{1,2,\ldots\}$ and assume (1.12) does not hold. Let $\mu = \mu_n$, $\delta = \delta_i$ given by (1.8), (1.10). There exists $r_0 > 0$ and C^∞ mappings

$$\rho :]-r_0,r_0[\times \mathbb{T} \to \mathbb{R},$$

$$h,k:]-r_0,r_0[\times \mathbb{T} \to U,$$

with $\mathbb{T} = \mathbb{R}/2\pi\mathbf{Z}$ such that

$$(L(r,\underline{x}), u(r,\underline{x}), v(r,\underline{x}))$$

is a solution of (0.1)-(0.2) for $(r,\underline{x}) \in]-r_0,r_0[\times \mathbf{T}$ with

$$L(r,\underline{x}) = \sqrt{\frac{\mu}{\delta}} + \frac{\mu^2 c(0)}{2\underline{\gamma} a} r^2 + r^3 \rho(r,\underline{x}),$$

$$u(r,\underline{x})(x) = a - \frac{\delta r}{\sqrt{d_1 d_2}} \cos n \left(\frac{\pi x}{L(r,\underline{x})} - \underline{x}\right) + r^2 h(r,\underline{x})(x),$$

$$v(r,\underline{x})(x) = \frac{b}{a} + \left(1 + \delta\sqrt{\frac{d_1}{d_2}}\right) \frac{r}{d_2} \cos n \left(\frac{\pi x}{L(r,\underline{x})} - \underline{x}\right) + r^2 k(r,\underline{x})(x),$$

where $\underline{\gamma}$, $c(0)$ are given by (2.4), (2.12).

3. SOME OTHER EXAMPLES

First consider the system (1.1) with $\Omega =]-1,1[\times]-1,1[$ and the boundary conditions

$$\frac{\partial u}{\partial n} = \frac{\partial v}{\partial n} = 0 \text{ on } \partial\Omega, \tag{3.1}$$

or

$$u = a,\ v = \frac{b}{a} \text{ on } \partial\Omega. \tag{3.2}$$

With a similar proof to the proof above, we obtain the following result:

PROPOSITION 3.1: Assume (0.3) holds; let $i \in \{1,2\}$ and n,m be positive integers (resp. even positive integers). Set $\delta = \delta_i$ and $L_{n_i} = (d_1 d_2)^{1/4}[(n^2 + m^2)/\delta]$. Assume that either $\delta = a^2$ or

$$\frac{a^2(n^2 + m^2)}{\delta^2} \text{ is not of the form } h^2 + k^2.$$

Then $(L_{n,m}, a, b/a)$ is a bifurcation point for the problem (1.1)-(3.1) (resp. (1.1)-(3.2)).

Consider now the system (0.1) with $\Omega =]-1,1[\times]-1,1[$, the boundary conditions (3.1) or (3.2) and take b as the parameter of bifurcation. In the same way we obtain the following result:

PROPOSITION 3.2: Let n,m be positive integers (resp. even positive integers). Assume either $a = (n^2 + m^2)\pi\sqrt{d_1 d_2}$ or

$$\frac{a^2}{(n^2 + m^2)\pi^2} \text{ is not of the form } (h^2 + k^2)d_1 d_2.$$

Set

$$b_{n,m} = \left(1 + (n^2 + m^2)\pi^2 \sqrt{\frac{d_1}{d_2}}\right)\left(1 + \frac{a^2}{(n^2 + m^2)\pi^2}\sqrt{\frac{d_1}{d_2}}\right).$$

Then, taking b as a parameter, $b_{n,m}$ is a critical value of order 2 and there is bifurcation at the point $(b_{n,m}, a, b_{n,m}/a)$ for the solutions of (0.1)-(3.1) (resp. (0.1)-(3.2)) where $\Omega =]0,1[\times]0,1[$.

Finally, consider (0.1) with $\Omega =]0,1[$ and the boundary conditions (3.1) or (3.2). For an integer n we set

$$E_1^n = \{\tfrac{n}{9}, \tfrac{n}{7}, \tfrac{n}{5}, \tfrac{n}{3}, 3n, 5n, 7n, 9n\},$$

$$E_2^n = E_1^n \cup \{\tfrac{n}{8}, \tfrac{n}{6}, \tfrac{n}{4}, \tfrac{n}{2}, 2n, 4n, 6n, 8n\}.$$

Then we have the following result:

PROPOSITION 3.3: Let n be a positive integer (resp. even positive integer). Assume that $\beta = a^{\frac{1}{2}}/\pi(d_1 d_2)^{1/4}$ is a positive integer and that $\beta/n \notin E_2^n$ (resp. $\beta \notin E_1^n \cap 2\mathbb{N}$). Set

$$b_n = \left(1 + n^2\pi^2 \sqrt{\frac{d_1}{d_2}}\right)\left(1 + \frac{a^2}{n^2\pi^2}\sqrt{\frac{d_1}{d_2}}\right).$$

Then, taking b as a parameter, b_n is a critical value of order 2 and there is bifurcation at the point $(b_n, a, b_n/a)$ for the solutions of (0.1)-(3.1) (resp. (0.1)-(3.2)) where $\Omega =]0,1[$.

REMARK 3.4: The case $\beta/n \in \{n/2, 2n\}$ (resp. n or β/n are odd integers) was studied in [1].

REMARK 3.5: Proofs of the Propositions 3.1, 3.2 and 3.3 will be given in [2].

References

[1] S. Amraoui and H. Labani, Etude de la bifurcation pour $A^2w - \lambda Aw + \alpha w = H(\lambda,\alpha,w,Aw)$. Publ. Math. Besançon. Analyse non linéaire, No. 9 (1985-86),

[2] S. Amraoui, Etude de la bifurcation pour le brussellator. Thèse Université de Besançon, in preparation.

[3] J.F. Auchmuty and G. Nicolis, Bifurcation analysis of nonlinear reaction-diffusion equations, II. Chemical oscillations, Bull. Math. Biology, 38, No. 4 (1976),

[4] M.G. Crandall and P.H. Rabinowitz, Bifurcation from simple eigenvalues. J. Funct. Anal. 8 (1971), 321-340.

[5] J. Esquinas and J. Lopez-Gomez, Optimal results in local bifurcation theory. J. Diff. Equations 71 (1988), 72-92.

[6] S.L. Hayle, Local solutions manifolds for nonlinear equations. Nonlinear Analysis. Theory Math. 4 No. 2 (1980), 283-295.

[7] J. Lopez-Gomez, Multiparameter local bifurcation, Nonlinear Analysis 10 No. 11 (1986), 1249-1259.

[8] J.B. McLeod and D.H. Sattinger, Loss of stability and bifurcation at a double eigenvalue. J. Funct. Anal. 14 (1973), 62-64.

[9] G. Meurant and J.C. Saut, Bifurcation and stability in a chemical system. J. Math. Anal. Appl. 59 (1977), 69-72.

[10] F. Peteuil, Etude de phénoménes de bifurcation sur un modèle du reaction-diffusion. Thèse 3ème cycle, Université Paris VI, 1983.

S. Amraoui
Département de mathématiques
Université de Franche Comté
route de Gray
25030 Besançon Cedex
France

M. ARTOLA

Existence and uniqueness for a diffusion system with coupling

0. INTRODUCTION

Let Ω be an open bounded set of R^N with boundary $\Gamma = \Gamma_1 \cup \Gamma_2$ (cap. $\Gamma_1 \neq 0$).

In what follows, only global existence and uniqueness theorems are given for the solution $u = [\tilde{u},v]$ $\tilde{u} = (u_1,\ldots,u_m)$ of the nonlinear system

$$\begin{cases} \dfrac{\partial u_r}{\partial t} - \operatorname{div} A^r(x,u_r)\nabla u_r - f_r(x,u_r,v) = 0, \quad 1 \leqq r \leqq m, \\ \\ \dfrac{\partial v}{\partial t} - \operatorname{div} \mathcal{B}(x,v)\nabla v - g[x,\tilde{u},v] = 0, \end{cases} \tag{I}$$

with boundaries conditions

$$\begin{cases} u = 0 \quad \text{on } \Gamma_1 \times]0,T[, \\ \\ \dfrac{\partial u}{\partial \nu_A} = h(\tilde{u},v) \text{ on } \Gamma_2 \times]0,T[, \end{cases} \tag{II}$$

$$\frac{\partial u}{\partial \nu_A} = (A^1(u_1)\nabla u_1.\nu,\ldots,A^m(u_m)\nabla u_m.\nu,\ \mathcal{B}(v)\nabla v.\nu),$$

where ν is the outward normal to Γ and initial data

$$u(x,0) = u^0(x) = [\tilde{u}^0(x),\ v^0(x)].$$

Proofs of uniqueness are given using special test functions in the case where the coupling terms satisfy some Lipschitz condition, and when the following properties are fulfilled:

$$g(x,\tilde{u},v) = - \sum_{k=1}^{m} f_k(x,u_k,v), \tag{IV}$$

and

$$\begin{cases} \forall k,\ \forall v,\ u_k \to f_k(x,u_k,v) \text{ is decreasing} \\ \forall k,\ \forall u_k,\ v \to f_k(x,u_k,v) \text{ is increasing.} \end{cases} \tag{V}$$

Condition (IV) is very important in various physical problems like Photonical Diffusion Theory (see also [12]). Stiff problems, invariant sets and other themes of my talk in Nancy are not considered here and will be published elsewhere.

The plan is as follows:

I. Statement of the problem.

II. Existence theorem.

III. Uniqueness theorems.

I. STATEMENT OF THE PROBLEM

1.1. Definitions

Let Ω be an open bounded set of the real N-dimensional Euclidean space R^n with sufficiently regular boundary $\partial\Omega = \Gamma$. Furthermore, we assume that $\Gamma = \Gamma_1 \cup \Gamma_2$ with Cap $\Gamma_1 > 0$.

$H = L^2(\Omega)$ denotes the usual Lebesgue space provided with the scalar product $(u,v) = \int_\Omega uv\, dx$ and the norm $|u| = (\int_\Omega |u|^2\, dx)^{\frac{1}{2}}$.

$H^1(\Omega)$ is the usual Sobolev space and we consider

$$V = \{v \in H^1(\Omega),\quad v|_{\Gamma_1} = 0\}, \tag{I.1}$$

which is a Hilbert space equipped with the norm $\|\ \|$ derived from the scalar product $((u,v)) = \int_\Omega \nabla u.\nabla v\, dx$. In what follows, we consider the spaces

$$\mathcal{V} = V^{m+1},\quad \mathcal{H} = H^{m+1},\ m \in \mathbb{N}\quad (m \geq 1). \tag{I.2}$$

If $\tilde{u} = (u_1,\dots,u_m)$ and if $u = [\tilde{u},v] \in \mathcal{H}$ (resp. $\mathcal{V}$) the norm in $\mathcal{H}$ (resp. $\mathcal{V}$) is given by

$$|u|_{\mathcal{H}} = (\sum_{i=1}^m |u_i|^2 + |v|^2)^{\frac{1}{2}}\quad (\text{resp. } \|u\|_{\mathcal{V}} = (\sum_{i=1}^m \|u_i\|^2 + \|v\|^2)^{\frac{1}{2}}.$$

Now $A^r(x,\lambda)$ $(r = 1,\ldots,m)$, $\mathcal{B}(x,\lambda)$ be m+1 $(N\times N)$ matrices with coefficients a^r_{ij}, b_{ij} $(r = 1,\ldots,m)$ $(i,j=1,\ldots,N)$ defined on $\Omega \times R$.

We recall that a function ϕ defined on $\Omega \times R^N$ is a Carathéodory function if

for all $\lambda \in R^N$, $x \to \phi(x,\lambda)$ is measurable on Ω,

for almost all $x \in \Omega$, $\lambda \to \phi(x,\lambda)$ is continuous on R^N.

Then we assume

(i) a^r_{ij}, $b_{ij} \in L^\infty(\Omega \times R)$ are Caratheodory functions from $\Omega \times R$ to R;

(ii) a^r_{ij}, $\xi_i\xi_j \geq \alpha\xi_j\xi_j$, b_{ij} $\xi_i\xi_j \geq \alpha$ $\xi_j\xi_j$ (I.3)

for all $\xi = (\xi_1,\ldots,\xi_N) \in R^N$ (α is a uniform constant);

and we define the functions f, h in the following way:

if $\lambda = [\tilde{\lambda},\mu]$ R^{m+1}, $\tilde{\lambda} = (\lambda_1,\ldots,\lambda_m)$, $x \in R^N$,

then

$$f(x,\lambda) = [\tilde{f}(x,\lambda),g(x,\lambda)],\ \tilde{f}(x,\lambda) = (f_1(x,\lambda_1,\mu),\ldots,f_m(x,\lambda_m,\mu)),$$
$$h(x,\lambda) = [\tilde{h}(x,\lambda),k(x,\lambda)],\ \tilde{h}(x,\lambda) = (h_1(x,\lambda_1,\mu),\ldots,h_m(x,\lambda_m,\mu)), \tag{I.4}$$

and we assume

(f_i,h_i) [resp. (g,k)] are Carathéodory's functions on $(\Omega \times R^2,\ \Gamma \times R^2)$ (resp. on $(\Omega \times R^{m+1},\ \Gamma \times R^{m+1})$). (I.5)

We recall that a Carathéodory function $\Phi(x,\lambda)$ operates from $L^p(\Omega)$ to $L^q(\Omega)$ if $\Phi(x,u(x))$ is in $L^q(\Omega)$, when u belongs to $L^p(\Omega)$, and in an obvious way we assume:

(i) f_i (resp. g) operates from $L^2(\Omega) \times L^2(\Omega)$

(resp. $[L^2(\Omega)]^{m+1}$) to $L^p(\Omega)$ with $p \geq 2N/N+2$;

(ii) h_i (resp. k) operates from $L^2(\Gamma_2) \times L^2(\Gamma)$ (I.6)

(resp. $(L^2(\Gamma_2))^{m+1}$ to $L^q(\Gamma_2)$ with $q \geq 2(N-1)/N$.

Now, if X is a real Banach space, $L^q(0,T;X)$, $1 \leq q < +\infty$ (resp. $C^0([0,T];X)$) is the space of (class of) functions ϕ which are measurable on (0,T) with values in X such that $\int_0^T |\phi(t)|_X^q \, dt < +\infty$ (resp. the space of continuous functions on [0,T] with values in X) provided with the natural norm. $L^\infty(0,T;X)$ is also defined in an obvious way.

It is well known that the space

$$W(0,T) = \left(u; u \in L^2(0,T;V), \frac{du}{dt} \in L^2(0,T;V')\right) \tag{I.7}$$

V' = dual of V (when H' is identified with H) is a Hilbert space equipped with the natural norm and we have

$$W(0,T) \subset C^0([0,T];H). \tag{I.8}$$

Moreover, because of the injective mapping $V \subset H$ is a compact mapping (we assume Γ sufficiently regular), we know that

$$\text{the identity mapping } W(0,T) \subset L^2(0,T;H) \text{ is compact.} \tag{I.9}$$

1.2 The Variational Problem (P)

We introduce the following forms on $V \times V$:

$$\begin{cases} a^r(w;\phi,\psi) = \int_\Omega A^r(x,w)\nabla\phi\cdot\nabla\psi \, dx, \\ w \in H, \quad \phi,\psi \in V, \; r = 1,\ldots,m, \end{cases} \tag{1.10}$$

$$\begin{cases} b(w;\phi,\psi) = \int_\Omega B(x,w)\nabla\phi\nabla\psi \, dx, \\ w \in H, \; \phi,\psi \in V. \end{cases}$$

Then putting

$$w = (w_1,\dots,w_{m+1}) \in H,\ \phi = (\phi_1,\dots,\phi_{m+1}),\ \psi = (\psi_1,\dots,\psi_{m+1}) \in V,$$

we can define

$$A(w;\phi,\psi) = \sum_{r=1}^{m} a^r(w_r;\phi_r,\psi_r) + b(w_{m+1};\phi_{m+1},\psi_{m+1}), \tag{I.11}$$

and we have

$$\begin{aligned} |A(w;\phi,\psi)| &\leqq M\|\phi\|, \|\psi\|_V \quad (M = \text{cte}), \\ A(w;\phi,\phi) &\geqq \alpha\|\phi\|_V^2, \end{aligned} \tag{I.12}$$

now giving us an initial data

$$u^0 = [\tilde{u}^0, v^0] \in H. \tag{I.13}$$

We are looking for $u = [\tilde{u},v]$ to be a solution of the following problem (the so-called problem (P)):

$$\begin{cases} u \in W(0,T), \quad \Omega_T = \Omega \times]0,T[; & \text{(I.14)} \\ \langle\frac{du}{dt}(\cdot),\psi\rangle^{(1)} + A(u(\cdot);u(\cdot),\psi) = \int_\Omega f(\cdot,u)\cdot\psi\, dx + \int_{\Gamma_2} h(\cdot,u)\cdot\psi\, d\Gamma_2 & \text{(I.15)} \\ \text{for all } \psi \in V. \ \text{(Equation (I.16) is taken with sense of } \mathcal{D}'\ (]0,T[).); & \\ u(0) = u^0 & \text{(I.16)} \end{cases}$$

<u>REMARK I.1</u>:

(i) If $\psi \in V$, then $\psi \in (L^r(\Omega))^{m+1}$, $r \leqq 2N/(N-2)$ $(N > 2)$ (for all $r < +\infty$ if $N = 2$) by the Sobolev imbedding theorem. From (I.6) - (I.14) $f[u(\cdot)] \in L^\infty(0,T;[L^p(\Omega)]^{m+1})$, $p \geqq 2N/(N+2)$. Then $p' \leqq 2N/(N-2)$ $(1/p + 1/p' = 1)$ and $\int_\Omega f(u)\cdot\psi\, dx$ exists.

(1) $\langle,\rangle$ means the bracket in the duality between V and V'.

(ii) In the same way, $\psi|_{\Gamma_2} \in H^{\frac{1}{2}}(\Gamma_2) \subset L^s(\Gamma_2)$, $s \leq 2(N-1)/(N-2)$ if $N > 3$ and for all $s < +\infty$ if $N = 3$. Then $h(u(\cdot)) \in L^\infty(0,T;(L^q(\Gamma_2))^{m+1})$ with $q \geq 2(N-1)/N$ and $\int_{\Gamma_2} h(u(\cdot))\cdot\psi \, dx$ exists.

(iii) Thus we can interpret problem (P) as

$$\begin{cases} \dfrac{\partial u_r}{\partial t} - \operatorname{div} A^r(u_r)\nabla u_r = f_r(u_r,v), \quad 1 \leq r \leq m, \\ \dfrac{\partial v}{\partial t} - \operatorname{div} \mathcal{B}(v)\cdot\nabla v = g[\tilde{u},v]. \end{cases} \tag{I.17}$$

$$\begin{cases} u = 0 \text{ on } \Gamma_1 \times]0,T[, \\ \dfrac{\partial u}{\partial \nu_A} = h(u) \text{ on } \Gamma_2 \times]0,T[, \end{cases} \tag{I.18}$$

$$u(x,0) = u^0(x) = [\tilde{u}^0(x), v^0(x)] \text{ a.e. in } \Omega. \quad \square \tag{I.19}$$

II. EXISTENCE THEOREM

THEOREM 1: With assumptions (I.3) to (I.6) there exists at least a solution for problem (P).

PROOF OF THEOREM 1:

Step 1. First, if $h \equiv 0$, let w be a given function in $L^2(0,T; \ )$ and let $\hat{u}$ be the unique solution $\in W(0,T)$ of the linear problem

$$\begin{cases} < \dfrac{\partial u}{\partial t},\psi> + A(w;\hat{u},\psi) = \displaystyle\int_\Omega f(w)\cdot\psi \, dx, \\ \hat{u}(0) = u^0. \end{cases} \tag{II.1}$$

If $h \not\equiv 0$ we should take care with the choice of w.

Let V_s be an interpolated space between V and H ($V_0 \equiv H$) such that $V \subset V_s$ with compact injective mapping and such that the trace of $w \in V_s$ lies in a convenient space $H^s(\Gamma_2)$ interpolated between $[H^{\frac{1}{2}}(\Gamma_2)]^{m+1}$ and $[L^2(\Gamma_2)]^{m+1}$. Such a choice is possible with $s < 1/2$ (see [10, Chap. 1]). Then we choose $w \in L^2(0,T;V_s)$ and we have $w|_{\Gamma_2} \in L^2(0,T;[L^2(\Gamma_2)]^{m+1})$ and from (I.6)

$h(\cdot,w)|_{\Gamma_2} \in L^q[\Gamma_2 \times]0,T[\ ]$.

Now $\hat{u}$ is the unique solution of

$$\begin{cases} \langle \frac{d\hat{u}}{dt},\psi\rangle + A(w;\hat{u},\psi) = \int_\Omega f(w)\cdot\psi\, dx + \int_{\Gamma_2} h(w)\cdot\psi\, d\Gamma_2, \\ \hat{u}(0) = u^0. \end{cases} \tag{II.1)'$$

Then we denote by T the mapping:

$$T: \quad \begin{matrix} w \to \hat{u} = T(w) \quad \hat{u} \text{ is a solution of (II.1)'}, \\ L^2(0,T;\ _2) \to W(0,T). \end{matrix} \tag{II.2}$$

Step 2. Since $V \subset V_S$ is compact the range of a bounded set in $W(0,T)$ is relatively compact in $L^2(0,T;V_s)$. In order to use the Schauder fixed-point theorem we have to show the continuity of T.

Let $\{w_n\}_{n\in\mathbb{N}}$ be a sequence of functions in $L^2(0,T;V_s)$ which converges toward w in the strong topology of $L^2(0,T;V_s)$ (a fortiori, $w_n \to w$) in the strong topology of $L^2(0,T;H)$ and $L^2(0,T;(L^{2s}(\Gamma_2))^{m+1})$). From (II.1)' if $\hat{u}_n = T(w_n)$ and $\hat{u} = T(w)$ then $\hat{v}_n = \hat{u}_n - \hat{u}$ satisfies

$$\begin{aligned} \langle \frac{\partial\hat{v}_n}{\partial t},\psi\rangle + A(w_n;\hat{v}_n,\psi) &= \int_\Omega [f(w_n)-f(w)]\psi\, dx \\ &+ \int_{\Gamma_2} (h(w_n)-h(w))\psi\, d\Gamma_2 + A(w;\hat{u},\psi) - A(w_n,\hat{u},\psi), \end{aligned} \tag{II.3}$$

$\forall\psi \in V$.

Taking $\psi = \hat{v}_n$ in (II.3) we derive from (II.3) by classical estimates for the solution of parabolic equations for all $t \in]0,T[$

$$\begin{aligned} &\frac{1}{2}|\hat{v}_n(t)|_V^2 + \frac{a}{2}\int_0^t \|\hat{v}_n(\sigma)\|_V^2\, d\sigma \\ &\leq C\Phi_n(T) + \|f(w_n)-f(w)\|^2_{L^2(0,T;L^p(\Omega))} + \|h(w_n)-h(w)\|^2_{L^2(0,T;L^q(\Gamma_2))} \end{aligned} \tag{II.4}$$

with

$$\Phi_n(T) = \sum_{ij,r} \int_{\Omega_T} |(a^r_{ij}(x,w_n)-a^r_{ij}(x,w)) \frac{\partial \hat{u}_r}{\partial x_j}|^2 \, d\Omega_T \;;$$

$$+ \sum_{ij} \int_{\Omega_T} |(b_{ij}(x,w_n)-b_{ij}(x,w)) \frac{\partial v}{\partial x_j}|^2 \, d\Omega_T. \tag{II.5}$$

From (I.6) and [9, Theorem 2.1, p. 22]

$$\lim_{n\to+\infty} \|f(w_n)-f(w)\|_{L^2(0,T:(L^p(\Omega))^{m+1},} = 0,$$

$$\lim_{n\to+\infty} \|h(w_n)-h(w)\|_{L^2(0,T:(L^q(\Gamma_2))^{m+1},} = 0. \tag{II.6}$$

On the other hand, since $w_n \to w$ in the strong topology of $L^2(0,T;H)$ $w_n \to w$ in measure. Then a^r_{ij}, b_{ij} being Carathéodory functions, the quantities $a^r_{ij}(\cdot,w_n)-a^r_{ij}(\cdot,w)$, $b_{ij}(\cdot,w_n)-b_{ij}(\cdot,w)$ converge toward 0 in measure and are uniformly bounded with respect to n ([9]).

Furthermore $|\partial u_r/\partial x_j|^2$ and $|\partial v/\partial x_j|^2 \in L^1(\Omega_T)$. Then using the following lemma (see [9] for the proof):

<u>LEMMA 1</u>: If $\chi_n \in L^\infty(\Omega_T)$, $\|\chi_n\|_{L^\infty(\Omega_T)} \leq C$ converges toward 0 in measure, then for all $\theta \in L^1(\Omega_T)$, $\int_{\Omega_T} \chi_n \theta \, d\Omega_T \to 0$ as $n \to +\infty$, we see that $\Phi_n(T) \to 0$ as $n \to +\infty$. Thus (from II.4) we have proved that $\int_0^T \|\hat{v}_n(t)\|^2_V \, dt \to 0$ as $n \to +\infty$, which implies $\hat{v}_n \to 0$ in $L^2(0,T;V_s)$. Thus T is continuous from $L^2(0,T;V_s) \to L^r(0,T;\ _s)$. □

III. <u>UNIQUENESS THEOREMS</u>

III.1. <u>Uniqueness Results for one Parabolic Equation</u>

First we recall (see [11]) the following definition: $\tau \to \chi(\tau)$ is an Osgood function if

$$\chi(\tau) \geq 0,\ \chi(0) = 0,\ \chi \text{ is increasing and}$$

$$\int_{\to 0} \frac{d\sigma}{\chi(\sigma)} = +\infty. \tag{III.1}$$

As a consequence of (III.1) we have

$$\text{for all } \eta > 0, \text{ there exists } \delta(\eta) > 0 \text{ such that} \quad \int_{\delta}^{\eta} \frac{d\sigma}{\chi(\sigma)} = 1. \tag{III.2}$$

Now we consider the problem

$$\begin{cases} u \in L^2(0,T;V), \; u' \in L^2(0,T;V'), \\ \langle \frac{\partial u}{\partial t}, \psi \rangle + \int_\Omega A(x,u) \nabla u \nabla \psi \, dx + \int_\Omega f(x,u)\psi \, dx = 0, \\ u(0) = u^0 \in H \\ \text{for all } \psi \in V \text{ in the } \mathcal{D}'(]0,T[) \text{ sense.} \end{cases} \tag{π}$$

We assume that $A(x,\lambda)$ ($N \times N$ matrix) satisfies (I.3) and

$$|A(x,\lambda)-A(x,\lambda^*)| \leqq \delta(|\lambda-\lambda^*|), \tag{III.3}$$

where

$$\tau \to [\omega(\tau)]^2 \text{ is an Osgood function.} \tag{III.4}$$

We also assume $(x,\lambda) \to f(x,\lambda)$ satisfies (I.5)-(I.6). Then

THEOREM III.1: If we have (I.3), (III.3), (III.4) and if

$$\lambda \to f(x, \;) \text{ is nondecreasing,} \tag{III.5}$$

or if

$$|f(x,\lambda)-f(x,\lambda^*)| \leqq K|\lambda-\lambda^*|, \tag{III.6}$$

then the solution of problem (π) is unique.

PROOF: In both cases (III.5), (III.6), with u and $\hat{u}$ being two solutions for problem (π), $w = u - \hat{u}$ satisfies:

$$< \frac{\partial w}{\partial t},\psi> + \int_\Omega A(x,u)\nabla w\cdot\nabla\psi\, dx + \int_\Omega (f(x,u)-f(x,\hat{u}))\psi\, dx =$$
$$= \int_\Omega (A(x,\hat{u})-A(x,u))\nabla\hat{u}\cdot\nabla\psi\, dx.$$

Thanks to (III.2), (III.3) we can introduce (see also [1], [3], [5], [6], [11]) the following approximation of the Heaviside function:

$$\forall\eta > 0,\ P_\eta(\tau) = \begin{cases} 0 & \text{if } \tau \leqq \delta(\eta) = \delta, \\ \displaystyle\int_\delta^\tau \frac{d\sigma}{[\omega(\sigma)]^2} & \text{if } \delta \leqq \tau \leqq \eta. \\ 1 & \text{if } \tau \geqq \eta\ . \end{cases} \tag{III.8}$$

Let w^+ be the positive part of w, then $P\,(w^+) \in V$ for all $\eta > 0$ and a.e. in $t \in]0,T[$. We can choose $\psi = P\,(w^+)$ in (III.7) and we get

$$\int_\Omega \int_\delta^{w^+(x,t)} P_\eta(\sigma)\, d\sigma\, dx + \frac{\alpha}{2} \int_0^t \int_{\substack{\Omega \\ \delta\leqq w^+\leqq\eta}} \frac{|\nabla w^+|^2}{[\omega(w^+)]^2}\, dx\, d\tau + \tag{III.9}$$
$$+ \int_0^t \int_{\substack{\Omega \\ \delta\leqq w}} (f(x,u)-f(x,\hat{u}))P_\eta(w^+)\, dx\, d\tau \leqq \frac{1}{2\alpha} \int_0^t \int_{\substack{\Omega \\ \delta\leqq w\leqq\eta}} |\nabla\hat{u}|^2\, dx\, d\tau,$$

since

$$\int_0^t \int_{\substack{\Omega \\ \delta\leqq w\leqq\eta}} |\nabla\hat{u}|^2\, dx\, d\tau \to 0 \text{ when } \eta \to 0, \tag{III.10}$$

then if we have

$$\begin{cases} \text{there exists } \rho > 0 \text{ such that} \\ E_\rho = \{(x,t) \in \Omega_T;\ w^+(x,t) \geqq \rho > 0\},\ \text{mes } E = |E_\rho| \neq 0, \end{cases} \tag{III.11}$$

and if we have (III.5), we obtain from (III.9)

$$0 \leqq \lim_{\eta\to 0} \int_{\Omega_T} \int_\delta^{w^+} P_\eta(\sigma)d\sigma) \leqq 0;$$

on the other hand,

$$\int_{\Omega_T}\int_\delta^{w^+} P_\eta(\sigma)d\sigma \geq \int_E\int_\delta^\rho P_\eta(\sigma)\, d\sigma\, dx\, dt \geq (\rho-\eta)|E_\rho| \neq 0,\ \eta < \rho, \tag{III.12}$$

and we get a contradiction. Thus $w^+ = 0$, a.e. on Ω_T. Now if f satisfies (III.6), from (III.9) we deduce:

$$\int_\Omega \int_\delta^{w^+(x,t)} P_\nu(\sigma)\, d\sigma\, dx \leq \frac{1}{2\alpha}\int_0^t \int_{\substack{\Omega \\ \delta\leq w^+<\eta}} |\nabla\hat{u}|^2\, dx\, d\tau + $$
$$+ K\int_0^t \int_{\substack{\Omega \\ w^+(x,\tau)\geq\delta}} w^+P_\eta(w^+)\, dx\, d\tau, \tag{III.13}$$

but $w^+P_\eta(w^+) = \int_\delta^{w^+} \sigma P'_\eta(\sigma)d\sigma + \int_\delta^{w^+} P_\eta(\sigma)d\sigma$ and $\int_\delta^{w^+} \sigma P'_\eta(\sigma)d\sigma \leq \int_\gamma^\eta \sigma P'(\sigma)d\sigma \leq \eta$ if $w^+ \geq \delta$.

Thus (III.13) gives

$$\int_\Omega \int_\delta^{w^+(x,t)} P_\eta(\sigma)d\sigma\, dx \leq \phi(\eta) + K\int_0^t\int_\Omega\int_\delta^{w^+(x,\tau)} P_\eta(\sigma)\, d\sigma\, dx\ \ d\tau,$$

$$\phi(\eta) = \frac{1}{2\alpha}\int_0^t \int_{\substack{\Omega \\ \delta\leq w^+<\eta}} |\nabla\hat{u}|^2 dx\, d\tau + \eta K|\Omega_T| \to 0 \text{ when } \eta\to 0. \tag{III.14}$$

Gronwall's inequality implies

$$\Phi(t) \leq \phi(\eta) + K\int_0^t \Phi(\tau)d\tau \leq \phi(\eta)e^{kt}, \text{ with } \Phi(t) = \int_\Omega\int_\delta^{w^+(x,t)} P_\eta(\sigma)d\sigma.$$

Now, integrating with respect to t on [0,T], we get

$$\int_{\Omega_T}\int_\delta^{w^+(x,t)} P_\eta(\sigma)d\sigma \leq \phi(\eta)\cdot C(T) \to 0 \text{ if } \eta \to 0 \tag{III.15}$$

(C(T) = cte which depends on T). If (III.11) occurs, according to (III.12)-(III.15), we get a contradiction. Thus $w^+ = 0$. In the same way, we get $w^- = 0$ and theorem (III.1) follows. □

III.2. The Case of Systems

We consider here two generalizations to problem $(\mathcal{P})$ (see Section I) when $A^r(x,\lambda)$ $(r = 1,\dots,m)$, $\mathcal{B}(x,\lambda)$ satisfy hypothesis (III.3) with (III.4).

III.2.1. First Extension: Uncoupled Systems

THEOREM III.2: We assume:

(i) $h(x,\lambda) \equiv 0$,

(ii) $f(u) = [f_k(x,u_k),\ 1 \leqq k \leqq m,\ g(x,v)]$, (III.16)

(iii) $f(u)$ is Lipschitzian(1) on $\mathbf{R}^{m+1}$.

Then the solution of problem $(\mathcal{P})$ is unique.

PROOF: Let P_η as in Section III.1 and let $u = (\hat{u},v)$, $\hat{u} = (\hat{u},\hat{v})$ be two solutions of problem (π). If $w^+ = [\tilde{w}^+,\tilde{v}^+]$, $\tilde{w}^+ = ((\tilde{u}_1-\hat{u}_1)^+,\dots,(\tilde{u}_m-\hat{u}_m)^+)$, $\tilde{v}^+ = (v-\hat{v})^+$ we define $P_\eta(w^+) = [P_\eta[(\tilde{u}_1-\hat{u}_1)^+],\ 1 \leqq i \leqq m,\ P_\eta(\tilde{v}^+)]$. Then it is easy to adapt the proof of theorem (III.7) with hypothesis (III.6) in that situation. □

III.2.2. Second Extension: Systems with Coupling

We consider now the more interesting case where

$$g(x,\lambda_1,\dots,\lambda_m,\mu) = -\sum_{k=1}^{m} f_i(x,\lambda_i,\mu), \tag{III.17}$$

$$k(x,\lambda_1,\dots,\lambda_m,\mu) = -\sum_{k=1}^{m} h_i(x,\lambda_i,\mu), \tag{III.17'}$$

are fulfilled in problem $(\mathcal{P})$.

As was said in the introduction, this type of condition expresses conservative laws in some physical applications. Moreover, we assume

---

(1) In the sense where f_k,g satisfy a Lipschitz condition like (III.6).

Moreover, we assume

$$\text{(i)} \quad \forall\mu,\ \lambda_i \to f_i(x,\lambda_i,\mu) \ (\text{resp. } h_i(x,\lambda_i,\mu)) \text{ is decreasing,}$$
$$\text{(ii)} \quad \forall\lambda_i,\ \mu \to f_i(x,\lambda_i,\mu) \ (\text{resp. } h_i(x,\lambda_i,\mu)) \text{ is increasing.} \tag{III.18}$$

If we put

$$F(u) = -f(u) = -[f_i,\ 1 \leqq i \leqq m, g],$$
$$H(u) = -h(u). \tag{III.19}$$

Then we have

<u>LEMMA III.1</u>: The operators $F(u)$ (resp. $H(u)$) are accretive in $(L^1(\Omega))^{m+1}$ (resp. $(L^1(\Gamma_2))^{m+1}$).

<u>PROOF</u>: We have to show (see [3], for instance)

$$I = \int_\Omega (F(\phi)-F(\hat{\phi}))\cdot S_0(\phi-\hat{\phi})\, dx \geqq 0,$$
$$\phi = [\phi_k,\psi],\ \hat{\phi} = (\hat{\phi}_k,\hat{\psi}) \in [L^1(\Omega)]^{m+1}, \tag{III.20}$$

where $S_0(\phi-\hat{\phi}) = [r_k,\ 1 \leqq k \leqq m, r]$ with $r_k = \text{sgn}(\phi_k-\hat{\phi}_k)$, $r = \text{sgn}(v-\hat{v})$, (sgn = "sign of x" = +1 if $x > 0$, 0 if $x = 0$, -1 if $x < 0$).

A simple computation shows

$$I = \sum_{k=1}^{m} \int_\Omega [f_k(x,\hat{\phi}_k,\psi)-f_k(x,\phi_k,\psi)](r_k-r)dx +$$
$$+ \sum_{k=1}^{m} \int_\Omega [f_k(x,\hat{\phi}_k,\hat{\psi})-f_k(x,\hat{\phi}_k,\psi)](r_k-r)dx = I_1 + I_2. \tag{III.21}$$

Now, we observe that each term I_j in (III.21) is positive. In fact, with assumption [(III.18)(i)]

$$[f_k(x,\hat{\phi}_k,\psi)-f_k(x,\phi_k,\psi)]r_k = |f_k(x,\hat{\phi}_k,\psi)-f_k(x,\phi_k,\psi)|$$

and $|r(x)| \leqq 1$ a.e. imply $I_1 \geqq 0$. Using assumption [(III.17(ii)] we get $I_2 \geqq 0$, (III.20) follows. □

THEOREM III.3: With assumptions (III.17), (III.17', (III.18) the solution of problem (P) is unique.

PROOF OF THEOREM III.3: Using the notations of Theorem III.2, we introduce the following approximation of sgn σ, $\sigma = \sigma^+ - \sigma^- \in R$:

$$s_\eta(\sigma) = P(\sigma^+) - P_\eta(\sigma^-) \text{ [we have: } s'_\eta(\sigma) \geqq 0,\ \forall\eta > 0,\ s'_\eta \in L^\infty(R)],$$

and we can take the following test function:

$$S_\eta(\sigma) = P_\eta(w^+) - P_\eta(w^-) = [s_\eta(w_i),\ 1 \leqq i \leqq m+1], \tag{III.22}$$

where $w_i = u_i - \hat{u}_i$, $w_{m+1} = v-\hat{v}$.

Now setting

$$\Phi_\eta(t) = \sum_{i=1}^{m+1} \int_\Omega [\int_0^{w_i(x,t)} s_\eta(\sigma)\, d\sigma]\, dx \tag{III.23}$$

from the system we obtain

$$\Phi_n(t) + \frac{\alpha}{2} \sum_{i=1}^{m+1} \int_{\Omega_t} s'_\eta(w_i)|\nabla w_i|^2\, dx\, d\tau \leqq \frac{1}{2\alpha} \sum_{i=1}^{m+1} \int_{\Omega_t} \chi_i|\nabla \hat{u}_i|^2 dx\, d\tau +$$

$$+ I_\eta(t), \tag{III.24}$$

$$I_\eta(t) = \sum_{i=1}^{m} \int_{\Omega_t} [f_i(u_i,v)-f_i(\hat{u}_i,\hat{v})][s\,(w_i)-s\,(w_{m+1})]\, dx\, d\tau,$$

where χ_i is the characteristic function of the set $\{(x,t) \in \Omega_T; \sigma \leqq |w_i| \leqq \eta\}$. Since, when $\eta \to 0$, $\int_0^\tau s_\eta(\sigma)d\sigma \to |\tau|$, $t \in R$, and from the fact that $t \to \Phi_\eta(t)$ is uniformly bounded with respect to η, into $C^0([0,T];(L^1(\Omega))^{m+1})$ we have

$$\lim_{\eta\to 0} \int_0^T \Phi_\eta(t)\, dt = \Phi = \sum_{i=1}^{m+1} \int_{\Omega_T} |w_i(x,t)|\, dx\, dt.$$

Now from (III.24) we deduce

$$\int_0^T \Phi_\eta(t)dt \leq TI_\eta(T) + \frac{T}{2\alpha} \sum_{i=1}^{m-1} \int_{\Omega_T} \chi_i |\nabla \hat{u}_i|^2 \, dx\, dt,$$

and we observe that:

$$\lim_{\eta \to 0} \int_{\Omega_T} \chi_i |\nabla u_i|^2 \, dx\, dt = 0.$$

On the other hand, from Lemma III.1, we have $\lim_{\eta \to 0} TI_\eta(T) \leq 0$. Hence $\Phi = 0$ and $|w_i| = 0$, $\forall i$. □

REMARK III.1:

1. Results of Sections III.1 and III.2 were obtained in August 1985 during a visit by the author to Stanford University (see also [2]). The results with assumption (III.6) seems a new result.

2. With assumption (III.5) the result of uniqueness for one equation is independently proved in another context in [6].

3. Naturally, the result of Theorem III.3 may be extended to L^1 theory for the accretive operator A in divergence form with $A(u) = -\mathrm{div}\, \mathcal{A}(u)\, \nabla u$ where $\mathcal{A}(u) = (A^r, \mathcal{B})$ as in problem (P)) and with $F(u)$ (resp. $\mathcal{H}(u)$) satisfying (III.17) and (III.18).

ACKNOWLEDGEMENT: I express my thanks to the referee for his careful reading of the manuscript.

REFERENCES

[1] Artola, M., Sur une classe de problèmes paraboliques quasi-linéaires, Boll. U.M.I. (6) 4-B (1986), 51-70.

[2] Artola, M., Approximation theorems for the solution of a nonlinear parabolic system with coupling. Pub. Dept. Math. Appliquées, Université de Bordeaux I, No. 8603.

[3] Benilan, Ph., Equations d'evolution dans un espace de Banach quelconque et applications. These, Orsay, 1972.

[4] Carrillo, J. and Chipot, M., On some nonlinear elliptic equations involving derivatives on the nonlinearity. Proc. Roy. Soc., Edinburgh, 100A (1985), 281-294.

[5] Chipot, M. and Michaille, Uniqueness results and monotonicity properties for strongly nonlinear elliptic variational inequalities. I.M.A. Oct. 87, 40347 Minneapolis.

[6] Chipot, M. and Rodrigues, J.F., Comparison and stability of solutions to a class of quasilinear parabolic problems (to appear.)

[7] Dautray, R. and Lions, J.L., Analyse mathématique et calcul numérique pour les sciences et les techniques, vol. III, collection CEA, Masson, 1985.

[8] Gagneux, G. and Guerfi, F., Approximations de la fonction d'Heaviside et résultats d'unicité pour une classe de problèmes quasi-linéaires stationnaires. Publ. Lab. Math. Appliquées, No. 1204, Université de Pau et des Pays de l'Adour, No. 87012.

[9] Kranoselski, Topological methods in the theory of nonlinear integral equations. Pergamon Press, 1964.

[10] Lions, J.L. and Magenes, F., Problèmes aux limites nonhomogènes et applications, vol. I, Dunod 1968.

[11] Piccinini, L.C., Stampacchia, G. and Vidossich, G., Equazioni differenziali ordinarie in R^N. Ed. Liguori, Napoli, 1978.

[12] Pierre, M., An L^1 method to prove global existence in some reaction-diffusion systems. Contributions to nonlinear partial equations. Vol. II, Pitman Research Notes in Math. Series No. 155, 1987.

[13] Trudinger, N.S., On the comparison principle for quasi-linear divergence structure equations. Arch. Rat. Mech. Anal. 57 (1974), 128-133.

M. Artola
Département de Mathématiques Appliquées
351, Cours de la Libération
Université de Bordeaux I
33 405 Talence
France

P. BÉNILAN AND H. LABANI

Existence of attractors for the Brusselator

We consider the following reaction-diffusion system, named brusselator and introduced by Lefever and Prigogine in [7]

$$\partial u/\partial t = d_1\Delta u - (b+1)u + u^2v + a \qquad \text{on } Q =]0,\infty[\times \Omega, \qquad (1)$$
$$\partial v/\partial t = d_2\Delta v + bu - u^2v$$

where d_1, d_2 a, b are strictly positive constants and Ω is a bounded open set in R^N, which is assumed to be smooth for simplicity. We will consider on the boundary $\partial\Omega$ Dirichlet conditions as well as Neumann conditions: more generally, we consider boundary conditions of the following form:

$$\lambda\ \partial u/\partial n + (1-\lambda)(u-\alpha) = 0, \qquad \text{on } \Sigma =]0,\infty[\times \partial\Omega, \qquad (2)$$
$$\lambda\ \partial v/\partial n + (1-\lambda)(v-\beta) = 0,$$

where λ, α, β are constants satisfying the basic assumptions

$$0 \leqq \lambda \leqq 1,\ \alpha \geqq 0,\ \beta \geqq 0. \qquad (3)$$

This system has been extensively studied; in particular, recently Hollis, Martin and Pierre [3] proved existence of a global bounded solution of (1), (2) with any bounded nonnegative initial data u(0), v(0). We state this result under the form

THEOREM 0 ([3]): For any $u_0, v_0 \in L^\infty(\Omega)^+$ there exists unique functions $u,v \in C^{1,2}(]0,\infty[\times \bar{\Omega})$ classical solutions of (1), (2) with the initial condition

$$u(0) = u_0, \; v(0) = v_0, \tag{4}$$

in the sense

$$\int_\Omega |u(t)-u_0| + |v(t)-v_0| \to 0 \text{ as } t \to 0.$$

Moreover, $u, v \in L^\infty(Q)^+$.

In other words, the problem (1), (2) defines a semigroup $\{S(t)\}_{t\geq 0}$ on the cone $E = \{(u,v) \in L^\infty(\Omega)^2;\ u \geq 0,\ v \geq 0\}$. In this article we prove the existence of a maximal attractor for this semigroup (see, for instance, [1] for the notion of attractor); we state the result precisely:

THEOREM 1: There exists a unique compact set M in the cone $L^\infty(\Omega)^{+2}$ satisfying:

(i) $S(t)M = M$ for any $t \geq 0$;

(ii) $\text{dis}(S(t)(u_0,v_0),M) \to 0$ as $t \to \infty$ uniformly for (u_0,v_0) bounded in $L^\infty(\Omega)^{+2}$.

A maximal attractor for the brusselator has already been studied: for instance, in [4], where the Neumann boundary conditions are considered ($\lambda = 1$), it is shown that the fractal dimension of such an attractor in $L^2(\Omega)$ is bounded from above and below, but its existence is not proved.

Remark that any solution $(\underline{u},\underline{v}) \in C^2(\bar{\Omega})$ of the elliptic problem

$$d_1\Delta\underline{u} - (b+1)\underline{u} + \underline{u}^2\underline{v} + a = d_2\Delta\underline{v} + b\underline{u} - \underline{u}^2\underline{v} = 0 \text{ on } \Omega,$$

$$\lambda\ \partial\underline{u}/\partial n + (1-\lambda)(\underline{u}-\alpha) = \lambda\ \partial\underline{v}/\partial n + (1-\lambda)(\underline{v}-\beta) = 0 \text{ on } \partial\Omega,$$

is in M; in particular, if $\alpha = a$ and $\beta = b/a$, then $(a,b/a) \in M$. It is also proved in [4], in the case of Dirichlet boundary conditions ($\lambda = 0$, with $\alpha = a$ and $\beta = b/a$), that M is reduced to $(a,b/a)$ provided that b/a is less than some positive constant depending only on d_1,d_2 and the first eigenvalue of the Laplace operator on Ω with Dirichlet boundary conditions.

Using compactness of the semigroup $S(t)$, the proof of Theorem 1 reduces to existence of an attractor A, that is, a bounded set in $L^\infty(\Omega)^{+2}$ satisfying

$\text{dist}(S(t)(u_0,v_0),A) \to 0$ as $t \to \infty$ uniformly for (u_0,v_0) bounded in $L^\infty(\Omega)^{+2}$.

The maximal attractor M will be then defined by

$$M = \bigcap_{t \geq 0} \text{cl}\Big(\bigcup_{t \geq \tau} S(t)(A)\Big).$$

In other words Theorem 1 reduces to

THEOREM 2: There exists a constant C such that for any R there exists T such that for any $(u_0,v_0) \in L^\infty(\Omega)^{+2}$ with $\|(u_0,v_0)\|_{L^\infty} \leq R$

$$\|S(t)(u_0,v_0)\|_{L^\infty} \leq C \text{ for any } t \geq T. \tag{5}$$

The proof of this Theorem 2 will be given in three steps.

Step 1 will prove the following:

PROPOSITION 3: There exists a constant C such that for any R there exists T and M such that for any $(u_0,v_0) \in L^\infty(\Omega)^{+2}$ with $\|v_0\|_{L^\infty} \leq R$, the solution (u,v) of (1), (2), (4) satisfies

$$\|v(t)\|_{L^\infty} \leq C \text{ for any } t \geq T,\ \|v(t)\|_{L^\infty} \leq M \text{ for any } t \leq T. \tag{6}$$

Step 2. Let us introduce the self-adjoint operator A in $L^2(\Omega)$ defined by

$$Aw = \Delta w \text{ on } D(A) = \{w \in H^2(\Omega);\ \lambda\ \partial w/\partial n + (1-\lambda)w = 0 \text{ on } \partial\Omega\}.$$

It generates a continuous sub-Markovian semigroup $e^{\tau A}$ on $L^2(\Omega)$. We will prove the following:

PROPOSITION 4: There exists a constant C such that for any R there exists T such that for any $(u_0,v_0) \in L^\infty(\Omega)^{+2}$ with $\|(u_0,v_0)\|_{L^\infty} \leq R$, the solution (u,v) of (1), (2), (4) satisfies

$$\|e^{d_1 A} u(t)\|_{L^\infty} \leq C \text{ for any } t \geq T. \tag{7}$$

Step 3 will prove end of proof of Theorem 2.

The organization of the paper is as follows: in Sections 1, 2, 3 we develop the proofs of Steps 1, 2, 3 above; in Section 4, we state an abstract result which will be developed in [5] and [2].

1. PROOF OF PROPOSITION 3

We consider separately the case $\lambda = 1$ (Neumann boundary conditions) and the case $\lambda < 1$. Indeed, remark that in this last case ($\lambda < 1$), the operator A is one-to-one from D(A) onto $L^2(\Omega)$; we will denote by G the Green operator $(-\Delta)^{-1}$ defined by

$$-\Delta Gf = f \text{ on } \Omega, \quad \lambda \partial Gf/\partial n + (1-\lambda)Gf = 0 \text{ on } \partial\Omega. \tag{8}$$

Recall that G is a positive operator.

PROOF OF PROPOSITION 3 IN THE CASE $\lambda < 1$: Let (u,v) be a solution of (1), (2) and set $w = v^2 - \beta^2$. Then $w \in C^{1,2}(]0,\infty[\times \bar{\Omega})$ and satisfies

$$\partial w/\partial t = 2v(d_2\Delta v + u(b-uv)) \leq d_2\Delta w + b^2/2 \text{ on } Q,$$

$$\lambda\partial w/\partial n + (1-\lambda)w = \lambda 2v\partial v/\partial n + (1-\lambda)(v^2-\beta^2) = -(1-\lambda)(v-\beta)^2 \leq 0 \text{ on } \Sigma.$$

By the comparison principle, using the semigroup e^{tA} introduced above, it follows that

$$w(t) \leq e^{tA}w(0) + \int_0^t e^{\tau A}b^2/2 \, d\tau. \tag{9}$$

Now

$$\int_0^t e^{\tau A} \, b^2/2 \, d\tau \leq b^2/2 \; G1.$$

Then

$$v(t)^2 \leqq \beta^2 + b^2/2 \, \|G1\|_{L^\infty} + \|e^{tA}1\|_{L^\infty} \max(\beta^2, \|v_0\|_{L^\infty}).$$

The proposition follows with

$$C = (\beta^2 + b^2/2 \, \|G1\|_{L^\infty})^{1/2}, \tag{10}$$

since $\|e^{tA}1\|_{L^\infty} \to 0$ as $t \to \infty$. □

In the case $\lambda = 1$, we use the ideas in [3], making them more precise. We begin with the following result stated without assumption on λ:

<u>LEMMA 6</u>: Any solution (u,v) of (1), (2) satisfies

$$u \geqq \gamma \, (1 - e^{-(b+1)t}) \text{ on } Q \text{ with } \gamma = \min(\alpha, a/(b+1)). \tag{11}$$

<u>PROOF OF LEMMA 6</u>: Set $m(t) = \min\{u(t,x);\ x \in \bar{\Omega}\}$. For any $t > 0$, the derivative from the right $m'(t)$ of $m(t)$ exists and there exists $x \in \bar{\Omega}$ such that

$$m'(t) = \partial u/\partial t(t,x), \ m(t) = u(t,x).$$

Moreover, using the Hopf maximum principle and the boundary conditions (2), if $m(t) < \alpha$ we may assume $x \in \Omega$ and we have $\Delta u(t,x) \geqq 0$. Then, using (1)

$$m'(t) \geqq -(b+1)m(t) + a \text{ for any } t > 0 \text{ with } m(t) < \alpha.$$

The lemma follows. □

<u>PROOF OF PROPOSITION 3 IN THE CASE $\lambda = 1$</u>: In this case Lemma 6 holds with $\gamma = a/(b+1)$ (we may choose α arbitrary). Set $M(t) = \max\{v(t,x);\ x \in \bar{\Omega}\}$. For any $t > 0$, the derivative from the right $M'(t)$ exists and there exists $x \in \bar{\Omega}$ such that

$$M'(t) = \partial v/\partial t(t,x), \ M(t) = v(t,x).$$

By the Hopf maximum principle again, we may choose $x \in \Omega$ and we have

$$M'(t) \leq u(t,x)(b-u(t,x)M(t)). \tag{12}$$

Now fix $\tau > 0$ and set $\gamma(\tau) = \gamma(1-e^{-(b+1)\tau})$. Looking to the variations of the function $r \to r(b-rM(t))$, it follows from (11) and (12) that

$$M'(t) \leqq \gamma(\tau)(b-\gamma(\tau)M(t)) \text{ for } t \geqq \tau \text{ according that } M(t) \geqq b/2\gamma(\tau), \tag{13}$$

and

$$M'(t) \leqq b/4M(t) \text{ for any } t \geqq 0. \tag{14}$$

By integration, we have then

$$M(t) \leq (b/\gamma(\tau))(1-(1/2)e^{-\gamma(\tau)^2(t-\tau)}) \text{ for } t \geq \tau, \tag{15}$$

and

$$M(t) \leq (M(0)^2 + bt/2)^{1/2}. \tag{16}$$

Then the proposition follows with

$$C > b(b+1)/a. \quad \square \tag{17}$$

REMARK 7: The above proof will work for general $\lambda \in [0,1]$, provided that $\alpha > 0$: the proposition will follow with $C > \max(b(b+1)/a, b/\alpha)$. Remark also that for such C, there exists T such that for any solution (u,v) of (1), (2)

$$\|v(t)\|_{L^\infty} \leq C \text{ for any } t \geq T$$

independently of the initial data.

2. PROOF OF PROPOSITION 4

Let C_0, and for fixed $R > 0$, T_0 and M_0 be the constants obtained in Proposition 3. We consider again separately the case $\lambda = 1$ and $\lambda < 1$. In the case $\lambda = 1$, we will use the following:

LEMMA 8: Assume $\lambda = 1$. Then for any solution (u,v) of (1), (2)

$$\int_\Omega u(t) + v(t) = e^{-t} \int_\Omega u(0) + v(0) + \int_0^t e^{s-t}(\int_\Omega v(s) + a)ds \tag{18}$$

for any $t \geq 0$.

PROOF OF LEMMA 8: Immediate since in the case $\lambda = 1$

$$d/dt \int_\Omega u(t) + v(t) = \int_\Omega a-u(t).$$

PROOF OF PROPOSITION 4 IN THE CASE $\lambda = 1$: For a solution (u,v) of (1), (2), (4) with $\|(u_0,v_0)\|_{L^\infty} \leq R$ we have by (18) and (6)

$$\|u(t)\|_{L^1} \leq (C_0+a+e^{-t}(R+e^{T_0}(M_0-C_0)^+))|\Omega| \quad \text{for } t \geq T_0. \tag{19}$$

Using the fact that $e^{d_1 A}$ applies $L^1(\Omega)$ into $L^\infty(\Omega)$, the proposition follows. □

In the case $\lambda > 1$ we will use the following lemma where G is the Green operator definec by (8):

LEMMA 9: Assume $\lambda < 1$, let (u,v) be a solution of (1), (2) and set $W(t) = G(u(t) + v(t))$. Then for any $t \geq 0$

$$\|W(t)\|_{L^2} \leq e^{-t} \|W(0)\|_{L^2} + \int_0^t e^{s-t} \|G(v(s)+a)+d_1\alpha+d_2\beta\|_{L^2} ds. \tag{20}$$

PROOF OF LEMMA 9: It is immediate from

$$dW/dt(t) = -d_1(u(t)-\alpha) - d_2(v(t)-\beta) + G(a-u(t))$$

$$\leq \min(d_1,d_2)AW(t)-W(t) + G(v(t)+a) + d_1\alpha + d_2\beta. \quad □$$

PROOF OF PROPOSITION 4 IN THE CASE $\lambda < 1$: For a solution (u,v) of (1), (2), (4) with $\|(u_0,v_0)\|_{L^\infty} \leq R$, using (20) and (6)

$$\|Gu(t)\|_{L^2} \leq (C_0+a)\,\|G1\|_{L^2} + (d_1\alpha+d_2\beta)|\Omega|^{1/2} + e^{-t}(R+e^{T_0}(M_0+a))\,\|G1\|_{L^2}. \qquad (21)$$

Using now the fact that Ae^{d_1A} applies $L^2(\Omega)$ into $L^\infty(\Omega)$, we obtain the proposition. □

REMARK 10: The estimate we gave in Lemma 9 is far from sharp: one main fact is that it is independent of (d_1,d_2) as well as the estimate following from Lemma 8. However, the constant C in Proposition 4 really depends on d_1: we have proved dependence in $d_1^{-N/2}$ in the first method we used for $\lambda = 1$, while it is in $d_1^{-1-N/2}$ in the second method. Remark also that we could use the second method in the case $\lambda = 1$, by taking

$$W(t) = (-A+\mu)^{-1}(u(t)+v(t)) \text{ with } 0 < \mu < (\max(d_1,d_2) - \min(d_1,d_2))^{-1}.$$

We could also use the first method in the case $\lambda < 1$, provided that $\alpha = \beta = 0$: indeed, then (18) will still be true with "=" replaced by "$\leq$".

3. PROOF OF THEOREM 2

We begin the following:

LEMMA 11: For any $1 < p < \infty$, there exists a constant C such that for any R there exists T such that for any $(u_0,v_0) \in L^\infty(\Omega)^{+2}$ with $\|(u_0,v_0)\|_{L^\infty} \leq R$, the solution (u,v) of (1), (2), (4) satisfies

$$\|u(t+1+.,.)\|_{L^p(]0,1[\times\Omega)} \leq C \text{ for any } t \geq T. \qquad (22)$$

PROOF OF LEMMA 11: For given R, let (u,v) be the solution of (1), (2), (4) with $\|(u_0,v_0)\|_{L^\infty} \leq R$ and set

$$w = d_1u + d_2v. \qquad (23)$$

We have

$$\partial w/\partial t = d_1\Delta w + d_1(-u+a) + (d_2-d_1)\partial v/\partial t \text{ on } Q,$$

$$\lambda\partial w/\partial n + (1-\lambda)(w-d_1\alpha - d_2\beta) = 0 \text{ on } \Sigma.$$

Then

$$w(t+1+\tau) = w_1(t,\tau) + d_1 w_2(t,\tau+1) + (d_2-d_1)(\partial/\partial\tau) w_3(t,\tau+1) \qquad (24)$$

with

$$w_1(t,\tau) = d_1\alpha + d_2\beta + e^{(1+\tau)d_1A}w(t),$$

and for $t \geqq 0$, $w_2(t)$, $w_3(t)$ are the solutions of

$$\partial w_i/\partial\tau = d_1\Delta w_i + f_i \text{ on } Q,\ \lambda\partial w_i/\partial n + (1-\lambda)w_i = 0 \text{ on } \Sigma,\ w_i(0,\cdot) = 0 \text{ on } \Omega$$

with $f_2(\tau) = a-u(t+\tau)$ and $f_3(\tau) = v(t+\tau)-v(t)$, respectively.

Since

$$\| e^{(1+\tau)d_1A} w(t) \|_{L^\infty} \leqq \| e^{d_1A} u(t) \|_{L^\infty} + \| v(t) \|_{L^\infty}$$

using Propositions 3 and 4, there exist C_0 (independent of R) and T_0 such that

$$w_1(t,\tau) \leqq C_0 \text{ for any } (t,\tau) \in [T_0,\infty[\times]0,\infty[. \qquad (25)$$

Also, it is clear that

$$w_2(t,\tau) \leqq a\tau \quad \text{for any } t \geqq 0,\ \tau \geqq 0. \qquad (26)$$

At last, using a classical result (see [6]), for any $1 < p < \infty$, there exists a constant $C(p)$ such that

$$\| (\partial/\partial\tau)w_3(t) \|_{L^p(]0,2[\times\Omega)} \leqq C(p) \| f_3 \|_{L^p(]0,2[\times\Omega)},$$

and then using Proposition 3,

$$\| (\partial/\partial\tau)w_3(t) \|_{L^p(]0,2[\times\Omega)} \leqq C(p,d_1,|\Omega|) \text{ for } t \geqq T_0. \qquad (27)$$

Collecting (23)-(27), clearly the lemma follows. □

PROOF OF THEOREM 2: For given R, let (u,v) be the solution of (1), (2), (4) with $\|(u_0,v_0)\|_{L^\infty} \leqq R$. Using (1),

$$u(t+2) \leqq \alpha \, \| e^{d_2 A} u(t+1) \|_{L^\infty} + a + \int_0^1 e^{\tau d_1 A} u(t+1+\tau)^2 v(t+1+\tau) d\tau .$$

Using Propositions 3 and 4, there exist C_0 (independent of R) and T_0 such that

$$u(t+2) \leqq C_0 (1 + \int_0^1 e^{\tau d_1 A} u(t+1+\tau)^2 d\tau) \text{ for } t \geqq T_0. \tag{28}$$

Now there exists a constant C(N) such that

$$\| e^{\tau d_1 A} w \|_{L^\infty} \leqq C(N) \, \|w\|_{L^q} \, \tau^{-N/2q} \text{ for } 0 < \tau < 1 \text{ and } 1 \leqq q \leqq \infty .$$

Fix $q < \infty$ with $q > 1 + (N/2)$; by (28) we have with some constant C depending only on C_0, N and q

$$u(t+2) \leqq C(1 + \int_0^1 \int_\Omega u(t+1+\tau,x)^{2q} \, dx \, d\tau) \text{ for } t \geqq T_0. \tag{29}$$

Using Lemma 11, Theorem 2 follows. □

4. AN ABSTRACT RESULT

The above proof may be developed to study the existence of global solutions and attractors for a system of two evolution equations

$$du/dt = d_1 A(u-\underline{u}) + f(u,v), \quad dv/dt = d_2 A(v-\underline{v}) + g(u,v), \tag{30}$$

on a measure set $(\Omega,\mathcal{B},\mu)$ with finite measure, where

(a) A is a self-adjoint operator in $L^2(\Omega)$ one-to-one with bounded inverse, infinitesimal generator of a sub-Markovian semigroup $\{e^{tA};\ t \geqq 0\}$ with $e^{tA}(L^2(\Omega)) \subset L^\infty(\Omega)$ for $t > 0$.

(b) f and g are locally Lipschitz maps from R^2 into R satisfying for any $u \geqq 0$, $v \geqq 0$

$$f(0,v) \geqq 0, \; g(u,0) \geqq 0, \tag{31}$$

$$f(u,v) + g(u,v) \leqq L(v), \tag{32}$$

$$f(u,v) \leqq L(v)(1+u)^{m(v)}, \tag{33}$$

with L and m nondecreasing continuous functions on R^+.

(c) d_1, d_2 are positive constants, $\underline{u}$, $\underline{v} \in L^\infty(\Omega)$ and satisfy

$$\int \underline{u}\, Aw \leqq 0, \; \int \underline{v}\, Aw \leqq 0 \text{ for any } w \in D(A), \; w \geqq 0. \tag{34}$$

By a bounded solution on $[0,T[$ of (30), we mean, of course, a couple of bounded functions $u,v : [0,T[\to L^\infty(\Omega)$ satisfying

$$u,v \in C([0,T[; L^2(\Omega)) \cap W^{1,1}_{loc}(]0,T[; L^2(\Omega)) \text{ and a.e. } t \in]0,T[,$$

$$u(t)-\underline{u} \in D(A), \; u'(t) = d_1 A(u(t)-\underline{u}) + f(u(t),v(t)),$$

$$v(t)-\underline{v} \in D(A), \; v'(t) = d_2 A(v(t)-\underline{v}) + g(u(t),v(t)).$$

Then we may state the following:

THEOREM 11: Assume, moreover, there is some continuous nonnegative function $\Gamma(R,t)$ on R^{+2} nondecreasing in R and nonincreasing in t such that for any bounded nonnegative solution (u,v) of (30) on some interval $[0,T[$

$$v(t) \leqq \Gamma(\|u(0) + v(0)\|_{L^\infty}, t) \text{ a.e. on } \Omega \text{ for any } t \in [0,T[. \tag{35}$$

(a) For any $u_0, v_0 \in L^\infty(\Omega)^+$ there exists a unique bounded solution of (30) on $[0,\infty[$; moreover, this solution is nonnegative. In other words (30) defines a continuous semigroups $\{S(t); t \geqq 0\}$ on $L^\infty(\Omega)^{+2}$.

(b) If the function Γ above satisfies

$$\lim_{R\to\infty} (\lim_{t\to\infty} \Gamma(R,t)) < \infty , \tag{36}$$

then there is an attractor for the semigroup $\{S(t)\}$.

(c) If, moreover, e^{tA} is compact in $L^2(\Omega)$ for some $t > 0$, then there is a maximal attractor for the semigroup $\{S(t)\}$.

This theorem will be proved completely in [5] and [2], where also other results of the same type will be given. Remark that it extends the results in [3]. Also, thanks to Proposition 3, it applies to the problem (1), (2), using, for operator A, the operator $A-d_1^{-1}I$ where A is the operator defined in the introduction. Other applications will be discussed in [5].

REFERENCES

[1] A.V. Babin and M.I. Vishik, Attractors of partial differential equations and estimates of their dimension. Russian Math. Survey 38, 4 (1983), 151-213.

[2] Ph. Bénilan and H. Labani, Existence d'attracteurs pour une classe de systèmes de réaction-diffusion, in preparation.

[3] S.L. Hollis, R.H. Martin and M. Pierre, Global existence and boundedness in reaction-diffusion system, SIAM J. Math. Anal. 18, No. 3 (1987), 744-761.

[4] S. Kouachi, Dimensions des ensembles fonctionnels invariants et attracteurs associés à des équations d'évolution semi-linéaires dans des espaces de Banach et applications, Thèse de 3ème cycle, Université Paris VII, 1985.

[5] H. Labani, Contribution a l'étude de systèms de réaction-diffusion, Thèse de l'Université de Franche-Comté, Besançon, 1988.

[6] O.A. Ladyzenskia, V.A. Solonnikov and R.N. Uralceva, Linear and quasilinear equations of parabolic type, Transl. Math. Mon., No. 33, Am. Math. Soc., 1968.

[7] R. Lefever and I. Prigogine, Symmetry breaking instabilities in dissipative systems II. J. Chemical Phys. 48, No. 4 1968.

Ph. Bénilan and H. Labani
Département de Mathematiques
Université de Franche Comté
route de Gray
25030 Besançon Cedex
France

L.C. EVANS

Some asymptotics for reaction-diffusion PDEs

1. INTRODUCTION

We will consider systems of reaction-diffusion PDEs of the general form

$$u_{k,t} = L_k u_k + f_k(u) \text{ in } \mathbb{R}^n \times (0,\infty) \qquad (k = 1,\ldots,m), \tag{1.1}$$

where the unknown is $u = (u_1,\ldots,u_m)$, $f = (f_1,\ldots,f_m)$ denotes the nonlinear coupling, and L_k is a second-order, uniformly elliptic operator of the form

$$L_k v = a_{ij}^k v_{x_i x_j} + b_i^k v_{x_i}.$$

To investigate the asymptotic behaviour of a solution u for large $x \in \mathbb{R}^n$ and $t > 0$, we perform an order ε^{-1} rescaling, and thereby shift our attention to the new functions

$$u^\varepsilon(x,t) \equiv u(x/\varepsilon, t/\varepsilon),$$

which solve the rescaled system

$$u_{k,t}^\varepsilon = L_k^\varepsilon u_k^\varepsilon + \frac{1}{\varepsilon} f_k(u^\varepsilon) \text{ in } \mathbb{R}^n \times (0,\infty) \quad (k=1,\ldots,m) \tag*{$(1.2)_\varepsilon$}$$

for

$$L_k^\varepsilon v \equiv \varepsilon\, a_{ij}^k\, v_{x_i x_j} + b_i^k\, v_{x_i}. \tag{1.3}$$

We inquire as to the limiting behaviour as $\varepsilon \to 0$ of solutions u^ε of $(1.2)_\varepsilon$. Such asymptotics can in general be exceedingly complex, so let us at once simplify by supposing that f vanishes precisely at finitely many points $\{a_k\}_{k=1}^N \subset \mathbb{R}^n$. Then rewriting $(1.2)_\varepsilon$ to read

$$f_k(u^\varepsilon) = \varepsilon u_{k,t}^\varepsilon - \varepsilon L_k^\varepsilon u_k^\varepsilon \qquad (k = 1,\ldots,m),$$

we may naively hope that as $\varepsilon \to 0$ the vector function u^ε will converge somehow to a limit u verifying

$$f(u) = 0 \text{ in } \mathbb{R}^n \times (0,\infty).$$

Then u would take on only the values $\{a_k\}_{k=1}^N$, and so we could write

$$u = \sum_{k=1}^N a_k \chi_{A_k},$$

the disjoint sets $\{A_k\}_{k=1}^N$ presumably subdividing $\mathbb{R}^n \times (0,\infty)$ into regions of different asymptotic behaviour of our solutions u^ε of $(1.2)_\varepsilon$ for small $\varepsilon > 0$. We may further attempt to envision the boundaries of the sets $\{A_k\}_{k=1}^N$ as somehow being evolving "wavefronts" separating regions where the solutions "switch" in the limit among the equilibria of the nonlinearity f. And if, to continue the fantasy, we could derive a direct analytic and/or geometric understanding of these wavefronts, we could then characterize the "complicated" large x- and t-asymptotics of solutions to the reaction-diffusion system (1.1) in terms of our "simple" characterization of wavefront propagation.

Now such an undertaking is obviously hopeless for general systems of the type (1.1): the complex interplay of the diffusion terms L_k with the nonlinear structure of f and its associated equilibria, limit cycles, etc. excludes any such simple characterization of the asymptotics. Nonetheless, Freidlin [3] has identified certain very specific systems amenable to an analysis of a kind as envisioned above. His probabilistic approach has been augmented recently by a purely PDE analysis undertaken by Souganidis and myself [2] for single equations, and by Barles, Souganidis, and myself [1] for systems of PDEs. The guiding idea in these studies is to focus upon the simplest case that f has just two equilibria, a_1 and a_2, the first of which is unstable and the second stable. Under various additional structural hypotheses, we construct a Hamiltonian H, which incorporates information concerning the diffusion effects and the instability of the point a_1. We next solve a corresponding Hamilton-Jacobi PDE:

$$J_t + H(DJ) = 0 \text{ in } \mathbb{R}^n \times (0,\infty). \tag{1.4}$$

It then turns out, roughly speaking, that we have

$$A_1 = \{J > 0\} \quad \text{and} \quad A_2 = \{J > 0\}.$$

In other words, to characterize the limiting asymptotic behaviour of solutions u^ε of the coupled parabolic system $(1.2)_\varepsilon$, we need only solve the single PDE (1.4). And since a simple, fairly explicit representation formula for the solution J is available, we have (for these very special cases) realized the program sketched out at the beginning.

My intention in this paper is to present, for the special cases that additional estimates are available, an extension and greatly simplified derivation of some convergence results from [1].

2. WAVEFRONT PROPAGATION

More precisely now, let us assume that $u^\varepsilon = (u_1^\varepsilon,\dots,u_m^\varepsilon)$ is a smooth solution of the system

$$\begin{aligned} u^\varepsilon_{k,t} &= L^\varepsilon_k u^\varepsilon_k + \frac{1}{\varepsilon} f_k(u^\varepsilon) \quad \text{in } \mathbb{R}^n \times (0,\infty) \quad (k = 1,\dots,m), \\ u^\varepsilon_k &= g_k \quad \text{on } \mathbb{R}^n \times \{0\}, \end{aligned} \tag{2.1$_\varepsilon$}$$

the uniformly elliptic operators L^ε_k defined by (1.3). We will for simplicity assume the coefficients a^k_{ij}, b^k_i to be constant. We additionally suppose $f = (f_1,\dots,f_m)$ is smooth, Lipschitz, and $g_k \geq 0$: consequently, our solution is unique and positive. We assume $G = \operatorname{spt} g_k$ $(k = 1,\dots)$ is smooth and bounded. Let us further hypothesize concerning the nonlinearity f that

(F1) $f(0) = 0$,

(F2) $f_k(\cdots u_{k-1}, 0, u_{k+1},\dots) > 0$ if $u_1,\dots,u_m \geq 0$ with $u_l > 0$ for some $l \neq k$,

and

(F3) $f_k(u) \leq 0$ if $u_k \geq \Lambda$, $u \in \Pi$,

for some constant Λ, where $\Pi \equiv \{u \in \mathbb{R}^m \mid u \geq 0\}$. Now write $C = DF(0)$. We suppose also

(F4) $c_{kl} > 0 \quad (1 \leqq k,l \leqq m)$,

and

(F5) $f_k(u) \leqq c_{kl}u_l \quad (k = 1,\ldots,m,\ u \in \Pi)$.

To characterize that region $A_1 \subset \mathbb{R}^n \times (0,\infty)$ where $u^\varepsilon \to 0$, let us define as follows an auxiliary Hamiltonian. Given $p \in \mathbb{R}^n$, set

$$B(p) \equiv \mathrm{diag}(\cdots,\ a^k_{ij}p_ip_j - b^k_ip_i,\ \ldots)$$

and write

$$A(p) \equiv B(p) + C.$$

Owing to (F4) the off-diagonal entries of $A(p)$ are nonnegative, whence the Perron-Frobenius theory asserts that $A(p)$ has a simple, real eigenvalue $\lambda^0 = \lambda^0(A(p))$ with Re $\lambda < \lambda^0$ for all other eigenvalues λ. We set

$$H(p) \equiv \lambda^0(A(p)). \tag{2.2}$$

Consider now the associated Hamilton-Jacobi PDE

$$\begin{aligned} J_t + H(DJ) &= 0 \quad \text{in } \mathbb{R}^n \times (0,\infty), \\ J &= 0 \quad \text{on } G \times \{0\}, \\ J &= +\infty \text{ on } (\mathbb{R}^n - G) \times \{0\}, \end{aligned} \tag{2.3}$$

which has a unique (viscosity) solution J. Then we have

<u>THEOREM</u>:

(i) $\lim_{\varepsilon\to 0} u^\varepsilon = 0$ uniformly on compact subsets of $\{J > 0\}$;

(ii) $\liminf_{\varepsilon\to 0} u^\varepsilon > 0$ uniformly on compact subsets of $\{J > 0\}$.

Since control-theoretic representation formulas are available for J (see [1]) we have obtained a fairly explicit characterization of the region $A_1 \equiv \{u^\varepsilon \to a_1 = 0\}$. And if additionally f has a unique stable equilibrium $a_2 \in \Pi$, then given additional structural assumptions, we can employ (ii) to characterize $A_2 \equiv \{u^\varepsilon \to a_2\}$.

PROOF:

1. It is not difficult using (F3) to derive the bounds

$$\sup_\varepsilon \|u^\varepsilon\|_{L^\infty} < \infty. \tag{2.4}$$

However, such estimates alone are not particularly useful.

2. We instead switch our attention to the new functions

$$v_k^\varepsilon = -\varepsilon \log u_k^\varepsilon \quad (k = 1,\dots,m).$$

From $(2.1)_\varepsilon$ we deduce

$$\begin{aligned} &v^\varepsilon_{k,t} - \varepsilon a^k_{ij} v^\varepsilon_{k,x_i x_j} - b^k_i v^\varepsilon_{k,x_i} + a^k_{ij} v^\varepsilon_{k,x_i} v^\varepsilon_{k,x_j} = \frac{-f_k(u^\varepsilon)}{u^\varepsilon_k} \quad \text{in } R^n \times (0,\infty), \\ &v_k^\varepsilon = -\varepsilon \log g_k \text{ on } G \times \{0\}, \\ &v_k^\varepsilon = +\infty \qquad \text{on } (R^n - G) \times \{0\}. \end{aligned} \tag{2.5$_\varepsilon$}$$

The idea now is to show that as $\varepsilon \to 0$, each function v_k^ε converges to I, the unique, viscosity solution of the Hamilton-Jacobi variational inequality

$$\begin{aligned} &\min(I_t + H(DI), I) = 0 \text{ in } R^n \times (0,\infty), \\ &I = 0 \text{ on } G \times \{0\}, \\ &I = +\infty \text{ on } (R^n - G) \times \{0\}. \end{aligned} \tag{2.6}$$

For this we obtain first using the maximum principle the estimate

$$\sup_{\varepsilon} \|v^{\varepsilon}\|_{L^{\infty}(Q)} \leqq C(Q) < \infty \tag{2.7}$$

for any bounded region $Q \subset \mathbb{R}^n \times (0,\infty)$ which is a positive distance away from $(\mathbb{R}^n - G) \times \{0\}$.

For the purpose of exposition let us additionally assume we have the bounds

$$\sup_{\varepsilon} \|Dv^{\varepsilon}, v_t^{\varepsilon}\|_{L^{\infty}(Q)} \leqq C(Q), \tag{2.8}$$

in which case we may assume (passing as necessary to a subsequence) that

$$v_k^{\varepsilon} \to v_k \quad (k = 1,\ldots,m), \text{ locally uniformly on } \mathbb{R}^n \times (0,\infty).$$

3. We first *claim*

$$v_1 = v_2 = \cdots = v_m \equiv v \quad \text{in} \quad \mathbb{R}^n \times (0,\infty). \tag{2.9}$$

To see this, observe from $(2.5)_{\varepsilon}$ and (2.8) that

$$\int_Q \zeta \frac{f_k(u^{\varepsilon})}{u_k^{\varepsilon}} \, dx \, dt = C(Q),$$

ζ denoting an appropriate cutoff function. Using (F2) we reach a contradiction should (2.9) fail.

4. Next we verify

$$\min(v_t + H(Dv), v) = 0 \quad \text{in} \quad \mathbb{R}^n \times (0,\infty) \tag{2.10}$$

is the viscosity sense. In view of (2.4) we easily check $v \geqq 0$. Now suppose ϕ is smooth, and $v - \phi$ has a strict local minimum at a point (x_0,t_0). Choose then $\psi = (\psi_1,\ldots,\psi_m)$ to be a positive eigenvector corresponding to the eigenvalue $H(D\phi(x_0,t_0))$, H defined by (2.2). Then

$$\min_{1\leq l\leq m} [v_l^\varepsilon + \varepsilon \log \psi_l] - \phi \text{ has a local minimum at } (x_\varepsilon, t_\varepsilon), \tag{2.11}$$

$$(x_\varepsilon, t_\varepsilon) \to (x_0, t_0) \text{ as } \varepsilon \to 0. \tag{2.12}$$

Fix an index $k \in \{1,\ldots,m\}$ such that (through a further subsequence of the ε's)

$$\min_{1\leq l\leq m} (v_l^\varepsilon + \varepsilon \log \psi_l - \psi)(x_\varepsilon, t_\varepsilon) = (v_k^\varepsilon + \varepsilon \log \psi_k - \phi)(x_\varepsilon, t_\varepsilon). \tag{2.13}$$

Then $v_k^\varepsilon + \varepsilon \log \psi_k - \phi$ has a minimum at $(x_\varepsilon, t_\varepsilon)$; whence $(2.5)_\varepsilon$ implies

$$0 \leqq \phi_t - \varepsilon a_{ij}^k \phi_{x_i x_j} - b_i^k \phi_{x_i} + a_{ij}^k \phi_{x_1} \phi_{x_j} + \frac{f_k(u^\varepsilon)}{u_k^\varepsilon}$$

$$\leqq \phi_t - b_i^k \phi_{x_i} + a_{ij}^k \phi_{x_i} \phi_{x_j} + c_{kl}\, e^{(v_k^\varepsilon - v_l^\varepsilon)/\varepsilon} + o(1) \text{ at } (x_\varepsilon, t_\varepsilon),$$

according to (F5). Using (2.12), (2.13) we then deduce

$$0 \leqq \phi_t + a_{ij}^k \phi_{x_i} \phi_{x_j} - b_i^k \phi_{x_i} + c_{kl} \frac{\psi_1}{\psi_k} + o(1) \text{ at } (x_0, t_0).$$

Since $A(D\psi(x_0,t_0))\psi = H(D\phi(x_0,t_0))\psi$, we may simplify this expression to find

$$0 \leqq \phi_t(x_0,t_0) + H(D\phi(x_0,t_0)).$$

Thus

$$\min(v_t + H(Dv), v) \geqq 0 \text{ in the viscosity sense.}$$

A similar argument shows that if $v - \phi$ has a strict local maximum at a point (x_0,t_0), and $v(x_0,t_0) > 0$, then

$$0 \geqq \phi_t(x_0,t_0) + H(D\phi(x_0,t_0)).$$

Thus

$\min(v_t + H(Dv), v) \leqq 0$ in the viscosity sense.

Assertion (2.10) is proved.

5. Since $(2.5)_\varepsilon$ implies further that $v = 0$ in $G \times \{0\}$, $v = +\infty$ on $(\mathbb{R}^n - G) \times \{0\}$, we see that v is a, and by uniqueness the, viscosity solution of (2.6): $v = I$. On the other hand, since the Hamiltonian H does not depend on x, it turns out that $V = I = \max(J,0)$, J solving (2.3).

6. Now since $u_k^\varepsilon = e^{(-v_k^\varepsilon/\varepsilon)} = e^{[-I+o(1)]/\varepsilon}$, we see that $u_k^\varepsilon \to 0$ uniformly on compact subsets of $\{I > 0\} = \{J > 0\}$. Assume now $(x_0,t_0) \in \{J < 0\}$. Then $I = 0$ near (x_0,t_0). Write $\phi(x,t) = |x-x_0|^2 + (t-t_0)^2$. Since $v_k^\varepsilon \to 0$ uniformly near x_0,t_0, $v_k^\varepsilon - \phi$ has a maximum at a point $(x_k^\varepsilon,t_k^\varepsilon)$, with $(x_k^\varepsilon,t_k^\varepsilon) \to (x_0,t_0)$ as $\varepsilon \to 0$. Using $(2.5)_\varepsilon$ we deduce

$$o(1) = \phi_t - \varepsilon a_{ij}^k \phi_{x_i x_j} - b_i^k \phi_{x_i} + a_{ij}^k \phi_{x_i} \phi_{x_j} - \frac{f_k(u^\varepsilon)}{u_k^\varepsilon} \text{ at } (x_k^\varepsilon,t_k^\varepsilon).$$

Thus

$$f_k(u^\varepsilon) \leqq o(1)\, u_k^\varepsilon \quad \text{at} \quad (x_k^\varepsilon,t_k^\varepsilon).$$

Since $f^k(u) \geqq c_{k1}u_1 - O(|u|^2)$, we deduce

$$u_k^\varepsilon(x_k^\varepsilon,t_k^\varepsilon) \geqq \delta > 0 \quad (1 = 1,\ldots,m)$$

for some constant $\delta > 0$. But then since

$$(v_k^\varepsilon-\phi)(x_k^\varepsilon,t_k^\varepsilon) \geqq (v_k^\varepsilon-\phi)(x_0,t_0),$$

we further discover

$$u_k^\varepsilon(x_0,t_0) \geqq u_k^\varepsilon(x_k^\varepsilon,t_k^\varepsilon) \geqq \delta > 0. \quad \square$$

REFERENCES

[1] G. Barles, L.C. Evans and P.E. Souganidis, Wavefront propagation for reaction-diffusion systems, in preparation.

[2] L.C. Evans and P.E. Souganidis, A PDE approach to geometric optics for certain semilinear parabolic equations, Indiana J. Math. J. (to appear).

[3] M.I. Freidlin, Functional Integration and Partial Differential Equations, Annals. of Math. Studies, Vol. 109, Princeton University Press, Princeton, 1985.

L.C. Evans
Department of Mathematics
University of Maryland
College Park, MD., 20742
U.S.A.

G. GAGNEUX, A.M. LEFEVERE AND M. MADAUNE-TORT

An implicit time-stepping scheme for polyphasic flows in porous media including coupled quasi-linear inequalities

SUMMARY. This contribution to the theoretical study of reservoir simulation develops a variational formulation of the standard black-oil model with coupled obstacles and gives an existence result for an implicit time-stepping scheme.

INTRODUCTION

We present an analytical approach to the pseudocompositional black-oil model, whose interest is to be an industrial tool of petroleum engineering for simulating a three-dimensional isothermal constrained polyphasic flow in porous media. In this simplified representation one has three components: water, which we will suppose to coincide exactly with the aqueous phase, the heavy hydrocarbon pseudocomponent only present in the oil phase and the light volatile pseudocomponent; depending on the thermodynamical conditions, the latter can eventually be in the oil phase and in the gas phase.

The governing equations are evolution equations of divergence form: they express conservation of mass of each component and behaviour laws such as the Darcy-Muskat law, and they take into account the fact there is no longer coincidence between the phases and the chemical components; the main unknowns are S_g, S_w, reduced saturations of gaseous and aqueous phases, P, an unknown of pressure dimension, and X_0^h, molar fraction of the heavy component in the oil phase which determines the composition of this phase.

Gibbs's law which gives the variance of a system at thermodynamical equilibrium, i.e. practically the number of independent thermodynamical variables of the problem, states that these unknowns are coupled by unilateral constraints of a quasi-variational type:

$$(P_1) \quad \begin{cases} S_g \geqq 0, \quad X_0^h \geqq C(P), \\ S_g(X_0^h - C(P)) = 0, \end{cases}$$

where $P \to C(P)$ is a thermodynamical equilibrium function whose table is given by the user.

1. MODELIZATION OF THE PHYSICAL PROBLEM

1.1. Physical data

Let Ω be a bounded domain in $\mathbb{R}^3$, representing the porous medium with a smooth boundary Γ, with n the outward normal vector to Γ, $[0,T]$ the time interval under consideration, and let $Q = \Omega \times]0,T[$. We consider the partition of Γ into the three open disjoint subsets Γ_e, Γ_s, Γ_L such that:

$$\Gamma = \Gamma_e \cup \Gamma_s \cup \Gamma_L \cup \partial\Gamma_L \ , \ \bar{\Gamma}_s \cap \bar{\Gamma}_e = \emptyset,$$

representing, respectively, the boundary of water injection (contact area with aquifer or injection wells) spanned by the nonnegative function f of $L^\infty(\Gamma_e)$, the production boundary subjected to an exterior pressure, and the impermeable part of the boundary.

For simplicity, and without loss of generality, the rock porosity is supposed here to be constant, the gravity is neglected and the permeability tensor, generally symmetrical and uniformly positive definite, is assumed to be the unit diagonal tensor. To take into account more general conditions, the reader is referred to [5], [14], [8], [15].

1.2. Conservation laws

Denoting by S_p the reduced saturation of phase p, p taking the values w, o, g, in order to represent the water, oil and gas phases, we write the mass conservation of each component c, c = w, h, ℓ (water, heavy and light components) according to [1], [5], [8], [9],

$$(E_1) \quad \begin{cases} \dfrac{\partial}{\partial t}\left(\sum\limits_p \phi S_p \rho_p \omega_p^c\right) + \operatorname{div}\left(\sum\limits_p \rho_p \omega_p^c Q_p\right) = 0, \\ \sum\limits_p S_p = S_w + S_o + S_g = 1 \text{ in } Q, \end{cases}$$

where for the phase p, we denote by ρ_p the density, by Q_p the filtration velocity vector and by ω_p^c the mass fraction of component c.

The vector Q_p, depending on the pressure $\mathcal{P}$, is given by the Darcy-Muskat law, according to the formula

$$(E_2) \qquad Q_p = -\frac{k_{r_p}}{\mu_p} (\nabla(\mathcal{P}-P_{c_p})),$$

where in phase p we denote by k_{r_p} the relative permeability, by μ_p the dynamic viscosity and by P_{c_p} the capillary pressure between oil and phase p.

Denoting by X_p^c the molar fraction of component c in phase p, we also have the following phase equilibrium relationships and constraints, in the strictly triphasic framework, i.e. the three components split into three distinct phases

$$X_g^w = 0, \quad X_o^w = 0, \quad X_w^w = 1,$$

$$(E_3) \qquad X_g^h = 0, \quad X_w^h = 0, \quad X_o^h + X_o^\ell = 1, \quad X_g^\ell = 1,$$

$$1 = K^\ell(P)X_o^\ell, \quad X_w^\ell = 0, \quad K^\ell \text{ being known.}$$

Moreover, we have the following relation between molar and mass fractions

$$(E_4) \qquad \omega_p^c = \frac{X_p^c M^c}{\sum_{c'} X_p^{c'} M^{c'}}, \text{ with } c' = w, h, \ell,$$

where we denote by M^c, the molar mass of component c.

1.3. <u>Simplifying hypotheses</u>

It seems mathematically advisable to choose as main unknowns S_w, S_g, X_o^h and $P = \mathcal{P} - P_{c_g}$, the pressure of gas phase, given these values at initial time.

For the simplication of equations we introduce, as usual, [1]:

$$d_p = \frac{\rho_p}{\mu_p}$$ inverse of kinematical viscosity of phase p,

$$d = k_{r_w} d_w + k_{r_o} d_o + k_{r_g} d_g$$ global mass mobility,

$$\nu_p = k_{r_p} \frac{d_p}{d}$$ fractional flow of phase p,

the main functional properties of which, in addition to natural hypotheses of continuity, are as follows:

$$\nu_p = \nu_p(X_o^h, S_w, S_g, P), \; 0 \leqq \nu_p \leqq 1,$$

ν_p reduces to zero when S_p vanishes.

It is essential to note for the sequel that the functions $[d\nu_w]$ and $[d\nu_g]$ do not depend on X_o^h. The capillary pressure functions are taken in accordance with petroleum engineers purposes (see [1], [3], [10] and [2, p. 205-209], for instance); the same holds true with respect to general behaviour laws (densities, triphasic relative permeabilities).

1.4. Development of representative model equations

In the domain Ω, at a fixed time t, we first write the conservation of mass for water described by

$$\frac{\partial}{\partial t} S_w + \text{div } Q_w = 0 \text{ in } Q,$$

which is expressed, for example, by

$$\frac{\partial}{\partial t} S_w - \text{div}\{[d\nu_w](\nabla P + \nabla P_w(S_w) - \nabla P_g(S_g))\} = 0 \text{ in } Q,$$

a quasi-linear parabolic equation (see [11]) containing a degenerate diffusion term in the main unknown S_w and a nonlinear transport term. Practically, the function $[d\nu_w]$ depends only on the variable S_w.

We next consider the global mass conservation of the three components, i.e.

$$\frac{\partial}{\partial t}(S_w + \rho_o(X_o^h)S_o + \rho_g(P)S_g) + \mathrm{div}(Q_w + \rho_o Q_o + \rho_g Q_g) = 0 \text{ in } Q,$$

usually referred to as "pressure equation" and further detailed. In order to obtain an equation between S_w, S_g, P independently of X_o^h, we introduce formally

$$\Omega^+(t) = \{x \in \Omega;\ S_g(t,x) > 0\},$$

$$\Omega^o(t) = \{x \in \Omega;\ S_g(t,x) = 0\}.$$

These a priori unknown areas correspond, respectively, to the effective presence of gas phase or its absence (i.e. to the cases of saturated oil and subsaturated oil). In fact, in $\Omega^+(t) \times]0,T[$, the molar fraction of the heavy component in the oil phase is determined unambiguously by knowledge of P according to Gibbs's law, i.e.

$$\frac{\partial}{\partial t}((1-S_w-S_g)[\rho_o\omega_o^h](C(P))) + \mathrm{div}([\rho_o\omega_o^h](C(P))Q_o) = 0 \text{ in } \Omega^+(t) \times]0,T[,$$

$$S_g = 0 \text{ in } \Omega^o(t) \times]0,T[.$$

Finally, we derive the equation satisfied by X_o^h by expressing the law of mass conservation of the heavy component in the whole domain Ω, i.e.

$$\frac{\partial}{\partial t}[S_o\rho_o(X_o^h)\ \omega_o^h(X_o^h)] + \mathrm{div}[\rho_o(X_o^h)\omega_o^h(X_o^h)Q_o] = 0 \text{ in } Q$$

related to the constraint

$$X_o^h \geqq C(P) \text{ in } Q.$$

Noting that the function

$$X_o^h \to [\rho_o\omega_o^h](X_o^h)$$

defined on [0,1] is increasing and vanishes for $X_o^h = 0$, we are led, for simplicity, to introduce the auxiliary function

$$C_o^h = S_o[\rho_o\omega_o^h](X_o^h)$$

and to rewrite the previous equation and the obstacle condition as a function of C_o^h. It must be noted that the knowledge of C_o^h determines unambiguously the value of X_o^h when S_o is nonzero, the determination of X_o^h being obviously purposeless in the absence of the oil phase.

The boundary condition used on Γ_e and Γ_L are Dirichlet and Neumann mixed conditions. On the production wells we admit that the partial mass flow-rates are in proportion to the respective mass mobilities, which yield non-linear Fourier boundary relations. A detailed discussion on boundary conditions usually encountered can be found in [1], [5], [8].

Finally it is advisable, in the pressure equation, and only there, in order to release the coupling, to suppose that the oil density does not depend on the phase composition. For the sake of simplicity we shall consider in this equation that $\rho_w = \rho_o = 1$. In the other equations, the dependence of ρ_o, as a function of X_o^h, characteristic of compositional models, is taken into account.

2. VARIATIONAL FORMULATION FOR THE IMPLICIT TIME-DISCRETIZED PROBLEM

2.1. Some notations

Following the usual notations of the Sobolev spaces, we introduce the sets

$$V = \{v \in H^1(\Omega);\ v = 0 \text{ in } \Gamma_e\},$$

$$K^+ = \{v \in V;\ v \geq 0 \text{ a.e. in } \Omega\},$$

$$C^+ = \{v \in L^2(\Omega);\ v \geq 0 \text{ a.e. in } \Omega\},$$

$$K_1 = \{v \in H^1(\Omega);\ v = 1 \text{ on } \Gamma_e\}.$$

For any system $(a_i)_{1\leq i\leq 3}$ of functions in $C^1(\bar{\Omega})$, we denote (summation convention of repeated indexes), by ℓ the function defined on Γ by $\ell(x) = a_i(x)\, n_i(x)$, and according to the usual notations of Bardos [12] and Mignot and Puel [13].

$$\tilde{W}(A) = \{u \in L^2(\Omega);\ Au \in L^2(\Omega),\ \ u/\Gamma \in L^2_\ell(\Gamma)\},$$

where $Au = a_i\ (\partial u/\partial x_i)$.

We consider a uniform time step h, $h > 0$, and we look for a sequence of 4-uplets (S_w^k, S_g^k, P^k, C^k), a priori claimed approximation of the 4-uplet (S_w, S_g, P, C_o^h) at time $t_k = kh$.

For every k in $\mathbb{N}$ and every function λ, we denote as λ_k the function defined by

$$\lambda_k(x) = \lambda(S_w^k, S_g^k, P^k, (X_o^h)^{k-1})$$

and $\tilde{\lambda}_k$ an interpolating term of class C^1 of λ_k on Ω. In the same way, we introduce

$$H_k = \{v \in L^2(\Omega);\ v \geqq S_o^k[\rho_o \omega_o^h](C(P^k)) \text{ a.e. in } \Omega\},$$

$$K_{r_o}(S_w,S_g,P) = \frac{1}{S_o} k_{r_o}(S_w,S_g,P) \text{ if } S_o \neq 0$$

$$= -\frac{\partial k_{r_o}}{\partial S_p}(S_w,S_g,P),\ p \in \{w,g\},\ \text{if } S_o = 0,$$

$$a_i^k = -\frac{1}{\mu_o((X_o^h)^{k-1})} K_{r_o}(S_w^k,S_g^k,P^k)\ [\frac{\partial P^k}{\partial x_i}\ \frac{\partial}{\partial x_i} P_g(S_g^k)],$$

$$A_k = \tilde{a}_i^k \frac{\partial}{\partial x_i},\ b_k = \frac{\partial \widetilde{a_i^k}}{\partial x_i},$$

the smooth family $\{\widetilde{a_i^k}\}$ being constructed such that $\partial \widetilde{a_i^k}/\partial x_i$ is uniformly bounded from below, i.e.

$$\text{there exists } \omega > 0, \text{ such that } \frac{\partial \tilde{a}_i^k}{\partial x_i} + \omega \geqq 0 \text{ in } \bar{\Omega}.$$

2.2. An existence result

In this functional framework, we can prove the existence of a solution to the related time-discretized system, according to a scheme which is fully implicit in the saturation and pressure unknowns and explicit in the determination of oil phase composition. In fact we have:

PROPOSITION: For any $\varepsilon > 0$ and for any $h > 0$, sufficiently small, for any initial data $(S_w^o, S_g^o, P^o, (X_o^h)^\circ)$ satisfying the natural hypotheses

$$0 \leqq S_w^o \leqq 1,\ 0 \leqq S_g^o \leqq 1,\ 0 \leqq S_w^o + S_g^o \leqq 1 \quad \text{a.e. in } \Omega,$$

$$P^o \in L^\infty(\Omega),\ 0 \leqq (X_o^h)^\circ \leqq 1 \quad \text{a.e. in } \Omega,$$

and setting $C^o = (1-S_w^o - S_g^o)\ [\rho_o\omega_o^h]\ ((X_o^h)^\circ)$, there exists at least a sequence (S_w^k, S_g^k, P^k, C^k), $k \in \mathbb{N}^*$, such that

$$S_w^k \in K_1,\ S_g^k \in K^+,\ P^k \in H^1(\Omega) \cap C^+,\quad C^k \in H_k \cap \tilde{W}(A_k),$$

$$0 \leqq S_w^k \leqq 1,\ 0 \leqq S_g^k \leqq 1,\ 0 \leqq S_w^k + S_g^k \leqq 1 \text{ a.e. in } \Omega,$$

solution of the strongly coupled system of variational equations and inequalities:

(i) $S_w^k \in K_1$, satisfying

$$\frac{1}{h}\int_\Omega (S_w^k - S_w^{k-1})v\ dx + \int_\Omega \{[d\nu_w]_k\ P_w'(S_w^k) + \varepsilon\}\ \nabla S_w^k \cdot \nabla v\ dx$$

$$+ \int_\Omega [d\nu_w]_k\ \nabla P^k . \nabla v\ dx + \int_{\Gamma_s} [d\nu_w]_k\ P^k v\ d\Gamma = 0, \quad \forall v \in V;$$

(ii) $S_g^k \in K^+$, satisfying

$$\frac{1}{h}\int_\Omega [\rho_o\omega_o^h]\ (C(P^k))\ (S_w^k + S_g^k - S_w^{k-1} - S_g^{k-1})(v-S_g^k)\ dx$$

$$+ \int_\Omega \omega_o^h(C(P^k))[d\nu_o]_k\ \nabla P_g(S_g^k)\cdot\nabla(v-S_g^k)dx + \varepsilon\int_\Omega \nabla(S_w^k+S_g^k)\nabla(v-S_g^k)\ dx$$

$$- \int_\Omega \omega_o^h(C(P^k))[d\nu_o]_k\ \nabla P^k\ \nabla(v-S_g^k)dx$$

$$\geqq \int_{\Gamma_s} \omega_o^h(C(P^k))[d\nu_o]_k\ P^k(v-S_g^k)d\Gamma,\ \forall v \in K^+;$$

(iii) $P^k \in H^1(\Omega) \cap C^+$, satisfying

$$\frac{1}{h}\int_\Omega S_g^k(\rho_g(P^k)-1)(v-P^k)dx + \int_\Omega d_k \nabla P^k . \nabla(v-P^k)dx$$

$$+ \int_{\Gamma_s} d_k P^k(v-P^k)d\Gamma \geq \int_{\Gamma_e} f(v-P^k)d\Gamma + \frac{1}{h}\int_\Omega S_g^{k-1}(\rho_g(P^{k-1})-1)(v-P^k)dx,$$

$$\forall v \in C^+ \cap H^1(\Omega);$$

(iv) $C^k \in H_k \cap \tilde{W}(A_k)$, satisfying:

$$\frac{1}{h}\int_\Omega (C^k-C^{k-1})(v-C^k)dx + \int_\Omega (A_k C^k + b_k C^k)(v-C_k)dx \geq 0, \ \forall v \in H_k$$

and the condition $C^k = S_o^k[\rho_o\omega_o^h](C(P^k))$ on the eventual part of Γ of nonzero measure where $\tilde{a}_i^k \cdot n_i < 0$.

Sketch of the proof: The proof for this artificial viscosity method follows from an application of a fixed-point theorem of the Kakutani-Ky-Fan type and the well-known results of F. Mignot and J.P. Puel on the strong solutions of first-order variational hyperbolic inequalities with obstacle. The strategy of calculation is as follows: given the 4-uplet (S_w^k, S_g^k, P^k, C^k), we can calculate $(X_o^h)^k$ and, with an implicit scheme for the saturations and pressure, we determine $(S_w^{k+1}, S_g^{k+1}, P^{k+1})$, X_o^h being stationary between the times t_k and t_{k+1}; then C^{k+1} (and so $(X_o^h)^{k+1}$) is obtained by resolution of the first-order hyperbolic inequality. We note that in the pressure equation, some diffusion terms treated as negligible were omitted. The effects of these terms can be taken into account by a classical predictor-corrector method.

2.3. Numerical stability for the time-discretized scheme

We introduce the following notation: for any sequence $(u_k)_{0 \leq k \leq N-1}$ of numerical functions defined on Ω, we denote by u_h the function defined on Q by

$$u_h(t,x) = \sum_{k=0}^{N-1} u^k(x)\, \chi^k(t),$$

where χ^k is the characteristic function of time interval $[kh,(k+1)h]$.

Then we have partial stability results as follows:

PROPOSITION: When the time-step h tends to zero, $(S_w)_h$, $(S_g)_h$, $(C)_h$ remain bounded in $L^\infty(Q)$, $(P)_h$ remains bounded in $L^\infty(0,T;L^6(\Omega))$.

REMARK: This modelization takes into account water injection with prescribed flow-rate. A variant, via an adequate change of functional spaces can be developed, by taking into consideration the presence of injection wells with prescribed pressure. For the description of practical states, the reader is referred to [1]-[8].

REFERENCES

[1] Marle, C., Cours de production, t.4. Les écoulements polyphasiques en milieu poreux, ed. Technip., Paris, 1972.

[2] Chavent, G. and Jaffre, J., Mathematical Models and Finite Elements for Reservoir Simulation, North-Holland, 1987.

[3] Ciligot-Travain, G., Les modèles de gisements, à paraître, ed. Technip., Paris.

[4] Bia, P. and Combarnous, M., Les méthodes thermiques de production des hydrocarbures; chapitre 1, Transfert de chaleur et de masse, Revue de l'Institut Français du Pétrole, mai-juin 1975, pp. 359-395.

[5] Gagneux, G., Sur les problèmes unilatéraux dégénérés de la théorie des écoulements diphasiques en milieu poreux, Thèse de Doctorat d'Etat 1982, Université de Besançon.

[6] Madaune-Tort, M., Perturbations singulières de problèmes aux limites du second ordre, hyperboliques et paraboliques non linéaires, Thèse de Doctorat d'Etat, avril 1981, Université de Pau.

[7] Gagneux, G., Une étude théorique sur la modélisation de G. Chavent des techniques d'exploitation secondaire des gisements pétrolifères, J. Mécan. Theor. and Appl. 2, No. 1 (1983), 33-56.

[8] Coore, B., Eymard, R. and Quettier, L., Applications of a thermal simulator to field cases, Proceedings of the 59th Annual Technical Conference, Society of Petroleum Engineers of AIME, 1984, Houston, Texas.

[9] Duvaut, G. and Lions, J.-L., Les Inéquations en Mécanique et en Physique, Dunod, Paris, 1972.

[10] Chavent, G. and Salzano, G., Un algorithme pour la détermination de perméabilités relatives triphasiques satisfaisant une condition de différentielle totale. Rapport de Recherche INRIA No. 355, janvier 1985.

[11] Lions, J.L., Quelques méthodes de résolution des problèmes aux limites non linéaires, Dunod, Gauthier-Villars, Paris, 1969.

[12] Bardos, C. Problèmes aux limites, pour les équations aux dérivées partielles du premier ordre à coefficients réels, Ann. Scient. Ec. Norm. Sup., 4ème série, 3 (1970), 185-233.

[13] Mignot, F. and Puel, J.P., Inéquations variationnelles et quasi-variationnelles hyperboliques du premier ordre, J. Math. Pures Appl. 55 (1976), 353-378.

[14] Gagneux, G., Lefévère, A.M. and Madaune-Tort, M. Une approche analytique d'un modèle black oil des écoulements triphasiques compressibles en ingénierie pétrolière, J. Mécan. Théor. Appl. 6, No. 4 (1987) 1-24.

[15] Gagneux, G., Lefévère, A.M. and Madaune-Tort, M., Modélisation d'écoulements polyphasiques en milieu poreux par un système de problèmes unilatéraux, M_2 AN-RAIRO, 22, No. 3 (1988) 389-415.

G. Gagneux, A.M. Lefévère and M. Madaune-Tort
Laboratoire de Mathématiques Appliquées
U.S.-CNRS 1204
Département de Mathématiques
Université de Pau
Avenue de l'Université
64000 Pau
France

M.A. POZIO AND A. TESEI

Global existence of solutions for a strongly coupled semilinear parabolic system

SUMMARY. We prove global existence results for a class of strongly coupled semilinear parabolic systems. Use of invariant rectangles techniques is made.

1. INTRODUCTION

We want to investigate the global existence of solutions of the following problem:

$$\begin{aligned} u_t &= a(x)\,\Delta u + \alpha(x)\,\Delta v + f(u,v) & & \\ v_t &= \beta(x)\,\Delta u + b(x)\,\Delta v + g(u,v) & &\text{in } (0,\infty)\times\Omega, \\ u &= v = 0 & &\text{in } (0,\infty)\times\partial\Omega, \\ u(0) &= u_0,\ v(0) = v_0 & &\text{in } \Omega. \end{aligned} \tag{1.1}$$

Here $\Omega \subset \mathbb{R}^N$ is an open bounded subset with smooth boundary $\partial\Omega$ and u_0, v_0 are given functions. Concerning the coefficients a, b, α, β we assume

$$\text{(C)} \quad \begin{aligned} &a,\ b,\ \alpha,\ \beta \in C^\sigma(\bar{\Omega}) \quad (\sigma \in (0,1)), \\ &a(x) \geq \underline{a} > 0,\quad b(x) \geq \underline{b} > 0 \text{ for any } x \in \Omega. \end{aligned}$$

Strongly coupled semilinear parabolic systems are not so widely investigated as those weakly coupled (corresponding to the case $\alpha \equiv \beta \equiv 0$ for problem (1.1)). Concerning local existence and uniqueness, very general results have been proved in [Am 1] (see also [CTV], [DPG]).

Conditions for global existence have been given in [Am 2], [Co 1], [Co 2], [Al], [Re], using a Lyapunov function to get a priori bounds. The investigation of the qualitative properties of the solutions for such problems is still at a preliminary level.

A general tool to obtain a priori estimates of solutions of parabolic problems is the maximum principle. This is usually applied to "*weakly coupled*" systems in the form of the "*invariant rectangles*" technique, which combines geometric properties of the field (f,g) with the maximum principle used for one equation each time.

In principle, the technique also applies to "*strongly coupled*" systems. In [CCS] systems of the form

$$w_t = D\Delta w + \tilde{f}(t,w) \tag{1.2}$$

were considered, where $w \equiv (w_1,\dots,w_n)$, and $D = D(x,w)$ is a matrix having real nonnegative eigenvalues. Denote by

$$\Sigma := \bigcap_{i=1}^{m} \{w \mid G_i(w) \leq 0\}$$

(where G_i is a smooth function, $i = 1,\dots,m$) a candidate for an invariant region of (1.2). For this to happen the following conditions have to be satisfied "*at each* $\underline{w_0} \in \partial\Sigma$", if D is positive definite (see [CCS]):

(i) dG_i is a left eigenvector of D;

(ii) if $dG_i = 0$, then $d^2G_i \geq 0$;

(iii) $dG_i \cdot \tilde{f} \leq 0$.

If D is diagonal, any rectangle "*with sides parallel to the axes*" is invariant for (1.2) if the field $\tilde{f}$ points inwards at its boundary. This gives the usual method of invariant rectangles for the weakly coupled case. Moreover, conditions (i)-(iii) are also sufficient for a wide class of systems [CCS].

However, we are interested in results which apply also to systems of the form (1.1) with the first two equations replaced by

$$\begin{aligned} u_t &= L_{11}u + L_{12}v + f(u,v), \\ v_t &= L_{21}u + L_{22}v + g(u,v) \text{ in } (0, +\infty) \times \Omega, \end{aligned} \tag{1.3}$$

where L_{ij} $(i,j = 1,2)$ are general uniformly elliptic operators. Thus we prefer to work with rectangles and ask for conditions such that the "field" $(\alpha\Delta v + f, \beta\Delta u + g)$ points inwards at the boundary of such rectangles. This happens if the following are true:

(a) the field (f,g) points strictly inwards at the boundary of the rectangle;

(b) the out-of-diagonal terms $\alpha\Delta v$, $\beta\Delta u$ are uniformly small in $[0,\infty)$ in some norm.

While (a) is a property of the field, (b) can be ensured if Δu, Δv are uniformly bounded and α, β are small in the same norm†. Therefore we need "*at the same time*" uniform estimates both of the solution and of its Laplacian. These depend on each other, while the estimate of (u,v) is obviously related to the size of the rectangle. Thus this will be invariant not for any solution of (1.1) having initial data in it, but only for those whose Laplacians are sufficiently small depending on the size of the rectangle. In order to prove global existence for arbitrary initial data in a suitable function space we seek conditions which imply that the problem has arbitrarily large contracting rectangles (see Section 2). It is apparent from the above remarks that the global solutions we prove to exist are uniformly bounded on $[0,\infty)$.

In order to prove global existence results "*for any initial data*" in a suitable function space, let us introduce the following assumptions concerning the field (f,g):

(H_0) $f,g \in C^2$;

(H_1) there exist k_U, $k > 0$ such that for any $U \geq k_U$ there is a rectangle

$$B(U) := [-U,U] \times [-kU, kU]$$

with

† Observe that the smallness of α, β is also required for the local existence results mentioned above [Am 1].

$$(\pm U)f(\pm U,v) < 0 \qquad \text{for any } v \in [-kU, kU],$$

$$(\pm kU)g(u,\pm kU) < 0 \qquad \text{for any } u \in [-U,U].$$

For $U \geqq k_U$ define

$$|f|_{B(U)} := \max |f|\Big|_{B(U)}, \quad |g|_{B(U)} := \max |g|\Big|_{B(U)};$$

$$\hat{f}_U := \min \{\min(-f)|_{\partial_1 B(U)}, \ \min f|_{\partial_3 B(U)}\},$$

$$\hat{g}_U := \min \{\min(-g)|_{\partial_2 B(U)}, \ \min g|_{\partial_4 B(U)}\},$$

where

$$\partial B(U) = \bigcup_{i=1}^{4} \partial_i B(U)$$

and

$$\partial_i B(U) := \{(u,v) \in \partial B(U) | \quad u = \pm U\} \qquad (i = 1,3),$$

$$\partial_i B(U) := \{(u,v) \in \partial B(U) | \quad v = \pm kU\} \qquad (i = 2,4).$$

Then assume

$$(H_2) \quad \lim_{U\to\infty} \frac{|f|_{B(U)}}{U} = \lim_{U\to\infty} \frac{|g|_{B(U)}}{U} = \infty;$$

(H_3) the following quantities are infinite of the same order as $U \to \infty$:

$$|f|_{B(U)}; \quad |g|_{B(U)}; \quad \hat{f}_U; \ \hat{g}_U;$$

$$\{|f_u|_{B(U)} + |f_v|_{B(U)}\}U; \ \{|g_u|_{B(U)} + |g_v|_{B(U)}\}U.$$

Requirement (H_3) seems to be crucial for the present approach. It arises naturally, as we can see discussing the case where a, b, α, β are constant. It can be weakened if either α or β are zero, or if we choose them depending on the initial data (see Section 3).

2. RESULTS AND PROOFS

For any integer $k \geqq 0$ and $\sigma \in (0,1)$ we denote by $C^{k+\sigma}(\bar{\Omega})$ the subspace of $C^k(\bar{\Omega})$ consisting of functions whose derivatives of order k are σ-Hölder continuous in $\bar{\Omega}$. Then

$$[u]_{k+\sigma} := \sum_{|m|=k} \sup_{\substack{x,y\in\Omega \\ x\neq y}} \frac{|D^m u(x) - D^m u(y)|}{|x - y|^\sigma} < \infty ;$$

$C^{k+\sigma}(\bar{\Omega})$ is a Banach space with norm

$$u \to |u|_{k+\sigma} := |u|_k + [u]_{k+\sigma}.$$

By $h^{k+\sigma}(\bar{\Omega})$ we denote the subspace of $C^{k+\sigma}(\bar{\Omega})$ consisting of functions whose derivatives of order k are little-Hölder continuous in $\bar{\Omega}$, namely

$$\lim_{z\to 0} \sup_{\substack{x,y\in\bar{\Omega} \\ 0<|x-y|\leqq z}} \frac{|D^m u(x) - D^m u(y)|}{z^\sigma} = 0 \qquad (|m| = k);$$

$h^{k+\sigma}(\bar{\Omega})$ is a Banach space under the $|\cdot|_{k+\sigma}$ - norm. We also set

$$h_0^{k+\sigma}(\bar{\Omega}) := \{u \in h^{k+\sigma}(\bar{\Omega}) | \Delta^p u = 0, \quad p = 0,1,\ldots,[k/2]\}.$$

Let us first mention a local existence and uniqueness result.

THEOREM 1: Assume (C), (H_0). Let moreover

$$|\alpha|_0 < \bar{\alpha}, \quad |\beta|_0 < \bar{\beta} \qquad (\bar{\alpha}, \bar{\beta} > 0).$$

Then for any $u_0, v_0 \in h_0^{2+2\theta}(\bar{\Omega})$ there exists $T > 0$ such that a unique solution of (1.1) exists in $[0,T]$. Here

$$u,v \in C(0,T;h_0^{2+2\theta}(\bar{\Omega})) \cap C^1(0,T;h_0^{2\theta}(\bar{\Omega})) \qquad (\theta \in (0,\sigma/2)).$$

The proof can be given combining the results in [CTV] with standard fixed-point arguments. In the following we denote by $[0,T^*)$ the "*maximal*" interval of existence of the solution of (1.1).

The global existence result mentioned in Section 1 can now be stated:

THEOREM 2: Let (C) and (H_0)-(H_3) be satisfied. Also assume that

$$|\alpha|_{2\theta} < \tilde{\alpha} \ , \ |\beta|_{2\theta} < \tilde{\beta} \qquad\qquad (\tilde{\alpha},\tilde{\beta} > 0).$$

Then for any $u_0, v_0 \in h_0^{2+2\theta}(\bar{\Omega})$ the solution of (1.1) is global ($\theta \in (0,\sigma/2)$).

Let us outline the main ideas of the proof. In the following we think of $\theta \in 0,\sigma/2)$ as fixed. Denote by U_p, V_p ($p = 0,\theta, 1,1 + \theta$) positive constants such that

$$\begin{aligned}
&U_0 > \max \{|u_0|_0, \tfrac{1}{k}|v_0|_0, k_U\}, && V_0 := kU_0;\\
&U_\theta > |u_0|_{2\theta}, && V_\theta > |v_0|_{2\theta};\\
&U_1 > |\Delta u_0|_0, && V_1 > |\Delta v_0|_0;\\
&U_{1+\theta} > |u_0|_{2+2\theta}, && V_{1+\theta} > |v_0|_{2+2\theta}.
\end{aligned} \tag{2.1}$$

Then define

$$\Theta := \sup \{t \in [0,T^*) |\ |u(t)|_0 < U_0, |v(t)|_0 < kU_0, |\Delta u(t)|_0 < U_1,$$

$$|\Delta v(t)|_0 < V_1,\ |u(t)|_{2p} < U_p,\ |v(t)|_{2p} < V_p (p = \theta,\ 1 + \theta)\}.$$

In $[0,\Theta)$ we have the following estimate.

LEMMA: There exist $P = P(U_0,U_\theta,V_\theta,V_{1+\theta})$, $Q = Q(kU_0,U_\theta,V_\theta,U_{1+\theta})$ such that

$$|\Delta u(t)|_0 \leqq P,$$

$$|\Delta v(t)|_0 \leqq Q, \tag{2.2}$$

for any $t \in [0,\Theta)$.

Similar bounds hold for $|u(t)|_{2p}$, $|v(t)|_{2p}$ ($p = \theta$, $1 + \theta$) for any $t \in [0,\Theta)$. These entail the following result.

PROPOSITION 1: Let

$$|\alpha|_{2\theta} < \alpha_1, \quad |\beta|_{2\theta} < \beta_1 \quad (\alpha_1, \beta_1 > 0). \tag{2.3}$$

Then there exist $U_p(U_0), V_p(U_0)$ such that

$$\Theta < T^* \Rightarrow (2.4) \begin{cases} |u(t)|_{2p} < U_p(U_0) \\ |v(t)|_{2p} < V_p(U_0) \end{cases} \quad (p = \theta,\ 1 + \theta),$$

with strict inequalities for any t in the *closed* interval $[0,\Theta]$ (the choice of U_0 depends on $|u_0|_{2+2\theta}$, $|v_0|_{2+2\theta}$).

In turn, this implies the following proposition.

PROPOSITION 2: Let the assumptions of Proposition 1 be satisfied. Moreover, assume U_1 and V_1 as in (2.1) and such that

(A_1) $\quad U_1 > \tilde{P}(U_0), \quad V_1 > \tilde{Q}(U_0),$

where

$$\tilde{P}(U_0) := P(U_0, U_\theta(U_0), V_\theta(U_0), V_{1+\theta}(U_0)),$$

$$\tilde{Q}(U_0) := Q(kU_0, U_\theta(U_0), V_\theta(U_0), U_{1+\theta}(U_0)).$$

Then

$$\Theta < T^* \Rightarrow (2.5) \begin{cases} |\Delta u(t)|_0 < U_1, \\ |\Delta v(t)|_0 < V_1, \end{cases}$$

with strict inequalities for any t in the "*closed*" interval $[0,\Theta]$.

REMARK: All the above functions P, Q, U_θ, V_θ, $U_{1+\theta}$, $V_{1+\theta}$, and the constants α_1, β_1 are explicitly known.

Concerning $|u(t)|_0$, $|v(t)|_0$ we have the following result:

PROPOSITION 3: Let

$$(A_2) \qquad |\alpha|_0 V_1 < \hat{f}_{U_0}, \qquad |\beta|_0 U_1 < \hat{g}_{U_0}.$$

Then

$$\Theta < T^* \Rightarrow (2.6)\begin{cases} |u(t)|_0 < U_0, \\ |v(t)|_0 < kU_0, \end{cases}$$

with strict inequalities for any t in the "*closed*" interval $[0,\Theta]$.

The proofs of the lemma and Proposition 1 are technical and will be omitted (see [PT]). Proposition 2 follows immediately from (2.2) and (A_1). Before proving Proposition 3 let us observe the following corollary.

COROLLARY: Let the assumption of Proposition 1 and (A_1)-(A_2) be satisfied. Then $\Theta = T^*$.

PROOF: Were $\Theta < T^*$, under the present assumptions the strict inequalities (2.4)-(2.6) would hold in the closed interval $[0,\Theta]$. Since u and v are continuous from $[0,T^*)$ to $h_0^{2+2\theta}(\bar{\Omega})$ (see Theorem 1), this would contradict the definition of Θ. Hence the conclusion.

PROOF OF PROPOSITION 3: If $\Theta < T^*$, we have by definition

$$|v(t)|_0 \leqq kU_0, \quad |\Delta v(t)|_0 \leqq V_1,$$

for any $t \in [0,\Theta]$. Then for any $t \in [0,\Theta]$ and $x \in \bar{\Omega}$ we get

$$\alpha(x)\Delta v(t,x) + f(U_0,v(t,x)) \leqq |\alpha|_0 V_1 - \min(-f)|_{\partial_1 B(U_0)}, \tag{2.7}$$

$$\alpha(x)\Delta v(t,x) + f(-U_0,v(t,x)) \geq - |\alpha|_0 V_1 + \min f|_{\partial_3 B(U_0)}. \tag{2.8}$$

The right-hand side of (2.7) is negative, that of (2.8) is positive by the first inequality in (A_2). Then the maximum principle proves that

$$-U_0 < u(t,x) < U_0$$

for any $t \in [0,\Theta]$ and $x \in \Omega$. The conclusion follows.

Now the proof of Theorem 2 reduces to show that α, β, U_0, U_1 and V_1 can be chosen in such a way that Proposition 1 holds and (A_1)-(A_2) are satisfied.

PROOF OF THEOREM 2: Define

$$\alpha_2 := \lim_{U_0 \to \infty} \frac{\hat{f}_{U_0}}{\tilde{Q}(U_0)}, \quad \beta_2 := \lim_{U_0 \to \infty} \frac{\hat{g}_{U_0}}{\tilde{P}(U_0)},$$

$$\tilde{\alpha} := \min\{\bar{\alpha},\alpha_1,\alpha_2\}, \quad \tilde{\beta} := \min\{\bar{\beta},\beta_1,\beta_2\}.$$

Observe that α_2,β_2 (hence $\tilde{\alpha},\tilde{\beta}$) are strictly positive by assumptions (H_2)-(H_3). Fix

$$|\alpha|_{2\theta} < \tilde{\alpha}, \quad |\beta|_{2\theta} < \tilde{\beta}.$$

By Theorem 1 for any $u_0,v_0 \in h_0^{2+2\theta}(\bar{\Omega})$ there exists $T = T(u_0,v_0) > 0$ (also depending on α,β) such that the solution of (1.1) exists in $[0,T]$. Choose U_0 so large that

(i) $|\Delta u_0|_0 < \tilde{P}(U_0)$, $|\Delta v_0|_0 < \tilde{Q}(U_0)$;

(ii) $|\alpha|_0 < \dfrac{\hat{f}_{U_0}}{\tilde{Q}(U_0)}$, $|\beta|_0 < \dfrac{\hat{g}_{U_o}}{\tilde{P}(U_0)}$.

Then choose $V_0 = kU_0$, $U_p = U_p(U_0)$ and $V_p = V_p(U_0)$ ($p = \theta$, $1 + \theta$) as in Proposition 1, and U_1, V_1 so as to satisfy (A_1)-(A_2). Then the conclusion follows.

It is clear from the above proof that U_0 and U_1, V_1 cannot be fixed independently (according to the remarks made in the Introduction).

3. REMARKS AND EXAMPLES

(a) Assumption (H_3) is not needed in the triangular case (for (1.1), if $\alpha \equiv 0$ or $\beta \equiv 0$). For instance, if $\beta \equiv 0$, Theorem 2 still holds with (H_3) replaced by the weaker condition

$$(H_3)' \quad \lim_{U\to\infty} \frac{\hat{f}_U}{|g|_{B(U)}} > 0, \qquad \lim_{U\to\infty} \frac{\hat{f}_U}{\{|g_u|_{B(U)} + |g_v|_{B(U)}\}U} > 0.$$

Clearly, (H_2) and $(H_3)'$ imply that $|f|_{B(U)}$ is an infinite stronger than $|g|_{B(U)}$ as U diverges. Let us mention that in the triangular case the present approach extends to some quasi-linear problems [PT].

(b) Assumption (H_3) is not necessary either, if α,β are allowed to vary with the initial data. In fact, let us state without proof the following result:

THEOREM 3: Assume (C), (H_0)-(H_1). Then for any $u_0, v_0 \in h_0^{2+2\theta}(\bar{\Omega})$ there exist $\hat{\alpha},\hat{\beta} > 0$ such that, if $|\alpha|_0 < \hat{\alpha}$, $|\beta|_0 < \hat{\beta}$, the solution of (1.1) is global ($\theta \in (0,\sigma/2)$).

(c) Changing the coefficient in (1.1) when u_0, v_0 change is clearly unsatisfactory. On the other hand, (H_3) is needed for the validity of Theorem 2. To discuss this point assume that the diffusion matrix

$$\begin{pmatrix} a & \alpha \\ \beta & b \end{pmatrix}$$

has "*constant entries*" and "*two simple positive eigenvalues*". The corresponding left eigenvectors give a parallelogram "close" to B(U), whose sides are rotated with respect to those of B(U) by angles which depend on α,β. This parallelogram being a candidate for an invariant region of (1.1), we ask whether the field (f,g) points inwards at its boundary under assumption

(H_1). This is true by uniform continuity in any compact region if α,β are sufficiently small – depending on the size of the region, hence on the ball where u_0,v_0 lie (this corresponds to Theorem 3). However, if f and g have a different growth at infinity, the field (f,g) cannot point inwards for a given choice of α and β – independently from the size of B(U) and of the corresponding parallelogram. This makes clear the role of (H_3) in Theorem 2.

(d) Let us give an example of (f,g) which satisfy assumptions (H_0)-(H_3). This is

$$f(u,v) = \sum_{k=0}^{(p-1)/2} v^{2k} P_{p-2k}(u), \quad g(u,v) = \sum_{k=0}^{(p-1)/2} u^{2k} Q_{p-2k}(v),$$

where p is odd,

$$P_{p-2k}(u) = \sum_{\ell=0}^{p-2k} a_{\ell,2k} u^{\ell}, \quad Q_{p-2k}(v) = \sum_{\ell=0}^{p-2k} b_{\ell,2k} v^{\ell},$$

and

$$a_{p-2k,2k} < 0, \quad b_{p-2k,2k} < 0.$$

It is easily seen that

$$|f|_{B(U)} \leqq C_1 U^p \leqq C_2 \hat{f}_u,$$

$$\max\{|f_u|_{B(U)}, |f_v|_{B(U)}\} \leqq CU^{p-1},$$

with suitable constants C_1, C_2, $C > 0$. Similar inequalities hold for g. Obviously, the same conclusion holds if we have lower order terms in f and g.

REFERENCES

[Am 1] Amann, H., Quasilinear parabolic systems under nonlinear boundary conditions, Arch. Rational Mech. Anal. 92 (1986), 153-192.

[Am 2] Amann, H., Global existence for semilinear parabolic systems, J. Reine Angew. Math. 360 (1985), 47-83.

[Al] Alikakos, N., Quantitative maximum principles and strongly coupled gradient-like reaction-diffusion systems, Proc. Royal Soc. Edinburgh Sect. A 94 (1983), 265-286.

[CCS] Chueh, K., Conley, C. and Smoller, J., Positively invariant regions for systems of nonlinear diffusion equations, Indiana Univ. Math. J. 26 (1977), 373-392.

[Co 1] Cosner, C., Pointwise a priori bounds for strongly coupled semilinear systems of parabolic partial differential equations, Indiana Univ. Math. J. 30 (1981), 607-620.

[Co 2] Cosner, C., Pointwise bounds for strongly coupled time-dependent systems of reaction-diffusion equations, SIAM J. Math. Anal. 15 (1984), 350-356.

[CTV] Cannarsa, P., Terreni, B. and Vespri, V., Analytic semigroups generated by nonvariational elliptic systems of second order under Dirichlet boundary conditions, J. Math. Anal. Appl. 112 (1985),56-103.

[DPG] Da Prato, G. and Grisvard, P., Equations d'évolution abstraites nonlinéaires de type parabolique, Ann. Mat. Pura Appl. 120 (1979), 329-396.

[PT] Pozio, M.A. and Tesei, A., Global existence of solutions for a strongly coupled quasilinear parabolic system, Nonlinear Analysis: TMA (to appear).

[Re] Redlinger, R., Pointwise a priori bounds for strongly coupled semilinear parabolic systems, Indiana Univ. Math. J. 36 (1987), 441-454.

M.A. Pozio
Seconda Università di Roma
Via O. Raimondo
I-00173 Roma
Italy

A. Tesei
Dipartimento di Matematica
"G. CASTELNUOVO"
Università di Roma "LA SAPIENZA"
Piazzale Delle Scienze
I-00185 Roma
Italy

3. ELLIPTIC EQUATIONS AND SYSTEMS

C.J. AMICK AND R.E.L. TURNER

Elliptic equations, internal waves and dynamical systems

1. INTRODUCTION

This is a report by R. Turner based on our joint research on internal waves in stratified fluids.

In the study of propagation of waves of permanent form in infinite domains the use of moving coordinates can often reduce a problem to a time-independent partial differential equation, typically elliptic. The setting of the new problem will be a domain $\mathbb{R} \times I$ where $\mathbb{R}$ denotes the real line and I a domain in a Euclidean space. In the example that motivated the present work, I is a finite interval in which a stream function for a fluid takes its values, while $\mathbb{R}$ is the domain of a horizontal, spatial variable. An unknown function w defined on $\mathbb{R} \times I$ describes an amplitude associated with a wave. One point of view in approaching such a problem is to treat x as a new "time" variable and $w(x,\cdot)$ as the "state" of a system at time x. One can then bring to bear the techniques of dynamical systems, in particular, the centre manifold reduction, thereby parametrizing all "small" solutions of the full system, typically infinite dimensional, in terms of solutions of an equation in a lower dimensional, centre manifold. In the most auspicious cases, the equations in the centre manifold are a finite-dimensional system of ordinary differential equations. This point of view was developed by Kirchgässner starting in the paper [11] and has been considerably expanded in recent years (see [2], [12]-[15], and their references), proving very successful as a tool for exhibiting all small solutions of wave problems and their dependence on physical parameters.

Progressing waves which are internal to a fluid, arising from density stratification, have been investigated analytically for several decades (see [3], [7], [8], [9], [10], [18] and their references). Experimental work and geophysical observations show striking examples of internal wave phenomena ([6], [16], [18]). In [3] we considered internal waves in a system consisting of two ideal fluids of constant density, filling a channel of finite vertical extent and infinite horizontal extent. A two-dimensional model based on the Euler equations was ultimately reduced to the nonlinear eigenvalue problem described in Section 2. The eigenvalue is essentially the wave speed and the unknown function w, a measure of streamline displacements. The analysis of solitary waves in [3] and the numerical work of Turner and VandenBroeck [17] strongly suggested the existence of solutions which we call "*surges*". These are smooth progressing fronts which propagate into an undisturbed fluid and leave a new parallel flow behind the front. In moving coordinates a surge corresponds to a heteroclinic orbit connecting different flows at $x = -\infty$ and $x = +\infty$. By using the dynamical systems approach we show that small surges do exist. In fact, through this approach one can give a complete picture of the collection of all small amplitude interfaces between two irrotational regimes of differing densities under the influence of gravity [4].

The quasi-linear elliptic equations arising in the present problem lead naturally, after some preliminary analysis, to functional equations in weighted Hölder spaces. Since beginning this work we have learned of Mielke's L^p approach [15] which presumably will overlap in applications with our work. The presence of an internal interface produces mappings of a more singular nature than are encountered in more classical free surface problems, for example, in [2] where water waves with surface tension are treated. In Section 3 we give some results on mappings in weighted Hölder spaces which are useful for the analysis of a general class of functional equations. The equations, treated in [5], include those arising in the internal wave problem as well as those from [2] and, we expect, will be useful in other wave problems. In Section 4 we state results for functional equations arising in the internal wave problem, including the existence of a centre manifold and its regularity properties. In Section 5 we discuss the final stage of analysis of small internal waves for the two fluid system. There is a conserved quantity in this flow which can be used to reduce the problem to the study of a first-order ordinary differential equation. This is exploited in [4]. Here we describe a method which does not depend on having a conserved quantity and which may be of interest for applications lacking such an invariant.

2. FORMULATION OF A WAVE PROBLEM

Consider an undisturbed system consisting of two inviscid, incompressible fluids of infinite horizontal extent, a lower one of density $\rho_- = 1$ and depth h and a lighter, upper fluid of density $\rho_+ = \rho$ and depth $1 - h$. The two fluids fill a closed channel of depth 1. In this system we seek steady, progressing waves which have phase velocity c under the influence of gravity g. The flow is assumed to be two-dimensional; that is, all variables are independent of one horizontal coordinate. In a Cartesian frame of reference rendering the flow steady, an x-axis is parallel to the channel walls and a z-axis is directed vertically upward so that gravity acts in the negative z direction. The origin of z is taken to be the level of the interface at $x = -\infty$. If the problem is posed with x and z as independent coordinates, then an unknown, free boundary separating the two fluids must be found as part of the solution. An alternative, semi-Lagrangian formulation will remove this difficulty (see [3] for details). As independent coordinates we choose x and a stream variable y which labels a streamline by its height at $x = -\infty$. The unknown is then the vertical deviation $w(x,y)$ of a streamline from its height at $x = -\infty$. To describe the problem in the new coordinates we define

$$f_1(\nabla w) = \frac{w_x}{1 + w_y} \quad \text{and} \quad f_2(\nabla w) = \frac{w_y}{1 + w_y} - \frac{|\nabla w|^2}{2(1 + w_y)^2} .$$

The problem becomes the following: Find an eigenvalue $\lambda = g/c^2$ and a function $w(x,y)$ satisfying

$$\frac{\partial}{\partial x}(f_1(\nabla w)) + \frac{\partial}{\partial y}(f_2(\nabla w)) = 0 \text{ in } T^- \cup T^+, \tag{2.1}$$

$$f_2(\nabla w^-) - \lambda w^- - \rho(f_2(\nabla w^+) - \lambda w^+) = 0 \text{ on } y = 0, \tag{2.2}$$

$$w(x,-h) = w(x,1 - h) = 0, \quad x \text{ in } \mathbb{R}. \tag{2.3}$$

Here $T^- = \mathbb{R} \times (-h,0)$, $T^+ = \mathbb{R} \times (0,1 - h)$, and $\pm$ are used to denote values or limits taken within $T^{\pm}$. We use a standard weak formulation for a continuous w which is C^1 in $\overline{T^{\pm}}$. The condition on the line $y = 0$ is merely the statement that the pressure is continuous across the fluid interface.

An important parameter is the bifurcation value

$$\lambda_d = \frac{1}{1 - \rho} \left(\frac{1}{h} + \frac{\rho}{1 - h}\right).$$

We shall be interested in solutions for which λ is near λ_d, and for which w is small together with its derivatives in L^∞. To that end, we rewrite the problem introducing

$$g_1(\nabla w) = w_x - f_1(\nabla w) \quad \text{and} \quad g_2(\nabla w) = w_y - f_2(\nabla w)$$

and express (2.1)-(2.2) as

$$\nabla w^\pm = \operatorname{div}(g_1(\nabla w^\pm),\ g_2(\nabla w^\pm)) \text{ in } T^\pm, \tag{2.4}$$

$$\rho(w_y{}^+ - \lambda_d w^+) - (w_y{}^- - \lambda_d w^-) = g(\nabla w^\pm, w, \lambda) \text{ on } y = 0, \tag{2.5}$$

where

$$g(\nabla w^\pm, w, \lambda) = \rho g_2(\nabla w^+) - g_2(\nabla w^-) + (\lambda_d - \lambda)(1 - \rho)w.$$

We write (2.3)-(2.5) symbolically as

$$Lw = MG(\nabla w^\pm, w, \lambda_d - \lambda), \tag{2.6}$$

where L and M are, respectively, Δ and div together with boundary conditions, and

$$G(\nabla w^\pm, w, \lambda_d - \lambda) = (g_1(\nabla w^+), g_1(\nabla w^-),\ g_2(\nabla w^+), g_2(\nabla w^-), g(\nabla w^\pm, w, \lambda)). \tag{2.7}$$

The approach to finding small solutions of (2.1)-(2.3) is to first solve the linear problem

$$Lw = MG(x,y) \tag{2.8}$$

with $G(x,y) = (g_1^+(x,y), g_1^-(x,y), g_2^+(x,y), g_2^-(x,y), g(x,y))$ given and then replace

$G(x,y)$ by $G(\nabla w^{\pm}, w, \lambda_d - \lambda)$ from (2.7) to obtain a fixed-point equation for w. However, the nullspace of L enters in a pivotal way and depends on the particular space in which the problem is posed. A formal separation of variables yields one solution

$$w_1(x,y) = (\xi_1 + \xi_2 x)t(y) \tag{2.9}$$

of $Lw = 0$ where

$$t(y) = \begin{cases} m(y + h), & y \leqq 0; \\ -mh(1 - h)^{-1}(y + h - 1), & y \geqq 0, \end{cases}$$

and m is chosen so that $\int_{-h}^{1-h} \rho_\infty(y)t^2(y)\, dy = 1$. Here $\rho_\infty = 1$ if $y \leqq 0$ and is ρ if $y \geqq 0$. There are, of course, other solutions of the form $w_k(x,y) = e^{\pm\nu_k x}t_k(y)$ where $0 < \nu_2 < \nu_3 < \ldots$ are eigenvalues and t_k, $k > 1$, eigenfunctions of a boundary value problem in y. It will be important later to allow slow exponential growth of functions in the x direction, but we explicitly exclude growth as rapid as $e^{\pm\nu_2 x}$, thereby rendering the nullspace of L two dimensional. The solution w of (2.1)-(2.3) is sought in the form

$$w(x,y) = Q(x)t(y) + R(x,y), \tag{2.10}$$

where the "remainder" R is orthogonal to t(y) at each x with respect to the weight ρ_∞ and will ultimately be of higher order than linear in Q and $\lambda_d - \lambda$.

3. SPACES AND MAPS

Let I denote a bounded open set in $\mathbf{R}^n$ or a point and let $T = \mathbf{R} \times I$. Thus when I is a point we merely identify T with $\mathbf{R}$. Let $C^j(T)$ denote the space of continuous real-valued functions on T with continuous derivatives through order j. For h in $C^0(T)$ and $0 < \alpha < 1$ we measure the Hölder constant of h "at x" by

$$(H_\alpha h)(x) = \sup_{\substack{|(x,y)-(x',y')|<1 \\ y \in I}} \frac{|h(x,y) - h(x',y')|}{|(x,y) - (x',y')|^\alpha}, \tag{3.1}$$

where $|\ |$ denotes the Euclidean distance in T. For each real number μ we define a norm

$$|h|_{j,\alpha,\mu} = \sum_{|\beta|\leq j} \sup_{x\in\mathbb{R},y\in I} e^{-\mu|x|}|\partial^\beta h(x,y)| + \sum_{|\beta|=j} \sup_{x\in\mathbb{R}} e^{-\mu|x|}(H_\alpha\partial^\beta h)(x) \qquad (3.2)$$

using the standard multi-index notation for derivatives, and let

$$C_\mu^{j,\alpha} = \{h \in C^j\colon |h|_{j,\alpha,\mu} < \infty\}. \qquad (3.3)$$

Let $C_\mu^{j,\alpha}(T,\mathbb{R}^n)$ denote $\mathbb{R}^n$-valued functions, normed by the supremum of the norms of the components; and C_μ^j, the space with the H_α estimate omitted from (3.2). For $\delta > 0$ let

$$C_{\mu,\delta}^{j,\alpha} = \{u \in C_\mu^{j,\alpha}(T)\colon |\partial^\beta u| \leq \delta\ |\beta| = k, 1 \leq k \leq j; H_\alpha\partial^\beta u \leq \delta, |\beta| = j\} \qquad (3.4)$$

and $C_{\mu,\delta}^{j,\alpha}(T,\mathbb{R}^n)$, the vector-valued version, or simply $C_{\mu,\delta}^{j,\alpha}$ when the context is clear. Note that $C_{\mu,\delta}^{j,\alpha}$ is a closed subset of $C_\mu^{j,\alpha}$ and that $u \in C_{\mu,\delta}^{j,\alpha}$ can be unbounded. We shall want to allow u to depend on a parameter $p = (p_1,\ldots,p_n)\in\mathbb{R}^n$ and use $C_{\mu,\delta}^{j,\alpha}(T \times \mathbb{R}^n)$ for a function $u(x,y,p)$ which is in $C_{\mu,\delta}^{j,\alpha}(T)$ uniformly for $p \in \mathbb{R}^n$. We use the standard notation L_s^q for symmetric q-linear maps in Banach spaces, D^k for a kth derivative of a map and $|D^k(\cdot)|$ for its norm (see [1]).

We shall be interested in composition of functions in the spaces $C_\mu^{j,\alpha}$ and $C_{\mu,\delta}^{j,\alpha}$ with smooth nonlinear maps from $\mathbb{R}^n$, where u takes values, into $\mathbb{R}$. The dependence on the parameter p is not central in the discussion and will sometimes be suppressed. We use C for a generic constant depending only on j, n, and other quantities indicated in the context.

<u>DEFINITION 3.1</u>: We call $g = g(u_1,\ldots,u_n,p_1,\ldots,p_n)$ *smooth* if, for some $M > 0$, of g satisfy

$$|D_u^k D_p^m g| \leq c_k < \infty;\quad k = 0,1,2,\ldots,M;\ m \leq M$$

for all $(u,p) \in \mathbb{R}^n \times \mathbb{R}^n$ and call g *flat* if it also satisfies

$g(0,\dots,0,p_1,\dots,p_n) = 0$ and $D_{u,p_1} g(0,\dots,0,0,p_2,\dots,p_n) = 0$ for all p.

<u>LEMMA 3.2</u>: Suppose g is smooth and assume $g(0,\dots,0,p_1,\dots,p_n) = 0$. Let M_g be the map defined by

$$M_g(u,p)(x,y) = g(u(x,y),p).$$

Assume $\mu \geqq 0$. Then $\mathbf{M}_g$ maps $C^{j,\alpha}_{\mu,\delta}(T,\mathbf{R}^n) \times \mathbf{R}^n$ into $C^{j,\alpha}_{\mu,\sigma}$ where

$$\sigma \leqq C \sup_{1 \leqq k \leqq j+1} c_k \delta^k.$$

<u>LEMMA 3.3</u>: Under the hypotheses of Lemma 3.2 the map M_g is Lipschitz continuous from $C^{j,\alpha}_{\mu,\delta}(T,\mathbf{R}^n) \times \mathbf{R}^n$ into $C^{j,\alpha}_{\mu,\sigma}$ with

$$|M_g(u,p) - M_g(u',p')| \leqq K^{j,\alpha}|u - u'| + K_1^{j,\alpha}|p - p'|,$$

where $K^{j,\alpha} \leqq C \sup_{1 \leqq k \leqq j+1} c_{k+1}\delta^k$, $K_1^{j,\alpha}$ is similarly bounded with c_k being replaced by derivative bounds for $D_p g(u,p)$, and the norms of functions are those in $C^{j,\alpha}_{\mu}$.

As is standard in an analysis leading to a centre manifold we shall use a truncation. Let $\eta:\mathbf{R} \to \mathbf{R}$ be an odd, nondecreasing, C^∞ function which is the identity for $-1 \leqq t \leqq 1$ and equal to 2 for $t \geqq 2$. Setting $\eta_r(t) = r\eta(t/r)$ one sees that η_r has derivative bounds $c_k \leqq Cr^{1-k}$, $k \geqq 0$. It is easily verified that if $g(u,p)$ is flat and $g(u,p,r)$ is defined to be $g(\eta_r(u_1),\dots,\eta_r(u_n),p_1,\dots,p_n)$, then $g(u,p,r)$ has derivative bounds $c_k \leqq C(r^{2-k} + |p_1|r^{1-k})$ where C depends on g and η.

At this point we indicate a difficulty that arises in the context of weighted spaces, where the analysis eventually requires some degree of smoothness in the resulting solutions. Let $g(u) = \eta_r^2(u)$ for $r > 0$. As a map of C^0_1 into itself the map M_g is not differentiable at $u = 0$. Were there a derivative, it would take each uniformly bounded function in C^0_1 to zero. Now let h be 0 for $x \leqq 0$, 1 for $x \geqq 1$, and linear on $0 \leqq x \leqq 1$. The function $h_k = h(x - k)$ has C^0_1 norm $|h_k| \leqq Ce^{-k} \to 0$ as $k \to \infty$. However, one verifies that the norm $|\eta_r^2(h_k)|$ does not approach zero faster than $|h_k|$. One can show that M_g is Lipschitz continuous from C_1^0 into itself. However, it is not Lipschitz from $C_\mu^{j,\alpha}$ into itself for $\mu > 0$ when either j or α is positive. The lack of regularity of M_g is attributable to the presence of multiplicative factors which can have exponential growth in expressions for derivatives. One can control these factors if one views the target for M_g as a space allowing more rapid growth than that of the domain and for applications, differentiability in this sense will suffice.

For fixed $(x,y) \in T$ and $q = 1,2,\ldots,D^q g = D^q_{(u,p)} g(u(x,y),p)$ is a symmetric q-linear map from $\mathbb{R}^n \times \mathbb{R}^n$ to $\mathbb{R}$. It induces a symmetric q-linear map on $C_\mu^{j,\alpha} \times \mathbb{R}^n$ defined by

$$M_{D^r g}(h_1,\pi_1)\ldots(h_q,\pi_q) = D^q g(u(x,y),p)(h_1(x,y),\pi_1)\ldots(h_q(x,y),\pi_q).$$

LEMMA 3.4: Let g be smooth from $\mathbb{R}^n \times \mathbb{R}^n$ to $\mathbb{R}$ and let q be a positive integer. Then

$$|M_{D^{q-1}g(u,p)} - M_{D^{q-1}g(u',p')} - M_{D^q g(u,p)}(u-u',p-p')|_{Z^{q-1}_{j+q+2}}$$

$$\leqq C(|u - u'|^2_{C^j} + |p - p'|^2),$$

where, for k real,

$$Z_k^q = L_s^q(C_\mu^{j,\alpha}(T,\mathbb{R}^n) \times \mathbb{R}^n,\ C_{k\mu}^{j,\alpha}(T,\mathbb{R})).$$

Further,

$$|M_{D^q g(u,p)} - M_{D^q g(u',p')}|_{Z^q_{j+q+2}} \leqq C(|u - u'|_{C_\mu^{j,\alpha}} + |p - p'|),$$

where C depends on g.

The next result is an immediate consequence of the previous lemma.

THEOREM 3.5: Suppose g is smooth and $\mu \geqq 0$. Then M_g defined in Lemma 3.2 is q times differentiable as a map from $C_\mu^{j,\alpha}(T,\mathbb{R}^n) \times \mathbb{R}^n$ to $C^{j,\alpha}_{(j+q+2)\mu}$ and the derivatives, defined by

$$D^k M_g = M_{D^k g}, \quad 1 \leqq k \leqq q,$$

are Lipschitz continuous.

DEFINITION 3.6: A transformation M defined on a function space over T is called translation invariant if

$$M(u_\tau) = (Mu)_\tau,$$

where for all $\tau \in \mathbb{R}$ and $(x,y) \in T$, $u_\tau(x,y) = u(x + \tau, y)$.

For the internal wave problem let $I^- = (-h,0)$, $I^+ = (0,1-h)$, $I = (-h,1-h)$ and

$$X^1_\mu = \{u \in C^0(\bar{T}): u^\pm \in C^{1,1/2}_\mu(T^\pm);\ u(x,-h) = u(x,1-h) = 0,\ x \in \mathbb{R}\},$$

the norm being the sum of those on $T^\pm$. The solution (Q,Q',R) is sought in

$$U_\mu = C^{1,1/2}_\mu(\mathbb{R}) \times C^{0,1/2}_\mu(\mathbb{R}) \times X^1_\mu \tag{3.5}$$

and the element G from (2.7) resides in

$$\Lambda_\mu = C^{0,1/2}_\mu(T^+) \times C^{0,1/2}_\mu(T^+) \times C^{0,1/2}_\mu(T^-) \times C^{0,1/2}_\mu(T^-) \times C^{0,1.2}_\mu(\mathbb{R}).$$

Finally, $U_{\mu,\delta}$ denotes a product analogous to (3.5) but with subscripts μ, δ on each factor. Note that smooth functions defined on these product spaces and their linear subspaces inherit the differentiability properties derived earlier.

For the remainder of the discussion we restrict μ to lie in the interval $(0,(\nu_2/2)^{1/2})$ where ν_2 is the smallest positive eigenvalue mentioned in Section 2. By using a partition of unity in the x direction, the Lax-Milgram lemma on a space of functions ρ_∞ orthogonal to t(y) at each x and known elliptic theory, one can derive the following result.

THEOREM 3.7: There exist bounded linear maps $L_1, L_2: \Lambda_\mu \to C^{0,1/2}_\mu(\mathbb{R})$ and $L_3: \Lambda_\mu \to X^1$ such that, given $G \in \Lambda_\mu$ and $(\xi_1,\xi_2) \in \mathbb{R}^2$, the weak form of (2.3)-

(2.5) has a unique solution $w(x,y) = Q(x)t(y) + R(x,y)$, with $(Q,Q',R) \in U_\mu$ satisfying

$$\begin{aligned} Q(x) &= \xi_1 + \beta \int_0^x Q', \\ Q'(x) &= \xi_2 + \beta^{-1} L_1 G|_0^x + \beta^{-1} \int_0^x L_2 G, \\ R(x,y) &= (L_3 G)(x,y), \quad (x,y) \in \bar{T}, \end{aligned} \tag{3.6}$$

where L_3 is translation invariant and

$$\int_{-h}^{1-h} \rho_\infty(y)t(y)(L_3G)(x,y)\, dy = 0, \quad x \in \mathbb{R}.$$

REMARKS: Here $\beta = 1$ and is included for a subsequent rescaling. The maps L_1, L_2 are given by explicit formulas and L_3 is also bounded from Λ_0 into X_0^1.

For use in the sequel we let Ω denote the triple $(Q_1 = Q,\ Q_2 = Q', R)$, $\xi = (\xi_1, \xi_2)$, $p = \lambda_d - \lambda$ and express the equation (3.6) as

$$\Omega = N_0(G,\xi,p). \tag{3.7}$$

4. NONLINEAR FUNCTIONAL EQUATIONS

If one replaces $G = (g_1^+(x,y),\ldots)$ by $G(\Omega) = (g_1(\nabla(Q_1(x)t(y) + R(x,y))^+),\ldots)$ in equation (3.7) and defines $N(\Omega,\xi,p)$ to be $N_0(G(\Omega),\xi,p)$, one obtains an equation $\Omega = N(\Omega,\xi,p)$ whose fixed points formally provide solutions of (2.3)-(2.5). If Q_1 is replaced by $\eta_r(Q_1)$; the partial derivative R_x by $\eta_r(R_x)$; and so on, one obtains an equation

$$\Omega = N(\Omega,\xi,p,r), \tag{4.1}$$

which shares "small" solutions with $\Omega = N(\Omega,\xi,p)$ and which has solutions in spaces of the type $C^{j,\alpha}_{\mu,\delta}$ for δ and r small.

One uses a contraction mapping to solve (4.1). For this it is useful to rescale letting $Q_1^* = Q_1$ and $Q_2^* = \beta^{-1}Q_2$ where $\beta > 0$. Assume this is done and then omit the stars. In the new nonlinear system resulting from (3.6) one can choose β small in the first equation to achieve contraction. In a term containing β^{-1}, the higher order vanishing of flat functions compensates for the large term β^{-1}. For simplicity we shall suppress the parameter β. The equations in (4.1) in which an initial value ξ_i is given have an Ω dependence of the form $F(\Omega)|_0^x$. In the equation for Q_1, F is merely convolution with a Heaviside function. Convolution with a more general kernel b(x) can be allowed provided it has limits $b(+\infty)$, $b(-\infty)$ which it approaches exponentially rapidly in such a way that $u \to (b * u)|_0^x$ induces a bounded map between suitable spaces with weight μ. Maps with convolutions arise naturally in [2] and are pursued in conjunction with general equations in [5] where it is shown that fixed points will exist in spaces of the type $U_{\mu,\delta}$ for suitably restricted parameters. Here, because of the particular structure of the resulting equations, one can take $\delta = r$; we state a result specific to (4.1).

<u>THEOREM 4.1</u>: For sufficiently small p, β and r, the map N in (4.1) takes $U_{\mu,r}$ into itself and is contractive in Ω. For each $\xi \in \mathbb{R}^2$ there is a unique solution $\Omega(\xi,p)$ which is Lipschitz continuous in ξ and p.

One can use the result of Section 3 to show that for a fixed x_0 the solution $\Omega(\xi,p)(x_0,\cdot)$ depends smoothly on parameters, for with x fixed the exponential growth plays no role. Supressing p and y and letting $\tilde{Q} = (Q_1,Q_2)$ one shows, as for ordinary differential equations (see [2]), that $\Omega(\xi)(x+x_0)$ and $\Omega(\tilde{Q}(\xi)(x_0))(x)$ both satisfy (4.1) with initial data $\tilde{Q}(\xi)(x_0)$ and so coincide. By setting $x = 0$ and then replacing x_0 by x one obtains the local dependence of R on $\tilde{Q}$. Further analysis of (4.1) leads to the following "centre manifold" result. Here $V = \{u \in C^0(\bar{I}): u^{\pm} \in C^{1,1/2}(\overline{I^{\pm}})\}$ and "smooth" means any finite order of differentiability desired.

<u>THEOREM 4.2</u>: For each ξ, the components of the solution of (4.1) are related by

$$R(x,\cdot) = W(\tilde{Q}(x),p), \tag{4.2}$$

where W is smooth from $\mathbb{R}^3$ to V and satisfies

$$|W(\xi,p)|_V \leqq C|\xi|(|\xi| + |p|). \tag{4.3}$$

Both R_x and R_y have a local dependence similar to (4.2) and bounds of type (4.3) in spaces analogous to V.

5. SMALL INTERNAL WAVES

If we suppose (4.1) has a solution for which p, Q and Q' are sufficiently small, then by (4.3) and its analogs for R_x and R_y, all the arguments occurring in truncations η_r will lie in (-r,r) and the truncations can be removed. One can use Theorem 4.2 to substitute for R in the first two equations of the system (4.1). Using estimates of the type (4.3) and rescaling to simplify coefficients one arrives at a second-order ordinary differential equation satisfied by Q, having the form

$$Q'' = pQ - eQ^2 + Q^3 + N^*(p,Q,Q'), \tag{5.1}$$

where

$$e = \frac{1}{h^2} - \frac{\rho}{(1 - h)^2}, \tag{5.2}$$

and N^* is a smooth expression in eQ^3, pQ^2, p^2Q, and higher order terms in p, Q and Q'.

Conversely, if for a fixed small β one can construct a solution of (5.1) with Q, Q' and $p = \lambda_d - \lambda$ sufficiently small, then $Q_1 = Q$, $Q_2 = \beta^{-1}Q'$ and $R(x,\cdot) = W(Q,\beta^{-1}Q',p)(x)$ will be a solution of (4.1). In turn, $w(x,y) = Q(x)t(y) + R(x,y)$ will satisfy (2.1)-(2.3). By examining classical work on waves one anticipates that a solution which decays to zero at $-\infty$ will behave like $\exp(-p^{1/2}|x|)$. This suggests a rescaling and use of an implicit function theorem (see [2], [12]). If we let $\sigma^2 = p$ and

$$q(x,\sigma) = \frac{1}{\sigma^s} Q(\frac{x}{\sigma}), \tag{5.3}$$

then for s = 2 the equation for q to lowest order in σ is

$$q'' = q - eq^2; \tag{5.4}$$

that is, the terms corresponding to Q^3 and N^* vanish at $\sigma = 0$. Equation (5.4) has a familiar solution in terms of sech^2. This leads to solitary wave solutions of (2.1)-(2.3) through an application of the implicit function theorem for small σ in a manner analogous to that described next for surges.

To extract the surge solution in the spirit of the last section one must use a different scaling. The work in [3] shows that a surge will have an amplitude proportional to e and so we look for e in the form $e = \sigma(\zeta_0 + \zeta(\sigma))$ where ζ is to vanish at $\sigma = 0$, and take $s = 1$ in (5.3) so that Q and e are of the same order. With the new scaling (5.1) becomes

$$q'' = [q - \zeta_0 q^2 + q^3] + \zeta(\sigma)q^2 + \tilde{N}(\sigma,q,q'), \tag{5.5}$$

where $\tilde{N}$ is smooth and vanishes when $\sigma = 0$. Denote the term in square brackets in (5.5) by F(q). If $\zeta_0 = 3/\sqrt{2}$, then the equation $q'' = F(q)$ has a "heteroclinic orbit" $q_0(x)$, increasing from $q_0(-\infty) = 0$ to $q_0(+\infty) = \sqrt{2}$, for which $\pi(x) = q_0'(x)$ is a positive, even function. For an implicit function theorem we require decay properties at ∞. Suppose now that $-1 < \mu < 0$, the -1 arising from $F_q(0) = F_q(\sqrt{2}) = 1$, and define

$$\Gamma_\mu^0 = \{q \in C^0(R): \lim_{x\to\infty} e^{\mp\mu x}[q(\pm x) - q(\pm\infty)] \text{ exist}\},$$

$$\Gamma_\mu^2 = \{q \in C^0(R): q' \in C_\mu^1\},$$

with appropriate norms. The point here is to have limits at $\pm\infty$ approached exponentially fast. Note that $q_0 \in \Gamma_\mu^2$, for it approaches its limits as $\pm\infty$ at a rate $\exp(-|x|)$ and the derivatives decay to zero at that rate. In this connection $F_q(q_0(x)) = 1 + V(x)$ with $V \in C_{-1}^0$. Consider the map

$$\Theta:(q,\zeta,\sigma) \to \begin{pmatrix} q'' - F(q) - \zeta q^2 - \tilde{N} \\ \int_{-\infty}^{\infty} (q - q_0)\pi dx \end{pmatrix}.$$

One verifies that Θ takes $(q_0,0,0)$ to zero and takes a neighbourhood of $(q_0,0,0)$ in $\Gamma_\mu^2 \times R \times R$ into $\Gamma_\mu^0 \times R$. The derivative $D\Theta$ with respect to (q,ζ) at $(q_0,0,0)$ takes the pair (ϕ,γ) to $(\phi'' - F_q(q_0)\phi - \gamma q_0^2, \int \phi\pi)$ and one must show this is one-to-one and onto.

Since π is a solution of $\psi'' - F_q(q_0)\psi = 0$ an integration by parts shows that $\gamma = 0$ for a null vector (ϕ,γ) and then $\phi = 0$ is the only solution Γ^2_μ which is perpendicular to π. To show $D\Theta$ is onto, one can split the problem according to odd and even parts. For the odd part one can add a multiple of π to a solution, so as to continue a solution starting at $x = +\infty$ with good asymptotic behaviour, as an odd function. For the even part, integration against π over $(0, +\infty)$ and a suitable choice of γ allow solvability. The implicit function theorem then yields solutions depending on σ (thus ρ and h are related for these solutions; see (5.2)) and the following results.

THEOREM 5.1: For σ in a neighbourhood of 0, (5.5) has a branch of solutions $(e(\sigma),q(\sigma))$ and (2.1)-(2.3), a corresponding branch of surges $w(x,y,\sqrt{\lambda_d - \lambda})$, having different limits at $x = -\infty$ and at $x = +\infty$.

C.J. Amick was supported by the National Science Foundation and the Sloan Foundation.

R.E.L. Turner was supported by the National Science Foundation and the Air Force OSR.

REFERENCES

[1] Abraham, R. and Robbin, J., Transversal Mappings and Flows, W.A. Benjamin, Inc., New York, 1967.

[2] Amick, C.J. and Kirchgassner, K., A theory of solitary water waves in the presence of surface tension, Arch. Rat. Mech. Anal. 105 (1989), 1-49.

[3] Amick, C.J. and Turner, R.E.L., A global theory of solitary waves in two-fluid systems, Trans. AMS 298 (1986), 431-481.

[4] Amick, C.J. and Turner, R.E.L., Small internal waves in two-fluid systems, CMS Report #89-4 Univ. of Wis.-Madison, Arch. Rat. Mech. Anal. (to appear).

[5] Amick, C.J. and Turner, R.E.L., Center manifolds in equations from hydrodynamics, in preparation.

[6] Apel, J.R., Byrne, H.M., Proni, J.R. and Charnell, R.L., Observations of oceanic internal and surface waves from the earth resources technology satellite, J. Geophys. Res. 80 (1975), 865-881.

[7] Benjamin, T.B., Internal waves of finite amplitude and permanent form, J. Fluid. Mech. 25 (1966), 241-270.

[8] Bona, J.L., Bose, D.K. and Turner, R.E.L., Finite amplitude steady waves in stratified fluids, J. Math. Pure Appl. 62 (1983), 389-439.

[9] Bowman, S.B., Mathematical aspects of wave motion in stratified fluids, Ph.D. thesis, Oxford, 1985.

[10] Kakutani, T. and Yamasaki, N., Solitary waves on a two-layer fluid, J. Phy. Soc. Japan 45 (1978), 674-679.

[11] Kirchgässner, K., Wave-solutions of reversible systems and applications, J. Diff. Equations 45 (1982), 113-127.

[12] Kirchgässner, K., Nonlinear wave motion and homoclinic bifurcation, Theoretical and Applied Mechanics, F. Niordson and N. Olhoff, eds., Elsevier Science Publishers B.V. North-Holland, 1985.

[13] Kirchgässner, K., Nonlinearly resonant surface waves and homoclinic bifurcation, Adv. Applied Mech. 26 (1988), 135-181.

[14] Mielke, A., A reduction principle for nonautonomous sytems in infinite-dimensional spaces, J. Diff. Equations 65 (1986), 68-88.

[15] Mielke, A., Reduction of quasi-linear elliptic equations in cylindrical domains with applications, Math. Meth. Appl. Sci. 10 (1988), 51-66.

[16] Osborne, A.R. and Burch, J.L., Internal waves in the Andaman Sea, Science 208 (1980), 451-460.

[17] Turner, R.E.L. and Vanden-Broeck, M.-M., Interfacial solitary waves, Physics of Fluids (to appear).

[18] Walker, L.R., Interfacial solitary waves in a two-fluid medium, Physics of Fluids, 16 (1973), 1796-1804.

[19] Yih, C.-S., Stratified Flows, Academic Press, New York, 1980.

C.J. Amick
Department of Mathematics
University of Chicago
Chicago
U.S.A.

R.E.L. Turner
Department of Mathematics
University of Wisconsin
480 Lincoln Drive
Madison, Wisconsin, 53706
U.S.A.

P. BARAS

Semilinear problem with convex nonlinearity

INTRODUCTION

The main goal of this article is to provide an approach which allows us to extend the necessary and sufficient condition given in [1] for the existence of a solution of semilinear equations of the following type:

$$\begin{aligned} Lu &= j(u) + f && \text{in } \Omega, \\ u &= 0 && \text{in } \Omega, \\ u &= 0 && \text{on } \partial\Omega, \end{aligned} \qquad (1)$$

where Ω is an open subset of R^n, L is an elliptic operator and j on $\Omega \times R$ into R is a nondecreasing convex function on u such that $j(x,0) = 0$ a.e. on Ω. As usual, $j(u)$ denotes the function $x \to j(x,u(x))$.

Let us recall that this criterium can be easily obtained in the following way: Let h be a suitable positive test function, we have

$$\int_\Omega f(x)h(x)\,dx = \int_\Omega (L^*h(x)u(x)-j(x,u(x))h(x))\,dx \leq \int_\Omega j^*(x,\frac{L^*h(x)}{h(x)})\,dx,$$

where $j^*(x,r) = \sup\{rz - j(x,z);\ z \geqq 0\}$ is the conjugate function of j. It is proved in [1] that there exists a weak solution of (1) if and only if

$$\int_\Omega f(x)h(x)\,dx \leq \int_\Omega j^*(x, \frac{L^*h(x)}{h(x)})\,dx \qquad (2)$$

for each h in a convenient subset.

In order to extend this result to semilinear equations involving a non-linear term depending also on the gradient of u, we establish a preliminary theoretical result on the abstract nonlinear equation

$$F(u) + f = 0, \qquad (3)$$

where F is a nonlinear operator into a linear space Y and f belongs to Y.

In the example (1), $F(u) = j(u) - Lu$ and (2) is nothing other than

$$(f,h) \leqq \sup \{-F(u),h); u \in D(F)\}, \tag{4}$$

where D(F) is a suitable set on which F(u) is well defined and where (f,h) is equal to the integral used above. Let Y^* be the algebraic dual of Y, we notice that (4) makes a sense for all h in Y^*. A natural generalization of the previous result is then to know if the existence of a solution of (3) and (4) for each h in a suitable subset of Y^* are equivalent.

Of course, we are unable to answer at this question in the large. The first part of this article is devoted to the abstract result which is in some sense an application of the Hahn-Banach theorem. We prove that, under two main hypotheses, that is, the convexity and, roughly speaking, the lower semicontinuity of $u \to (F(u)\cdot h)$ for each h in a convenient subset Z^+ of Y^*, (4) and the existence of a "weak subsolution" of (3) are equivalent. By a "weak subsolution", we understand u such that $(F(u) + f\cdot h) \leqq 0$ for all h in Z^+.

In the second part of this article, we derive from the above result a necessary and sufficient condition on f to give the existence of a solution of the following problem:

$$\begin{aligned} & u \in W^{1,1}_{loc}(\Omega), \quad j(u,\nabla u) \in L^1_{loc}(\Omega), \\ & Lu \geqq j(u,\nabla u) + f \qquad \text{in } \mathcal{D}'(\Omega), \\ & u \geqq 0 \qquad\qquad\qquad \text{in } \Omega. \end{aligned} \tag{5}$$

where j is a Carathéodory function which mainly satisfies:

(i) $u \to j(u,\nabla u)$ is a convex function;

(ii) there exist some constants $a > 0$, $b > 0$, $d,\gamma > 1$, $\sigma > 1$, such that

$$j(x,r,p) \geqq a|r|^\gamma + b|p|^\sigma - d \quad \text{for all } r \text{ in } \mathbf{R},\ p \text{ in } \mathbf{R}^n \text{ and for a.e. } x \text{ in } \Omega.$$

This last hypothesis can be easily improved by using some technical refinements. For example, if j does not depend of p, (ii) must be satisfied with b = 0. Surprisingly, L may be any first- or second-order operator and no ellipticity is required. However, when L is an elliptic operator and j depends only on r and is nondecreasing in r, we can deduce by using the usual tools that the existence of a solution of (5) implies the existence of a solution of (1).

THEORETICAL PRELIMINARIES

Let X and Y be linear spaces and X^*, Y^* their respective algebraic dual spaces. Let D(F) be a subset of X^* and F be a function in D(F) into Y. For each (k,h) in $X \times Y^*$, we define a function L(k,h) by

$$L(k,h) = \{(u,k)-(F(u),h);\ u \in D(F)\}, \tag{6}$$

where (,) is the product of duality between a space and its dual. We can easily see that L is a positively homogeneous function on $X \times Y^*$ into $\mathbb{R} \cup \{+\infty\}$. It is not possible to compute L in the large, so we restrict L on subspaces K and Z of X and Y^*, respectively, and we need the following important hypothesis:

$$\left\| \begin{array}{l} \text{For each } k \text{ in } K, \text{ there exist } h \text{ and } h' \text{ in } Z \\ \text{such that } L(k,h) \text{ and } L(-k,h') \text{ are finite.} \end{array} \right\| \tag{7}$$

Let f be given in Y, we define a function ϕ on K into $\mathbb{R} \cup \{\pm\infty\}$ by

$$\phi(k) = \inf\{L(k,h)-(f,h);\ h \in Z\}. \tag{8}$$

PROPOSITION 1: Let f be given in Y and suppose (7). Then the following assertions are equivalent:

$$(f,h) \leq L(0,h) \text{ for all } h \text{ in } Z. \tag{9}$$

$$\phi(k) \text{ is finite for all } k \text{ in } K \text{ and is a positively homogeneous function on } K. \tag{10}$$

for each k_0 in K, there exists u in X^* such that: (11)

(i) $(u,k_0) = \phi(k_0)$

(ii) $(u,k) + (f,h) \leqq L(k,h)$ for all (k,h) in $K \times Z$.

PROOF: The hypothesis (7) implies that ϕ takes its values in $\mathbb{R} \cup \{-\infty\}$. ϕ is then a positively homogeneous convex function on K, we have $\phi(0) \leqq \phi(k) + \phi(-k)$ and $\phi(0)$ belongs to $\{-\infty, 0\}$.

Since (9) means that $\phi(0) \geqq 0$, (9) and (10) are equivalent.

If we set k = 0 in (11)(ii), we obtain that (11) implies (9).

Suppose (9). By the Subdifferentiability Theorem [2], we know that for each k_0 in K, there exists u in X^* and a in R such that $a + (u,k) \leqq \phi(k)$ for all k in K and $a + (u,k_0) = \phi(k_0)$. Choosing $k = \lambda k_0$ with $\lambda > 0$, we obtain that $a \leqq \lambda a$ for all $\lambda > 0$ and thus a = 0 which proves (11). □

Notice that (9) is the same condition as (4). Since (11)(ii) means that for each h in Z, we have:

$$\sup\{(u,k)-L(k,h);\ k \in K\} + (f,h) \leqq 0.$$

We obtain that if h is such that

$$(F(u),h) = \sup\{(u,k)-L(k,h);\ k \in K\} \tag{12}$$

the best result that we can hope to deduce from (9) is the existence of a solution u in D(F) of the following weak inequality:

$$(F(u) + f,h) \leqq 0 \tag{13}$$

for all h in Z such that (12) holds. But (12) implies that $u \to (F(u),h)$ is the restriction on D(F) of a convex lower semicontinuous function on X^* endowed with the topology $\sigma(X^*,K)$. So the following functional frame may seem quite natural.

Given K and Z satisfying (7), we suppose that there exist a convex subset $D(F^0)$ of X^*, a function F^0 on $D(F^0)$ into Y and a subset Z^+ of Z such that

$$\left\| D(F) \subset D(F^0), \quad (F^0(u),h) = (F(u),h) \text{ for all } (u,h) \text{ in } D(F) \times Z^+ \right\| . \tag{14}$$

Let φ_h be the function defined by $\varphi_h(u) = (F^0(u),h)$ if u belongs to $D(F^0)$, $\varphi_h(u) = +\infty$ elsewhere. We shall use the following assumption where X^* is topogized by $\sigma(X^*,K)$:

$$\left\| \begin{array}{l} \varphi_h \text{ is a convex function on } X^*; \\ \text{if for all h in } Z^+, \text{ lim inf } \varphi_h(u) \text{ is finite, then u belongs to } D(F^0) \\ \text{and for each h in } Z^+, \ \varphi_h(u) \leqq \text{lim inf } \varphi_h(u). \end{array} \right\| \tag{15}$$

PROPOSITION 2: If F^0 and Z^+ satisfy (14), (15) and u in X^* is a solution of (11) then u belongs to $D(F^0)$ and

$$(F^0(u) + f,h) \leqq 0 \quad \text{for all h in } Z^+. \tag{16}$$

PROOF: Let h be fixed in Z^+. We have with (14)

$$L(k,h) \leqq \varphi_h^*(k) = \sup\{(u,k) - \varphi_h(u);\ u \in X^*\},$$

then we deduce from (11) that

$$(f,h) + \varphi_h^{**}(u) \leqq 0,$$

where $\varphi_h^{**}(u) = \sup\{(u,k) - \varphi_h^*(k);\ k \in K\} = \text{lim inf } \varphi_h(u)$. Next, we use (15) to conclude. □

In order to work with some usual topological vector space, we introduce a subspace K_0 of K which is a separated locally convex vector space. K_0' denotes its topological dual space and X_0' is the subspace of X^* such that each restriction of an element of X_0' on K_0 defines a continuous linear function. X_0' is topologized by the topology inherited from the strong topology of K_0'. We need the following assumption on the topology of K_0 and its relations with f in Y and F:

$$\left\| \begin{array}{l} K_0 \text{ is reflexive and } \phi \text{ is bounded above} \\ \text{on a neighbourhood of } 0 \text{ in } K_0. \end{array} \right\| \tag{17}$$

Let F^0 and Z^+ satisfy (14), we shall use the following assumption:

$$\left\| \begin{array}{l} D(F^0) \subset X_0', \ \varphi_h \text{ is a convex function on } X_0'; \\ \text{if for all } h \text{ in } Z^+, \ \liminf \varphi_h(u) \text{ is finite, then } u \text{ belongs to} \\ D(F^0) \text{ and for each } h \text{ in } Z^+, \ \varphi_h(u) \leqq \liminf \varphi_h(u). \end{array} \right\| \tag{18}$$

In this last hypothesis, φ_h is restricted on X_0' and the lim inf is taken in the strong topology of X_0'. We have the following result which summarizes the above proposition.

<u>THEOREM 1</u>: Let f be given in Y. Suppose that (7), (9), (14), (17) and (18) hold. Then for each k_0 in K, there exists u in $D(F^0)$ such that

$$(u,k_0) = \inf\{L(k_0,h) - (f,h);\ h \in Z\}, \tag{19}$$

$$(u,k) + (f,h) \leqq L(k,h) \text{ for all } (k,h) \text{ in } K \times Z, \tag{20}$$

$$(F^0(u) + f,h) \leqq 0 \qquad \text{for all } h \text{ in } Z^+. \tag{21}$$

<u>PROOF</u>: Equations (19) and (20) follow from Proposition 1 and we have just to prove the last point. We deduce from (20) that $(u,k) \leqq \phi(k)$ for all k in K, so using the second part of the hypothesis (17), we obtain that u belongs to X_0'. On the other hand, the first part of the hypothesis (17) implies that the topological dual of X_0' (which is the dual space of K_0') is identifiable with K_0 and so the topology $\sigma(X_0',K_0)$ is the weak topology of X_0'. The proof is then the same as in Proposition 2 when we substitute X_0' and K_0 in X^* and K. The crucial point is that we have again

$$\sup\{(u,k) - \varphi_h^*(k);\ k \in K_0\} = \liminf \varphi_h(u),$$

where here

$$\varphi_h^*(k) = \sup\{(u,k) - \varphi_h(u);\ u \in X_0'\},$$

which is a consequence of the Hahn-Banach theorem. □

Let Z_f be the subset of Z defined by

$$Z_f = \{h \in Z \text{ such that } L(0,h) \text{ is finite}\} \tag{22}$$

and consider the following hypothesis which is more precise than (14)

$$\left\| \begin{array}{l} D(F) \subset D(F^0),\ (F^0(u),h) = (F(u),h) \text{ for all } (u,h) \text{ in } D(F) \times Z_+ \\ \text{and for each } (u,h) \text{ in } D(F^0) \times Z_+, \text{ there exists a sequence } u_n \\ \text{in } D(F) \text{ such that } (F^0(u),h) = \lim(F(u_n),h). \end{array} \right\| \tag{23}$$

COROLLARY 1: Suppose that (7), (18) and (23) hold with $Z^+ = Z_f$. Let f be given in Y satisfying (17). The following assertions are equivalent:

(i) f satisfies (9);

(ii) there exists u in $D(F^0)$ such that

$$(F^0(u) + f,h) \leqq 0 \quad \text{for all } h \text{ in } Z_f.$$

PROOF: By Theorem 1, we already know that (i) implies (ii). For all h in Z_f, we have $(f,h) \leqq -(F^0(u),h)$ and so $(f,h) \leqq \sup\{-(F^0(u),h);\ u \in D(F^0)\}$. We easily deduce from (23) that $\sup\{-(F^0(u),h);\ u \in D(F^0)\} = L(0,h)$ for all h in Z^+. Using $Z^+ = Z_f$, we obtain that f satisfies (9).

SEMILINEAR SECOND-ORDER EQUATIONS

Given Ω an open subset of $\mathbb{R}^n$, and L a second-order differential operator

$$Lu = \sum_{i,j} \frac{\partial}{\partial x_i}\, a_{i,j} \frac{\partial u}{\partial x_j} + \sum_j b_j \frac{\partial u}{\partial x_j},$$

where a_{ij} and b_j belong to $L^1_{loc}(\Omega)$, we denote by A, B the matrices $(a_{i,j})$

and (b_j), and we write

$$Lu = \mathrm{div}(A\nabla u) + B\cdot\nabla u.$$

Let j be a Carathéodory function on $\Omega \times \mathbf{R} \times \mathbf{R}^n$ into $\mathbf{R} \cup \{+\infty\}$. We suppose that there exist $a > 0$, $b > 0$, $d,\gamma > 1$, $\sigma > 1$, such that

$$j(x,r,p) \geqq a|r|^{\gamma} + b|p|^{\sigma} - d \text{ for all } r \text{ in } \mathbf{R},\ p \text{ in } \mathbf{R}^n \text{ and a.e. on } \Omega. \quad (24)$$

Let f be given in $\mathcal{D}'(\Omega)$ and consider the equation

$$Lu = j(u,\nabla u) + f \text{ in } \mathcal{D}'(\Omega). \quad (25)$$

To apply the above result, we take in a first stage,

$$X = (W^{1,\infty}(\Omega))',\ Y = \mathcal{D}'(\Omega),\ D(F) \equiv \{u \in C^{\infty}(\Omega),\ j(u,\nabla u) \in L^1_{loc}(\Omega)\}$$

and we set $F(u) = j(u,\nabla u) - Lu$. To u in D(F) corresponds the element still denoted u in X^* defined by $(u,k) = ((u,k))$ where ((,)) is the product of duality between $W^{1,\infty}(\Omega)$ and X.

Let $L^1_c(\Omega)$ be the set of functions in $L^1(\Omega)$ with a compact support. We begin by some a priori estimates on the function L(k,h) defined in (6). Let (k_1,k_2) be in $L^1_c(\Omega) \times (L^1_c(\Omega))^n$, we define $k = k_1 - \mathrm{div}\, k_2$ in X by

$$u \to \int_{\Omega} (k_1(x)u(x) + (k_2\cdot\nabla u(x)))\, dx.$$

Let h be in $\mathcal{D}(\Omega)$, $h \geqq 0$, by integration by parts, we have

$$L(k,h) = \sup\{\int_{\Omega} (uk_1+(k_2\cdot\nabla u)+(A\nabla u\cdot\nabla h)+(B\cdot\nabla u)h-j(u,\nabla u)h)\, dx;\ u \in D(F)\},$$

we deduce from (24)

$$L(k,h) \leqq \sup\{\int_{\Omega} (uk_1+(k_2\cdot\nabla u)+(A\nabla u\cdot\nabla h)+(B\cdot\nabla u)h-a|u|^{\gamma}h-b|\nabla u|^{\sigma}h+dh)dx\}.$$

For all u in $C^{\infty}(\Omega)$, we have

$$\int_\Omega ((k_2\cdot\nabla u)+(A\nabla u\cdot\nabla h)+(B\cdot\nabla u)h-b|\nabla u|^\sigma h)\,dx \leqq c(\sigma,b)\int_\Omega \frac{|k_2+A^t\nabla h+Bh|^{\sigma'}}{h^{\sigma'-1}}\,dx,$$

$$\int_\Omega (uk_1-a|u|^\gamma h)dx \leqq c(\gamma,a)\int_\Omega \frac{|k_1|^{\gamma'}}{h^{\gamma'-1}}\,dx,$$

where $1/\gamma + 1/\gamma' = 1$, $1/\sigma + 1/\sigma' = 1$ and $c(\alpha,r)$ is a finite constant for $r > 0$ and $\alpha > 1$. Since $a > 0$ and $b > 0$, we obtain

$$L(k,h) \leqq c(\gamma,a)\int \frac{|k_1|^{\gamma'}}{h^{\gamma'-1}}\,dx + c(\sigma,b)\int_\Omega \frac{|k_2+A^t\nabla h+Bh|^{\sigma'}}{h^{\sigma'-1}}\,dx + \int_\Omega dh\,dx.$$

Let J and K be the functions given by

$$J(h) = \int_\Omega \frac{|A^t\nabla h+Bh|^{\sigma'}}{h^{\sigma'-1}}\,dx + \int_\Omega dh\,dx,$$

and

$$K(k,h) = \int \frac{|k_1|^{\gamma'}}{h^{\gamma'-1}}\,dx + \int_\Omega \frac{|k_2|^{\sigma'}}{h^{\sigma'-1}}\,dx.$$

Of course, all the integrals occurring in these estimations may be infinite. At last, we define

$$Z = \{h \in \mathcal{D}(\Omega) \text{ and } J(|h|) < \infty\}\ ,$$

$$K = \{k=k_1-\operatorname{div} k_2, (k_1,k_2) \in L^1_c(\Omega) \times (L^1_c(\Omega))^n,\ \exists h \in Z/K(k,|h|) < \infty\}\ .$$

Now we can summarize the natural conditions on the coefficients of L

$a_{i,j}$ and b_j belong to $L^{\sigma'}_{loc}(\Omega)$. (26)

We have the following result about Z and K:

<u>LEMMA 1</u>: Suppose (26), then

(i) $\{\xi^m, \xi \in \mathcal{D}(\Omega),\ \xi \geq 0\} \subset Z$ for $m \geq \max\{2\gamma', \sigma'\}$. For each compact subset C of Ω, $\exists h_c$ in $\mathcal{D}(\Omega) \cap Z$, $h_c \geq 0$ and $h_c = 1$ on C.

(ii) $\exists c > 0$ such that $L(k,h) \leq c(J(h)+K(k,h))$ for all (k,h) in $K \times Z$. (27)

(iii) $\mathcal{D}(\Omega) \subset K$ and if C is a compact subset of Ω, $\exists M$ depending only on C such that

$$L(\zeta,h_c) \leq M\left(\int_\Omega |\zeta|^{\gamma'} dx + 1\right) \text{ for all } \zeta \text{ in } \mathcal{D}(\Omega),\ \mathrm{supp}(\zeta) \subset C. \tag{28}$$

(iv) K and Z satisfy (7).

<u>PROOF</u>: (i) When we compute $J(\xi^m)$ with $m \geq \max\{2\gamma', \sigma'\}$, the denominator disappears and one easily verifies with (26) that the numerator belongs to $L^1(\Omega)$. Take $h_c = \xi^m$ with ξ in $\mathcal{D}(\Omega)$, $\xi \geq 0$, $\xi = 1$ on C.

(ii) Equation (27) can be easily deduced from the inequalities obtained above with the Hölder inequality.

(iii) Let ζ be in $\mathcal{D}(\Omega)$ with $\mathrm{supp}(\zeta) \subset C$. We have

$$K(\zeta,h_c) = \int_\Omega |\zeta|^{\gamma'} dx$$

and so $\zeta \in K$. By (27), we have $L(\zeta,h_c) \leq c(J(h_c) + K(\zeta,h_c))$ which can be rewritten as (28).

(iv) We have $K(k,h) = K(-k,h)$, so the result follows from the definition of K. □

Next, we set

$$X = K,\ K_0 = \mathcal{D}(\Omega),\ Z^+ = \{h \in Z,\ h \geq 0\},\ Y = Z^* = \text{the algebraic dual of } Z,$$

$$D(F^0) = \{u \in W^{1,1}_{loc}(\Omega),\ j(u,\nabla u) \in L^1_{loc}(\Omega)\},$$

$$(F^0(u),h) = \int (-(A\nabla u \cdot \nabla h) - (B \cdot \nabla u)h + j(u,\nabla u)h)dx \text{ for all } h \text{ in } Z.$$

Let us observe that we have by the Hölder inequality

$$|(A\nabla u\cdot\nabla h)+(B\cdot\nabla u)h| \leqq |\nabla u|^{\sigma}|h| + c\,\frac{|A^{t}\nabla h+Bh|^{\sigma'}}{|h|^{\sigma'-1}}\quad dx,$$

and by (24)

$$|\nabla u|^{\sigma}|h| \leqq j(u,\nabla u)|h|,$$

so $(F^{0}(u),h)$ is well defined whenever h belongs to Z.

Notice that X_0' is identifiable with a subspace of $\mathcal{D}'(\Omega)$ and its topology is those of $\mathcal{D}'(\Omega)$. Now, we suppose

$$u \to j(u,\nabla u) \text{ is convex on } D(F^{0}) \text{ into } L^{1}_{loc}(\Omega). \tag{29}$$

LEMMA 2: Let f be in $\mathcal{D}'(\Omega)$ and let K_0, K, Z, Z^{+}, $D(F^{0})$ and F^{0} be given above. Then

(i) assertions (14), (17) hold:

(ii) suppose (29) then (18) holds.

PROOF: (i) We obtain (14) by integration by parts. It is well known that $\mathcal{D}(\Omega)$ is reflexive and so we have to prove the second part of (17). We deduce from the definition of ϕ and from Lemma 1 that for each compact subset C of Ω there exists M(C) such that for all ζ in $\mathcal{D}(\Omega)$ with $\mathrm{supp}(\zeta) \subset C$, we have

$$\phi(\zeta) \leqq M(C)\left(\int_{\Omega} |\zeta|^{\gamma'}\, dx + 1\right) - (f,h_C),$$

which proves that ϕ is bounded on a neighbourhood of 0 in $\mathcal{D}(\Omega)$.

(ii) For u in $D(F^{0})$ and h in Z^{+}, we have

$$\varphi_h(u) = \int_{\Omega} (-(A\nabla u\cdot\nabla h)-(B\cdot\nabla u)h+j(u,\nabla u)h)\, dx,$$

so by (29) φ_h is a convex function. We deduce from the Hölder inequality

$$\int_\Omega ((A\nabla u\cdot\nabla h)+(B\cdot\nabla u)h)\,dx \le c(\varepsilon)\int_\Omega \frac{|A^t\nabla h+Bh|^{\sigma'}}{h^{\sigma'-1}}\,dx + \varepsilon\int_\Omega |\nabla u|^\sigma h\,dx,$$

and so

$$\int (a|u|^\gamma h+(b-\varepsilon)|\nabla u|^\sigma h)\,dx \le \varphi_h(u)+c(\varepsilon)\int_\Omega \frac{|A^t\nabla h+Bh|^{\sigma'}}{h^{\sigma'-1}}\,dx - \int_\Omega dh\,dx.$$

Let u be in $\mathcal{D}'(\Omega)$ such that lim inf $\varphi_h(u)$ is finite for each h in Z^+. We deduce from this estimate that u and ∇u can be obtained as the weak limit in $L^\gamma(\Omega;\, h\,dx)$ and in $L^\sigma(\Omega;\, h\,dx)$ of sequences u_n and ∇u_n where u_n belongs to $D(F^0)$ and $\varphi_h(u_n)$ converges to lim inf $\varphi_h(u)$. We can build a sequence still denoted u_n such that the convergence holds in the strong topology and almost everywhere in supp(h). Then we deduce from the Fatou theorem that $j(u,\nabla u)h$ belongs to $L^1(\Omega)$. Applying this result with $h = h_C$, where C described an exhaustive sequence of compact in Ω, we obtain that u belongs to $W^{1,1}_{loc}(\Omega)$ and so u belongs to $D(F^0)$. By the Fatou theorem, we have also $\lim\ \varphi_h(u_n) \ge \varphi_h(u)$ and so we obtain (18). □

Now we introduce the last assumptions

for each u in $D(F^0)$, $\exists u_n$ in D(F) such that $j(u_n,\nabla u_n) \to j(u,\nabla u)$ in $L^1_{loc}(\Omega)$. (30)

for each open subset U in Ω, $\exists u_n$ in D(F) such that

$$\lim \int_U j(u_n,\nabla u_n)\,dx = +\infty \text{ and } \lim\sup \int_{C/U} j(u_n,\nabla u_n)dx < +\infty \qquad (31)$$

for all compact subset C in Ω.

THEOREM 2: Suppose that j satisfies (24), (29) and $a_{i,j}$ and b_j belong to $L^{\sigma'}_{loc}(\Omega)$. Let f be in $\mathcal{D}'(\Omega)$, consider the following assertions:

$$(f,h) \le \sup\{\int_\Omega (-(A\nabla u\cdot\nabla h)-(B\cdot\nabla u)h+j(u,\nabla u)h)dx;\ u\in C^\infty(\Omega)\} \quad \text{for all } h \text{ in } Z^+, \qquad (32)$$

there exists u in $D(F^0)$ such that

$$\operatorname{div}(A\nabla u)+B\cdot\nabla u \geqq j(u,\nabla u) + f \text{ in } \mathcal{D}'(\Omega) \tag{33}$$

for each k_0 in K, there exists u in $D(F^0)$ such that

$$\int_\Omega (k_{01}(x)u(x) + (k_{02}\cdot\nabla u(x))\, dx = \inf\{L(k,h)-(f,h);\ h \in Z^+\}$$

and (34)

$$\int_\Omega (k_1(x)u(x)+(k_2\cdot\nabla u(x))dx \leqq L(k,h)-(f,h) \text{ for all } (k,h) \text{ in } K \times Z,$$

then (32) and (34) are equivalent and imply (33). Moreover, if we suppose (30), (31) all these assertions are equivalent.

PROOF: First prove that (21) and (33) are equivalent. Let ξ be in $\mathcal{D}(\Omega)$, let $\xi \geqq 0$, and let p_n be a sequence of function in $C^\infty([0, +\infty))$ such that $p_n(r)/r^m$ is a constant on $(0,1/n)$ and $p_n(r)/r = 1$ on $(2/n, +\infty)$; one easily verifies that the functions $p_n(\xi)$ belong to Z^+ and so we can apply (21) with $h = p_n(\xi)$. On the other hand, we can choose p_n such that $p_n'(r) \leqq 1$ for all $r \geqq 0$, and so it is possible to take the limit in (21) when n goes to infinity. We obtain (33). It is clear that (33) implies (21). Now we deduce from Lemmas 1 and 2 that the hypotheses of Theorem 1 are fulfilled and so (32) and (34) are equivalent and imply (33).

We easily deduce from (24) and (30) that (23) holds. Show that (29) implies $Z^+ = Z_f$ (recall that Z_f is defined in (22)). Let h be in Z_f and suppose h nonnegative, we can find $\varepsilon > 0$ and U an open subset of Ω such that $h < -\varepsilon$ on U. Using (31) with C = supp(h) and (24), we obtain that $\lim (F(u_n)\cdot h) = -\infty$ and so $L(0,h) = +\infty$. The result follows from Corollary 1.

REMARK 1: To get the result claimed in the Introduction about a nonnegative solution, we put $j(x,r,p) = +\infty$ if $r \leqq 0$ and we apply Theorem 1.

REFERENCES

[1] P. Baras and M. Pierre, Critères d'existence de solutions positives pour des équations semi-linéaires non monotones, Annales de l'I.H.P. **2**, No. 3 (1985), 185-212.

[2] J.R. Giles, Convex Analysis with Application in Differentiation of Convex Functions, Pitman

P. Baras
INPG Grenoble
TIM3 - Tour des Mathématiques
BP 68 38402 St Martin d'Hères Cedex
France

J. BLAT AND J.M. MOREL

Elliptic problems in image segmentation and their relation to fracture theory

1. INTRODUCTION

Mumford and Shah [11] have suggested that a global variational method can be used for boundary detection in images, motivated by the failure of present local methods for giving psychologically reasonable contours in some non-pathological cases. This method is related to a Markovian model due to S and D. Geman [4]. By an "image" we mean a real function f on a bounded open set R in $\mathbb{R}^2$ (in general, a rectangle or a disk); f(x,y) is the "grey-level" at the point (x,y). An "image segmentation" is a pair (u,C) where C is a set of piecewise C^1 curves, which we call "contours", and u a real function which is regular on the connected components of $R \setminus C$. The curves of C should be the boundaries of the homogeneous areas in the image and u a sort of mean or, more generally, a regularized version of f in the interior of such areas.

According to Mumford and Shah u and C should be obtained by minimizing the functional

$$E(u,C) = \int_{R\setminus C} |\nabla u|^2 + a \int_R (u - f)^2 + b\cdot \text{length } (C).$$

Minimizing this functional corresponds to a quadratic approximation of f by a piecewise "flat" function u. The set of the contours of the smooth zones of u have to be as simple as possible, and this should be achieved by putting their length as a term of the functional. Indeed, Mumford and Shah conjecture that the contours thus obtained have the following geometric property: either the points of C are regular or the singular points are of three types, namely, triple points where three branches meet with 120° angles, boundary points where C meets the boundary of R at a 90° angle, and crack-tips where C comes to an abrupt end. While it is clear that in each component of $R \setminus C$, u is a solution of a linear elliptic equation with homogeneous Neumann boundary condition, there are some other Euler-Lagrange equations derived from the minimality of the contours:

At a point s where C is a C^2 manifold, Mumford and Shah give in [11] the following relation, which relies on one-dimensional elasticity theory:

$$\mathrm{curv}(s) = \text{jump at } s \text{ of } (|\nabla u|^2 + a(u-f)^2).$$

At a "crack-tip" of C we shall derive in Section 2 of this article a formula which expresses its stability considered as a sort of a crack in fracture mechanics. We follow the methods of Knowles for the computation of the energy release rate in a propagating crack [8]. As an application, we give in Section 2 an example of an image in a disk whose segmentation consists of a radius of the disk.

2. EULER-LAGRANGE EQUATIONS AT A CRACK-TIP

2.1. Statement of the problem and notation

For simplicity, we are going to consider an image on an open disk R, with a radial crack C_l on the X-axis of variable length l and starting from the boundary. We take u_l as verifying the minimum of $E(u,C_l)$ and thus the solution of the elliptic equation

$$-\Delta u_l + au_l = af \quad \text{in } R\setminus C_l,$$

$$\partial u_l/\partial n = 0 \quad \text{on } \partial R \text{ and } C_l.$$

We find the necessary conditions for a local minimum with respect to l, i.e. the lengthening or shortening of the contour.

We consider thus

$$E(l) = \int_{R\setminus C_l} |\nabla u_l|^2 + a\int_R (u_l-f)^2 + b\cdot l$$

and the conditions for local minimum will be conditions on the derivative with respect to l of this functional. Only the first two terms of the functional are going to be considered, as the last one is easy; for the rest of the article

$$E(l) = \int_{R\setminus C_l} |\nabla u_l|^2 + a\int_R (u_l-f)^2.$$

We set the origin at the point where the crack meets the boundary of the image, and denote by D_r a small mobile disk of radius r centred at the crack-tip; S_r will denote $R \setminus D_r$. The derivatives with respect to 1 will be noted E', u', etc. Our regularity assumptions on f are that $f \in L^\infty$ and its derivative f_x is in L^2. In fact, it is not difficult to see that the weakest regularity hypothesis for our results to hold is that f is in the space SBV of Special Bounded Variation functions introduced by de Giorgi and Ambrosio [2]. We suppose that $u(x,y,1) = u_1(x,y)$ is continuously differentiable with respect to 1 pointwise in the interior of $R \setminus C_1$. (In the case of the particular shape considered here, this property is easily proved.) Our main theorem is the following formula for E'.

THEOREM 1: Under the hypotheses stated above:

$$E' = \lim_{r \to 0} \left(\int_{\partial D_r} \left(u_1 \frac{\partial u_x}{\partial n} - u_x \frac{\partial u_1}{\partial n}\right) - a \int_{D_r \setminus C_e} (f u_{1x} - f_x u_1) \right).$$

Note that all the terms appearing in it are computable from known local behaviour of the functions u_1 and f. When f is regular enough the second term has limit zero and then the above formula becomes the same given by Gurtin [5] and Knowles [8].

In order to prove the main theorem we need some estimates which we collect in

LEMMA 1:

(a) $u_1 \in H^1 \cap L^\infty$.

(b) $\int_{\partial D_r} u_1 \frac{\partial u_1}{\partial n} \to 0$ as $r \to 0$.

(c) $\int_{D_r} |v'|^2 \leq Cr^2$ as $r \to 0$ where $v' = u' - u_x$.

(d) $\int_{D_r} |\nabla v'|^2 = O(r)$ as $r \to 0$.

2.2 Development of the main theorem

For clarity of exposition, we give first a derivation of the formula for E' using the estimates in Lemma 1. We know that u_1 is a solution of

$$\text{Min}\ \{\int_{R\smallsetminus C_1} |\nabla u_1|^2 + a \int_R (u_1-f)^2 : u \in H^1\}.$$

This linear variational problem has a (unique) solution satisfying the equation

$$-\Delta\ u_1 + a\ u_1 = a\ f \quad \text{in } R\smallsetminus C_1,$$

$$\partial u_1/\partial n = 0 \qquad \text{on } \partial R \text{ and } C_1.$$

From the definition of E(1)

$$E(1) = \lim_{r\to 0} \int_{S_r\smallsetminus C_1} (|\nabla u_1|^2 + a\ (u_1-f)^2).$$

The domain $S_r\smallsetminus C_1$ is regular enough to integrate by parts; then we use the equation satisfied by u_1 and get

$$E(1) = \lim_{r\to 0} (-a \int_{S_r\smallsetminus C_1} f(u_1-f) + \int_{\partial D_r} u_1 \frac{\partial u_1}{\partial n}).$$

Now we use (b) from Lemma 1 and thus

$$E(1) = -a \int_R f\ (u_1-f).$$

Formally, we can write from that

$$E'(1) = -a \int_{R\smallsetminus C_1} f\ u_1'.$$

Note that u' satisfies, at least in the distributional sense,

$$-\Delta u_1' + a\ u_1' = 0 \text{ in } R\smallsetminus C_1,$$

$$\partial u_1'/\partial n = 0 \text{ on } \partial R \text{ and } C_1.$$

However, this formula for E' is not really useful, as u_1' is difficult to compute. Our aim is to obtain a condition which can be checked for given u_1 and f. In order to do that we define

$$E_r'(1) = - a \int_{S_r \setminus C_1} f\, u_1'.$$

This formula makes sense since u_1' is clearly C^∞ on $S_r \setminus C_1$ by the equation satisfied by u_1'. Using again integration by parts, the equations satisfied by u_1 and u_1' and "*dropping the subscript 1 for convenience of notation*", we get

$$E_r'(1) = - \int_{S_r \setminus D_1} u'(-\Delta u + au) = - \int_{S_r \setminus C} u(-\Delta u' + au') +$$

$$+ \int_{\partial(S_r \setminus C)} (u' \frac{\partial u}{\partial n} - u \frac{\partial u'}{\partial n}) = \int_{\partial D_r} (u \frac{\partial u'}{\partial n} - u \frac{\partial u}{\partial n}) =$$

$$= \int_{\partial D_r} (u \frac{\partial u_x}{\partial n} - u_x \frac{\partial u}{\partial n}) + \int_{\partial D_r} (u \frac{\partial (u'-u_x)}{\partial n} - (u'-u_x) \frac{\partial u}{\partial n}).$$

Following Knowles, we set

$$v(r,\theta,1) = u(-1 + r \cos \theta,\ r \sin \theta, 1).$$

The underlying idea is that v should be more regular (with respect to 1) than u. Indeed by considering v we are eliminating the effect of the moving crack-tip by changing the origin of coordinates. For instance, if the region is an infinite strip and the crack is parallel to the boundary, v is obviously stationary.

Then we have

$$u' - u_x = v'$$

and thus

$$\int_{\partial D_r} (u \frac{\partial u_x}{\partial n} - u_x \frac{\partial u}{\partial n}) + \int_{\partial D_r} (v \frac{\partial v'}{\partial n} - v' \frac{\partial v}{\partial n}) = E_r'(1).$$

Hence we get formally the following formula

$$\int_{\partial D_r} (u \frac{\partial u_x}{\partial n} - u_x \frac{\partial u}{\partial n}) + \int_{\partial D_r} (v \frac{\partial v'}{\partial n} - v' \frac{\partial v}{\partial n}) - a \int_{D_r \setminus C} f\, u' = E'(1).$$

To see that the formula is correct, we estimate the last two terms in it, by using again integration by parts on $(D_r \setminus D_s) \setminus C_1$ for $0 < s < r$ and get

$$\int_{\partial D_r \setminus \partial D_s} (v \frac{\partial v'}{\partial n} - v' \frac{\partial v}{\partial n}) - a \int_{D_r \setminus D_s} f\, u' = a \int_{D_r \setminus D_s} (fv' + f_x v - fu')$$

$$= a \int_{D_r \setminus D_s} (f_x v - fu_x)$$

(where we have dropped everywhere "$\setminus C_1$" for convenience of notation, a convention that we keep from now on).

One sees then that

$$K(r) = \int_{\partial D_r} (v \frac{\partial v'}{\partial n} - v' \frac{\partial v}{\partial n}) - a \int_{D_r} (fu_x - f_x v - fu')$$

does not depend on $r > 0$, and by substitution in E'

$$\int_{\partial D_r} (u \frac{\partial u_x}{\partial n} - u_x \frac{\partial u}{\partial n}) + K(r) + a \int_{D_r} (fu_x - f_x u) = E'(1).$$

We now show that $K(r) = 0$ and the limit as $r \to 0$ of the last term in E' is 0 when f is regular enough and we get our formula:

$$E' = \lim_{r \to 0} (\int_{\partial D_r} (u \frac{\partial u_x}{\partial n} - u_x \frac{\partial u}{\partial n})).$$

In order to prove the latter statements, we call $K(r) = k$ and integrate between 0 and r, we get

$$\int_0^r k = \int_{D_r} (vv'_r - v'v_r) - \int_0^r a \int_{D_r} (fv' + f_x v).$$

Using Schwarz's inequality,

$$k^2r^2 \leqq \int_{D_r} |\nabla v|^2 \int_{D_r} |v'|^2 + \int_{D_r} |v|^2 \int_{D_r} |\nabla v'|^2 + r^2(a \int_{D_r} (|fv| + |f_x v|)).$$

Due to our estimates on v and v' collected in Lemma 1, we see that the right-hand side of this inequality is $O(r)^2$ and thus $k = 0$, and with the same estimates we can prove when f is regular, e.g. f bounded and f_x in L^2 that

$$\lim_{r \to 0} \int_{D_r} (fu_x - f_x u) = 0$$

which ends the proof of our main theorem. □

The following section is devoted to the proof of remaining points.

2.3. Justification of the development

PROOF OF LEMMA 1(a): Obviously, $u_1 \in H^1$ by the formulation of the minimization problem. In order to prove that u_1 is bounded, we see that for each $u \in H^1$ the value of E(1) decreases if we replace u by u^* defined as follows:

$$u^*(x,y) = \sup f \quad \text{if } u(x,y) > \sup f,$$

$$u^*(x,y) = \inf f \quad \text{if } u(x,y) < \inf f,$$

$$u^*(x,y) = u(x,y) \quad \text{otherwise.}$$

This shows that $|u_1(x,y)| \leqq |f|_\infty$.

PROOF OF LEMMA 1(b): This part of the lemma is essentially contained in Lemma 2 below.

PROOF OF LEMMA 1(c) and (d): As u = v by definition, from part (a) of Lemma 1 we have that

$$\int_{D_r} v_1^2 \leqq Cr^2 \quad \text{and} \quad \int_{D_r} |\nabla v_1|^2 \to 0 \text{ as } r \to 0. \qquad \square$$

We intend to prove the same type of estimates for v_1', following the lines of Knowles [8]. We begin by stating

LEMMA 2: (Knowles [8]: Assume that $\phi: D_p \setminus C_1 \to \mathbb{R}$ is C^2 in the interior of the region and up to the boundary and

(i) for each r such that $(0 < r \leqq p)$ $(1/r)\int_{\partial D_r} \phi^2 \leqq C$;

(ii) $-\Delta\phi = 0$ on $D_p \setminus C_1$;

(iii) $\partial\phi/\partial n = 0$ on both sides of C_1.

Then:

(a) $E(r) = \int_{D_r} |\nabla\phi|^2$ is finite for $0 < r \leqq p$;

(b) $E(r) = \int_{\partial D_r} \phi\phi_r \geqq 0$;

(c) $E(r) \to 0$ as $r \to 0$.

This lemma is proved by getting a differential inequality for E(r). It is easily seen that similar conclusions hold if we replace $-\Delta$ by $-\Delta + a$ and add to the integrand in E(r) the term $a\phi^2$.

In Knowles [8] where a homogeneous problem (i.e. with f = 0) is studied, the procedure to get the estimates for v_1' and $\nabla v_1'$ is to see that as $(v_1 - v_{1'})/(1-1')$ verifies the hypotheses of Lemma 2, the inequality of the conclusion (b) is verified; this inequality is preserved when taking the limit $1' \to 1$ and from this inequality for v_1' one gets the estimates of (i) in the lemma and using the lemma again we have the estimate for $\nabla v_1'$.

In our more complex situation we argue as follows: We define $W_{1,1'}$ as the solution obtained by minimization of the inhomogeneous problem corresponding to the finite increment quotient $(f_1 - f_{1'})/(1-1')$ and by $V_{1,1'}$ the difference $(v_1 - v_{1'})/(1-1') - W_{1,1'}$. We show that the type of estimates of (c) and (d) of Lemma 1 hold for $w_1 = \lim_{1' \to 1} W_{1,1'}$. Since $V_{1,1'}$ and its limit V_1 are solutions of homogeneous problems and the arguments of Knowles described above allow us to get the estimates for V_1. As $v_1' = V_1 + w_1$, the desired result follows immediately.

More formally, we define for a sufficiently small p

$$D_p \setminus C = \{(r,\theta): 0 < r \leqq p,\ 0 < \theta < 2\pi\};$$

and

$$f_{1,1'}(r,\theta) = (f(-1 + r\cos\theta, r\sin\theta) - f(-1' + r\cos\theta, r\sin\theta))/(1-1')$$

We take as $W_{1,1'}$ the solution of the problem

$$\mathrm{Min}\{\int_{D_p \setminus C} (\tfrac{1}{2}|\nabla\phi|^2 + \tfrac{1}{2}a\phi^2 - a\, f_{1,1'}\phi): \phi \in H^1\}.$$

<u>LEMMA 3</u>: For each r such that $0 < r \leqq p$ we have

$$\frac{1}{r}\int_{\partial D_r} w_1^2 \leqq C.$$

<u>PROOF</u>: In order to prove this inequality, we shall see that it is satisfied by each $W_{1,1'}$ and that the inequality is preserved when taking limits. Integrating by parts the equation satisfied by $W_{1,1'}$ we have

$$\int_{D_r} |\nabla W_{1,1'}|^2 - a\int_{\partial D_r} W_{1,1'}(W_{1,1'})_r = a\int_{D_r} f_{1,1'}W_{1,1'}$$

and this integration is justified because of part (c) of the lemma of Knowles. Then we have that

$$a\int_{\partial D_r} W_{1,1'}(W_{1,1'})_r + a\int_{D_r} f_{1,1'}\, W_{1,1'} \geqq 0$$

and thus, by integration

$$(\frac{1}{r}\int_{\partial D_r} W_{1,1'}^2 + \int_{D_r} (r - |x|)f_{1,1'}W_{1,1'})' \geqq 0.$$

If we integrate this expression between r and r_0 with $r_0 > r$ fixed,

$$\frac{1}{r}\int_{\partial D_r} W_{1,1'}^2 \leqq C_{1'} - \int_{D_r} (r - |x|) f_{1,1'} W_{1,1'},$$

where $C_{1'}$ is the value for r_0. Now we note that $f_{1,1'} \to f_x$ in L^2, $W_{1,1'}$ tends uniformly to w_1 off the origin, $C_{1'}$ has a limit because of these facts and from the variational formulation we get

$$\int_{D_r} |\nabla W_{1,1'}|^2 + a \int_{D_r} W_{1,1'}^2 \leqq a \, \|f_{1,1'}\|_{L^2} \|W_{1,1'}\|_{L^2}.$$

We can fix some element of the sequence to be minimized and thus it is clear that $W_{1,1'}$ is bounded in L^2 (and in H^1). Hence we can take limits as $1' \to 1$ and we have

$$\frac{1}{r}\int_{\partial D_r} W_1^2 \leqq C,$$

where C depends on the L^2 bounds of f_x and w_1 .

Integrating from 0 to r, we have for w_1 an estimate similar to that of (c) from Lemma 1. Using now the results from Knowles for V_1, the proof of this part of the lemma is finished. As $W_{1,1'}$ is bounded in H^1, extracting a subsequence, we see that the limit is in H^1 and using again the results from Knowles for the homogeneous problem, Lemma 1 is completely proved. □

3. A CASE STUDY

Here we give an example of a smooth picture having a crack with a free tip in its segmentation. According to Mumford and Shah, an important aim of using global methods in segmentation is to obtain psychologically more reasonable contours and thus it is interesting to give concrete examples.

We take an eyebrow-like picture in the form of a function $f(r,\theta)$ defined on a disk D of radius 1, which is black for small positive θ, white for small negative θ, and keeps smooth as θ increases from 0 to 2π; we call C the radius $\{(r,0): r \in [0,1]\}$.

Our almost explicit f and u are constructed as follows: It is well known (Leguillon and Sánchez-Palencia [9]) that a harmonic function v on $D \setminus C$ verifying homogeneous Neumann conditions on C has the following Fourier expansion:

$$v = \sum_{n\in\mathbf{Z}} a_n r^{(n+1)/2} \cos(\tfrac{n+1}{2})\theta, \quad r \in [0,1], \ \theta \in (0,2\pi).$$

(Note that $-\Delta v + v = v$ in $D\setminus C$.) We start from this v to construct our f and u.

For f to be in H^1 we take only $n \geq 0$, and u is going to behave as the first term of the expansion, which has the highest "elastic energy" near the origin. Set

$$u_1(r,\theta) = Ar^{1/2} \cos(\theta/2) + Br^{3/2} \cos(3\theta/2)$$

and define f by

$$\begin{aligned} -\Delta f &= -u_1 && \text{in } D\setminus C, \\ \partial f/\partial n &= -\partial u_1/\partial n && \text{on } (D\setminus C). \end{aligned}$$

Then we set

$$u = u_1 + f$$

and thus

$$\begin{aligned} -\Delta u + u &= -u_1 + u_1 + f = f && \text{in } D\setminus C, \\ \partial u/\partial n &= 0 && \text{on } \partial(D\setminus C). \end{aligned}$$

Note that f appears as a regularized version of $-u_1$ and is indeed eyebrow-like.

Now we determine A and B such that the necessary conditions for minimization hold for u and the "elastic energy" corresponding to (u,C) is less than that of $(f,\emptyset)$.

Since u_1, f and u verify an antisymmetry condition with respect to C, the condition given in the Introduction holds as the curvature and the jump are zero. The formula in Theorem 1 reads as

$$1 = \lim_{r\to 0} \int_{\partial D_r} u\, u_{xr} - u_x u_r = \lim_{r\to 0} \frac{1}{4} A^2 r^{-1} \int_0^{2\pi} 2r \cos^2 \frac{\theta}{2}\, d\theta = \frac{A^2}{2} \pi .$$

Thus if $A = (2/\pi)^{1/2}$, the necessary condition in Theorem 1 holds. It is not easy to decide whether (u,C) in fact realizes a minimum of energy.

Anyway, we prove that for B large enough $E(f,\emptyset) < E(u,C)$ and thus a segmentation with a crack is likely to appear. In fact, one has

$$\int_D |\nabla f|^2 = - \int_D f u_1 - \int_{\partial D} f \frac{\partial u_1}{\partial n} ,$$

and

$$\int_{D\setminus C} |\nabla u|^2 + \int_D (u - f)^2 = - \int_D f(u - f) = - \int_D f u_1 .$$

Thus

$$E(u,C) - E(f,\emptyset) = \int_{\partial D} f \frac{\partial u_1}{\partial n} + 1 .$$

It is easily seen that the first term of this difference behaves like $-cB^2$ with $c > 0$ when B is large.

The interpretation of this result is that as the image f gets a higher contrast, it becomes interesting to tear it and produce a "crack".

ACKNOWLEDGEMENTS: We thank Thierry Gallouët and Pierre Baras for useful conversations. After completing this work we learned that Carriero, Leaci, Pallara and Pascali [1] have obtained related results in the case without forcing term ($f = 0$). Mumford and Shah are preparing a new paper where, among many other results, significant parts of the preceding study will be treated.

This work has been partially supported by U.S. Army under Contract DAJA45-88-C-0009.

REFERENCES

[1] Carriero, M., Leaci, D., Pallara, E. and Pascali, E., Euler conditions for a minimum problem with discontinuity surfaces, preprint Universite degli studi di Lecce, Italy.

[2] De Giorgi, E. and Ambrosio, L., Un nuovo tipo di funzionale del calcolo delle variazioni, Atti Acad. Naz. Lincei Rend. Cl. Sci. Fis. Mat. Natur. (to appear).

[3] Fonseca, L. and Tartar, L., The gradient theory of phase transitions for systems with two potential wells, Carnegie Mellon Univ. (to appear).

[4] Geman, S. and Geman, D., Stochastic relaxation, Gibbs distributions and the Bayesian restoration of images, IEEE PAMI 6 (1984).

[5] Gurtin, M.E., On the energy release rate in quasi-static elastic crack propagation. Journal of Elasticity, 9, No. 2 (1979).

[6] Grisvard, P., Elliptic Problems in Nonsmooth Domains, Pitman, Advanced Publishing Program, 1985.

[7] Henrot, A. and Pierre, M., Existence d'états d'équilibre en formage électromagnétique, Colloque International Problèmes Elliptiques et Paraboliques Nonlinéaires, Université de Nancy 1, Mars 1988.

[8] Knowles, J.K., A note on the energy release rate in quasi-static elastic crack propagation, SIAM J. Appl. Math., 41 (1981), 401-412.

[9] Leguillon, D. and Sánchez-Palencia, E., Computation of Singular Solutions in Elliptic Problemes and Elasticity. Wiley and Sons, 1987.

[10] Modica, L., Gradient theory of phase transitions with boundary contact energy. Ann. Inst. Poincaré, 4, No. 5 (1987), 487-512.

[11] Mumford, D. and Shah, J., Boundary detection by minimizing functionals, IEEE Conference on Computer Vision and Pattern Recognition, San Francisco, 1985.

J. Blat
Dept. Matemàtiques i
Institut d'Estudis Avaneats
(CSIC-UIB)
Universitat Illes Balears
07071 Palma de Mallorca
Spain

J.M. Morel
CEREMADE
Université de Paris Dauphine
Place de Lattre de Tassigny
75775 Paris Cedex 16
France

L. BOCCARDO, J.I. DIAZ, D. GIACHETTI AND F. MURAT

Existence of a solution for a weaker form of a nonlinear elliptic equation

ABSTRACT. Consider the nonlinear elliptic equation

$$-\mathrm{div}(A \text{ grad } u) - \mathrm{div}(\phi(u)) = f \text{ in } \Omega,\ u \in H_0^1(\Omega), \tag{1}$$

where A is a $(L^\infty(\Omega))^{N\times N}$ coercive matrix, $f \in H^{-1}(\Omega)$ and $\phi \in (C^0(R))^N$; no growth restriction is assumed on ϕ; thus the term $\mathrm{div}(\phi(u))$ cannot be understood in the distributional sense.

In this paper we prove the existence of a solution of

$$\begin{aligned}&[-\mathrm{div}(A \text{ grad } u)]h(u) - \mathrm{div}(\phi(u))h(u)) + \phi(u)h'(u)\text{grad } u\\ &= fh(u) \text{ in } \Omega,\ \forall h \in C_c^1(R),\ u \in H_0^1(\Omega)\end{aligned} \tag{2}$$

where in contrast with (1) every term has a meaning in $\mathcal{D}'(\Omega)$ if $u \in H_0^1(\Omega)$. Equation (2) is a weaker form of the original problem, obtained in a formal way through a pointwise multiplication of (1) by h(u).

SUNTO. Consideriamo l'equazione ellittica non lineare

$$-\mathrm{div}(A \text{ grad } u) - \mathrm{div}(\phi(u)) = f \text{ in } \Omega,\ u \in H_0^1(\Omega), \tag{1}$$

con A matrice coerciva a coefficienti limitati, $f \in H^{-1}(\Omega)$e $\phi \in (C^0(R))^N$; non facciamo su ϕ nessuna ipotesi di crescenza; in tal caso il termine $\mathrm{div}(\phi(u))$ non è, a priori, una distribuzione.

In quest' articolo proviamo l'esistenza di una soluzione del problema

$$\begin{aligned}&[-\mathrm{div}(A \text{ grad } u)]h(u) - \mathrm{div}(\phi(u))h(u)) + \phi(u)h'(u)\text{grad } u\\ &= fh(u) \text{ in } \Omega,\ \forall h \in C_c^1(R),\ u \in H_0^1(\Omega)\end{aligned} \tag{2}$$

dove invece ogni termine ha un senso in $\mathcal{D}'(\Omega)$. L'equazione (2) è una forma indebolita del problema originale, ottenuta in modo formale moltiplicando puntualmente (1) per h(u).

RESUMEN. Consideramos la ecuación elíptica no lineal siguiente

$$-\mathrm{div}(A \text{ grad } u) - \mathrm{div}(\phi(u)) = f \text{ en } \Omega,\ u \in H_0^1(\Omega), \tag{1}$$

siendo A una matriz coerciva con coeficientes en L^∞, $f \in H^{-1}(\Omega)$ y $\phi \in (C^0(R))^N$; no hacemos ninguna hipóthesis sobre el crecimiento de ϕ en el infinito; por lo que el término $\mathrm{div}(\phi(u))$ no puede ser definido como una distribución.

En este artículo demostramos la existencia de una solución de

$$\begin{aligned}&[-\mathrm{div}(A \text{ grad } u)]h(u) - \mathrm{div}(\phi(u))h(u)) + \phi(u)h'(u)\text{grad } u\\ &= fh(u) \text{ en } \Omega,\ \forall h \in C_c^1(R),\ u \in H_0^1(\Omega)\end{aligned} \tag{2}$$

donde todos los términos están bien definidos en $\mathcal{D}'(\Omega)$. La ecuación (2) es una forma débil del problema original obtenida de manera formal al multiplicar (1) por h(u) en todo punto.

RÉSUMÉ. Considérons l'équation elliptique non linéaire

$$-\mathrm{div}(A \text{ grad } u) - \mathrm{div}(\phi(u)) = f \text{ dans } \Omega,\ u \in H_0^1(\Omega), \tag{1}$$

où A est une matrice coercive à coefficients L^∞, $f \in H^{-1}(\Omega)$ et $\phi \in (C^0(R))^N$; nous ne faisons aucune hypothèse sur la croissance de ϕ à l'infini; le terme $\mathrm{div}(\phi(u))$ n'a donc a priori aucune raison d'être une distribution.

Nous démontrons dans cet article l'existence d'une solution de

$$\begin{aligned}&[-\mathrm{div}(A \text{ grad } u)]h(u) - \mathrm{div}(\phi(u))h(u)) + \phi(u)h'(u)\text{grad } u\\ &= fh(u) \text{ dans } \Omega,\ \forall h \in C_c^1(R),\ u \in H_0^1(\Omega)\end{aligned} \tag{2}$$

où, contrairement à (1), chaque terme a un sens dans $\mathcal{D}'(\Omega)$ dès que $u \in H_0^1(\Omega)$. L'équation (2) est une forme affaiblie du problème original obtenue de façon formelle en multipliant ponctuellement (1) par h(u).

1. INTRODUCTION AND MAIN RESULTS

This paper investigates the existence of a solution for the following non-linear elliptic problem:

$$-\mathrm{div}(A \text{ grad } u) - \mathrm{div}(\phi(u)) = f \text{ in } \Omega, \tag{1.1}$$

$$u = 0 \text{ on } \partial\Omega. \tag{1.2}$$

Here Ω denotes a bounded open subset of $\mathbb{R}^N$, and A is a $N \times N$ coercive matrix with components in $L^\infty(\Omega)$, i.e. there exists $\alpha \in \mathbb{R}$, $\alpha > 0$, such that

$$A \in (L^\infty(\Omega))^{N \times N}, \tag{1.3}$$

$$A(x)\xi\xi \geq \alpha|\xi|^2, \quad \forall\xi \in \mathbb{R}^N, \text{ a.e. } x \in \Omega; \tag{1.4}$$

the right-hand side f of equation (1.1) is assumed to satisfy

$$f \in H^{-1}(\Omega). \tag{1.5}$$

Finally, let ϕ be a continuous function defined on $\mathbb{R}$ with values in $\mathbb{R}^N$, i.e.

$$\phi \in (C^0(\mathbb{R}))^N. \tag{1.6}$$

The main feature of the problem under consideration is that *no growth restriction is assumed on* ϕ.

It is natural to seek a solution u of (1.1), (1.2) which belongs to $H_0^1(\Omega)$ since the right-hand side f of (1.1) belongs to $H^{-1}(\Omega)$. But when u is only in $H_0^1(\Omega)$ there is no reasonable ground for $\phi(u)$ to be in $(L^1(\Omega))^N$ since no growth restriction is assumed on ϕ. Hence $\mathrm{div}(\phi(u))$ may be ill defined, even as a distribution.

This obstacle is bypassed by solving some weaker problem, obtained through pointwise multiplication of the original equation (1.1) by h(u) where h belongs to $C_c^1(\mathbb{R})$, the class of the $C^1(\mathbb{R})$ functions with compact support.

THEOREM 1.1: Assume that (1.3), (1.4), (1.5) and (1.6) hold true. Then there exists a solution u of

$$u \in H_0^1(\Omega), \tag{1.7}$$

$$[-\operatorname{div}(A \operatorname{grad} u)]h(u) - \operatorname{div}(\phi(u)h(u)) + \phi(u)h'(u) \operatorname{grad} u$$
$$= fh(u) \text{ in } \mathcal{D}'(\Omega), \ \forall h \in C_c^1(R). \quad \square \tag{1.8}$$

In the equation (1.8) every term is meaningful in the distributional sense; indeed, for h in $C_c^1(R)$ and u in $H_0^1(\Omega)$, h(u) belongs to $H^1(\Omega)$; thus for f in $H^{-1}(\Omega)$ the product fh(u) is the distribution defined by

$$\langle fh(u),\varphi\rangle_{\mathcal{D}'(\Omega),\mathcal{D}(\Omega)} = \langle f,\varphi\, h(u)\rangle_{H^{-1}(\Omega),H_0^1(\Omega)}, \quad \forall\varphi \in \mathcal{D}(\Omega);$$

the same holds true for [- div(A grad u)] h(u) since -div(A grad u) belongs to $H^{-1}(\Omega)$. Further because ϕh and $\phi h'$ belong to the class $C_c^0(R)$ of continuous functions with compact support, $\phi(u)$ h(u) and $\phi(u)$ h'(u) belong to $(L^\infty(\Omega))^N$ for any measurable u, which implies that $-\operatorname{div}(\phi(u)h(u))$ and $\phi(u)h'(u)$ grad u are respectively a distribution (in $W^{-1,\infty}(\Omega)$) and a $L^2(\Omega)$ function.

Equation (1.8) follows formally from (1.1) by multiplying by h(u) since

$$[-\operatorname{div}(\phi(u))]h(u) = -\operatorname{div}(\phi(u)h(u)) + \phi(u)h'(u) \operatorname{grad} u. \tag{1.9}$$

Note, however, that in contrast with the right-hand side, the left-hand side of (1.9) does not make sense when $h \in C_c^1(R)$. Thus (1.8) is to be viewed as a weaker form of (1.1).

The original equation (1.1) will be recovered whenever $h(u) \equiv 1$ (which does not belong to $C_c^1(R)$!) can be used in (1.8); such is not usually the case in general, except when stronger (regularity) requirements are met by u.

<u>THEOREM 1.2</u>: Assume that (1.3), (1.4), (1.5) and (1.6) hold true and define $\tilde{\psi} \in C^1(R)$ by

$$\tilde{\psi}(t) = \int_0^t |\phi(s)|ds. \tag{1.10}$$

Let u be a solution of (1.7), (1.8) such that:

$$\phi(u) \in (L^1_{loc}(\Omega))^N, \tag{1.11}$$

$$\tilde{\psi}(u) \in L^1_{loc}(\Omega). \tag{1.12}$$

Then u is a (usual weak) solution of the original problem (1.1), (1.2). □

We do not know if (1.12) is a necessary condition for Theorem 1.2; (1.11) seems to be necessary to lend a distributional meaning to $\mathrm{div}(\phi(u))$.

Consider now a solution u of (1.1). Formal multiplication of (1.1) by u and integration by parts yields

$$\int_\Omega A \operatorname{grad} u \operatorname{grad} u \, dx + \int_\Omega \phi(u) \operatorname{grad} u \, dx = \langle f,u\rangle. \tag{1.13}$$

Define $\tilde{\phi} \in (C^1(R))^N$ as

$$\tilde{\phi}(t) = \int_0^t \phi(s)ds.$$

Then, formally, $\mathrm{div}(\tilde{\phi}(u)) = \phi(u) \operatorname{grad} u$ and since $\tilde{\phi}(0) = 0$

$$\int_\Omega \phi(u) \operatorname{grad} u \, dx = \int_\Omega \mathrm{div}(\tilde{\phi}(u))dx = \int_{\partial\Omega} \tilde{\phi}(0)n \, ds = 0; \tag{1.14}$$

thus

$$\int_\Omega A \operatorname{grad} u \operatorname{grad} u \, dx = \langle f,u\rangle. \tag{1.15}$$

Let us stress that most of the operations performed before are purely formal. However, relation (1.15) (and even an extension of (1.15)) can be proved whenever u is a solution of (1.7), (1.8).

<u>THEOREM 1.3</u>: Assume that (1.3), (1.4), (1.5) and (1.6) hold true and that u is a solution of (1.7), (1.8). Then

$$\int_\Omega s'(u) A \operatorname{grad} u \operatorname{grad} u \, dx = \langle f,s(u)\rangle_{H^{-1}(\Omega),H^1_0(\Omega)} \tag{1.16}$$

for any Lipschitz continuous, piecewise $C^1(R)$ function s such that $s(0) = 0$. □

Note that Theorem 1.3 holds true for any solution of (1.7), (1.8), and not only for the solution that we will construct by approximation in the proof of Theorem 1.1; for the latter, (1.16) follows immediately from Theorem 2.1 below.

Existence results for (1.1), (1.2) were proved by Boccardo and Giachetti [4], [5] under a smoothness assumption on the right-hand side of (1.1) as well as a convenient growth condition on ϕ: indeed a regularity result permits constructing a solution u in $L^{p^*}(\Omega)$ whenever f lies in $W^{-1,p}(\Omega)$, $p \geq 2$; the condition $|\phi(t)| \leq c(t^{p^*}+1)$ then implies that $\phi(u)$ belongs to $(L^1(\Omega))^N$, and that u is a solution of the original equation (1.1). The uniqueness of the solution for equations of the type of (1.1) was investigated by Carrillo and Chipot [9], Carrillo [8] and Chipot and Michaille [10]; in these papers the solution u belongs to $L^\infty(\Omega)$ or ϕ is assumed to grow at most linearly at infinity.

The weaker form (1.8) of the problem (1.1) is very similar to the idea of "*renormalized solution*" introduced by Di Perna and Lions in their important papers [12], [13] when investigating the existence of solutions for the Boltzmann equation. It is also reminiscent of the introduction by Benilan et al. [1] of the space $T^{1,p}(\Omega)$ in the study of the existence and uniqueness of a solution for $-\text{div}(|\text{grad } u|^{p-2} \text{ grad } u) = f$ with f in $L^1(\Omega)$. Finally, it should be mentioned that this weaker form is related to the idea of entropy solutions for scalar nonlinear hyperbolic equations of the Burgers' type.

Our proof of Theorem 1.1 starts with an approximation ϕ^ε of ϕ which is bounded on R. In this case the above performed formal operations are licit and (1.15) provides a $H_0^1(\Omega)$ bound for the corresponding solution u^ε. The key point is then to prove that u^ε is actually a compact sequence for the strong topology of $H_0^1(\Omega)$; this is achieved through the use of nonlinear (with respect to u^ε) test functions, in a spirit closely related to Bensoussan, Boccardo and Murat [2]. Passing to the limit to obtain Theorem 1.1 is then easy.

The proof of Theorem 1.2 consists in observing that assumptions (1.11), (1.12) allow us to pass to the limit in (1.8) for a sequence of functions h_ε that converges to 1. In a similar manner the proof of Theorem 1.3 consists in approximating s(u) by s(u) $h_\varepsilon(u)$ with h_ε converging to 1. (Use of such functions h (u) has already been made in Boccardo, Murat and Puel [7].)

Extensions of the present work to general Leray-Lions operators as well as to parabolic equations will be given in our forthcoming paper [3].

2. PROOF OF THEOREM 1.1

Define T_m to be the truncation to level $m > 0$, i.e.

$$T_m(t) = \begin{cases} t & \text{if } |t| \leq m, \\ m \frac{t}{|t|} & \text{if } |t| \geq m, \end{cases} \tag{2.1}$$

and ϕ^ε, to be the following approximation ϕ^ε of ϕ:

$$\phi^\varepsilon(t) = \phi(T_{1/\varepsilon}(t)). \tag{2.2}$$

Consider the nonlinear elliptic equation

$$\begin{aligned} &-\mathrm{div}(A \text{ grad } u^\varepsilon) - \mathrm{div}(\phi^\varepsilon(u^\varepsilon)) = f \text{ in } \Omega, \\ &u^\varepsilon \in H_0^1(\Omega). \end{aligned} \tag{2.3}$$

Since ϕ^ε lies in $(C^0(\mathbf{R}) \cap L^\infty(\mathbf{R}))^N$, a simple application of Schauder's fixed point theorem in $L^2(\Omega)$ implies that (2.3) has (at least) one solution.

Recall now the following:

LEMMA 2.1: Let θ be a $(L^\infty(\mathbf{R}))^N$, piecewise continuous function and v belong to $H_0^1(\Omega)$. Define the Lipschitz continuous, piecewise $(C^1(\mathbf{R}))^N$ function $\tilde{\theta}$ by:

$$\tilde{\theta}(t) = \int_0^t \theta(s) \, ds. \tag{2.4}$$

Then

$$\begin{aligned} &\tilde{\theta}(v) \in (H_0^1(\Omega))^N, \\ &\text{grad } v = 0 \text{ a.e. on the set } \{x \in \Omega | v(x) = c\} \text{ for any } c \in \mathbf{R}, \\ &\mathrm{div}(\tilde{\theta}(v)) = \theta(v) \text{ grad } v \text{ in } \Omega, \end{aligned} \tag{2.5}$$

$$\int_\Omega \theta(v) \text{ grad } v \, dx = 0. \qquad \square \tag{2.6}$$

This lemma is a classical result (see e.g. Kinderlehrer and Stampacchia [11, p. 54] or Boccardo and Murat [6, Theorem 4.2]). The assertion (2.6) follows from Stokes' theorem

$$\int_\Omega \theta(v) \text{ grad } v \, dx = \int_\Omega \text{div}(\tilde{\theta}(v))dx = 0,$$

since $\tilde{\theta}(v) \in (H_0^1(\Omega))^N$.

Multiplication of (2.3) by u^ε and integration by parts yields in view of (2.5), (2.6)

$$\langle -\text{div}(\phi^\varepsilon(u^\varepsilon)), u^\varepsilon \rangle = \int_\Omega \phi^\varepsilon(u^\varepsilon) \text{ grad } u^\varepsilon \, dx = 0, \tag{2.7}$$

since ϕ^ε lies in $(C^0(R) \cap L^\infty(R))^N$ [contrast (2.7), which is licit since ϕ^ε belongs to $(C^0(R) \cap L^\infty(R))^N$ with (1.14), which is formal since ϕ only belongs to $(C^0(R))^N$]. Thus we have

$$\int_\Omega A \text{ grad } u^\varepsilon \text{ grad } u^\varepsilon \, dx = \langle f, u^\varepsilon \rangle \tag{2.8}$$

and the coersiveness (1.4) implies that

$$\| u^\varepsilon \|_{H_0^1(\Omega)} \leq \frac{1}{\alpha} \| f \|_{H^{-1}(\Omega)}.$$

At the possible expense of extracting a subsequence (still denoted by ε), we conclude that

$$u^\varepsilon \longrightarrow u \text{ in } H_0^1(\Omega) \text{ weakly and a.e. in } \Omega \text{ as } \varepsilon \to 0. \tag{2.9}$$

We shall prove the following:

<u>THEOREM 2.1</u>: The subsequence u^ε tends *strongly* to u in $H_0^1(\Omega)$. □

The proof of Theorem 2.1 is based on two lemmas.

LEMMA 2.2: Define for $k > 0$ the set

$$E_k^\varepsilon = \{x \in \Omega \,|\, |u^\varepsilon(x)| \geqq k\}. \tag{2.10}$$

Then for any fixed $k > 0$

$$\limsup_{\varepsilon \to 0} \int_{E_k^\varepsilon} |\text{grad } u^\varepsilon|^2 \, dx \leqq \frac{1}{\alpha} \langle f, u - T_k(u)\rangle. \quad \square$$

PROOF: Consider the test function

$$v^\varepsilon = u^\varepsilon - T_k(u^\varepsilon) \in H_0^1(\Omega).$$

Defining χ_k as

$$\chi_k(t) = \begin{cases} 0 & \text{if } |t| < k, \\ 1 & \text{if } |t| \geqq k, \end{cases} \tag{2.12}$$

we obtain by virtue of Lemma 2.1:

$$\text{grad } v^\varepsilon = \chi_k(u^\varepsilon) \text{ grad } u^\varepsilon.$$

Application of Lemma 2.1 to the $(L^\infty(R))^N$, piecewise continuous function $\theta(s) = \phi^\varepsilon(s)\chi_k(s)$ yields

$$\langle -\text{div}(\phi^\varepsilon(u^\varepsilon)), v^\varepsilon \rangle = \int_\Omega \phi^\varepsilon(u^\varepsilon)\chi_k(u^\varepsilon) \text{ grad } u^\varepsilon = 0.$$

Thus multiplication of (2.3) by v^ε and integration by parts yields:

$$\int_\Omega \chi_k(u^\varepsilon) \, A \text{ grad } u^\varepsilon \text{ grad } u^\varepsilon \, dx = \langle f, u^\varepsilon - T_k(u^\varepsilon)\rangle. \tag{2.13}$$

Since the right-hand side of (2.13) tends to $\langle f, u - T_k(u)\rangle$ as $\varepsilon \to 0$, the coerciveness assumption (1.4) implies Lemma 2.2.

LEMMA 2.3: Define for $i > 0$ and $j > 0$ the set

$$F_{ij}^{\varepsilon} = \{x \in \Omega \mid |u^{\varepsilon}(x) - T_j(u(x))| \leq i\}. \tag{2.14}$$

Then for any fixed $i > 0$ and $j > 0$

$$\limsup_{\varepsilon \to 0} \int_{F_{ij}^{\varepsilon}} |\mathrm{grad}(u^{\varepsilon}-T_j(u))|^2 \, dx \leq$$

$$\leq \frac{1}{\alpha} \langle f, T_i(u-T_j(u))\rangle - \frac{1}{\alpha} \int_{\Omega} A \,\mathrm{grad}\, T_j(u) \,\mathrm{grad}\, T_i(u-T_j(u))dx. \quad \square \tag{2.15}$$

PROOF: Consider the test function

$$w^{\varepsilon} = T_i(u^{\varepsilon}-T_j(u)) \in H_0^1(\Omega),$$

and define

$$X^{\varepsilon} = \langle -\mathrm{div}(\phi^{\varepsilon}(u^{\varepsilon})), w^{\varepsilon}\rangle = \int_{\Omega} \phi^{\varepsilon}(u^{\varepsilon}) \,\mathrm{grad}\, w^{\varepsilon} \, dx;$$

we claim that

$$X^{\varepsilon} \to 0 \text{ as } \varepsilon \to 0 \text{ for any fixed } i > 0 \text{ and } j > 0. \tag{2.16}$$

Indeed by Lemma 2.1

$$\mathrm{grad}\, w^{\varepsilon} = \begin{cases} \mathrm{grad}\,(u^{\varepsilon}-T_j(u)) & \text{on } F_{ij}^{\varepsilon}, \\ 0 & \text{on } \Omega \setminus F_{ij}^{\varepsilon}. \end{cases} \tag{2.17}$$

Since

$$|u^{\varepsilon}(x)| \leq |u^{\varepsilon}(x) - T_j(u(x))| + |T_j(u(x))| \leq i+j \text{ on } F_{ij}^{\varepsilon}$$

we have

$$\phi^{\varepsilon}(u^{\varepsilon}(x)) = \phi(T_{1/\varepsilon}(u^{\varepsilon}(x))) = \phi(T_{i+j}(u^{\varepsilon}(x))) \text{ on } F_{ij}^{\varepsilon} \text{ for } 1/\varepsilon \geq i+j.$$

Thus whenever $1/\varepsilon \geqq i + j$

$$X^\varepsilon = \int_\Omega \phi^\varepsilon(u^\varepsilon) \text{ grad } w^\varepsilon \, dx = \int_{F_{ij}^\varepsilon} \phi^\varepsilon(u^\varepsilon) \text{ grad } w^\varepsilon \, dx$$

$$= \int_\Omega \phi(T_{i+j}(u^\varepsilon)) \text{ grad } w^\varepsilon \, dx,$$

which tends to

$$X = \int_\Omega \phi(T_{i+j}(u)) \text{ grad } T_i(u-T_j(u))dx, \tag{2.18}$$

since w^ε tends to $T_i(u-T_j(u))$ weakly in $H_0^1(\Omega)$ while $\phi(T_{i+j}(u^\varepsilon))$, which is bounded in $(L^\infty(\Omega))^N$ and converges a.e., tends to $\phi(T_{i+j}(u))$ strongly in $(L^2(\Omega))^N$ through application of Lebesgue's dominated convergence theorem. Applying Lemma 2.1 to the $(L^\infty(R))^N$ piecewise continuous function $\theta(t) = \phi(T_{i+j}(t)) [1 - \chi_i(t-T_j(t))]\chi_j(t)$ proves that the right-hand side of (2.18) is zero, hence (2.16).

Multiplying now (2.3) by w^ε, integrating by parts and using (2.17) and (2.16) we obtain

$$\lim_{\varepsilon\to 0} \int_{F_{ij}^\varepsilon} A \text{ grad } u^\varepsilon \text{ grad}(u^\varepsilon-T_j(u))dx = \langle f,T_i(u-T_j(u))\rangle. \tag{2.19}$$

Substracting to both sides of (2.19) the quantity

$$\lim_{\varepsilon\to 0} \int_{F_{ij}^\varepsilon} A \text{ grad } T_j(u) \text{ grad } (u^\varepsilon-T_j(u))dx$$

$$= \lim_{\varepsilon\to 0} \int_\Omega A \text{ grad } T_j(u) \text{ grad } T_i(u^\varepsilon-T_j(u))dx$$

$$= \int_\Omega A \text{ grad } T_j(u) \text{ grad } T_i(u-T_j(u))dx$$

and using the coerciveness (1.4) complete the proof of Lemma 2.3. □

PROOF OF THEOREM 2.1: From

$$|u^\varepsilon(x)| \geqq |u^\varepsilon(x)-T_j(u(x))| - |T_j(u(x))| \geqq i-j \text{ on } \Omega\setminus F_{ij}^\varepsilon,$$

We deduce that $\Omega \setminus F^{\varepsilon}_{ij} \subset E^{\varepsilon}_{i-j}$. Then:

$$\int_{\Omega} |\mathrm{grad}(u^{\varepsilon}-u)|^2\, dx \leq \int_{F^{\varepsilon}_{ij}} |\mathrm{grad}(u^{\varepsilon}-u)|^2\, dx + \int_{E^{\varepsilon}_{i-j}} |\mathrm{grad}(u^{\varepsilon}-u)|^2\, dx$$

$$\leq 2\int_{F^{\varepsilon}_{ij}} |\mathrm{grad}(u^{\varepsilon}-T_j(u)|^2 dx + 2\int_{F^{\varepsilon}_{ij}} |\mathrm{grad}(T_j(u)-u)|^2 dx \qquad (2.20)$$

$$+ 2\int_{E^{\varepsilon}_{i-j}} |\mathrm{grad} u^{\varepsilon}|^2 dx + 2\int_{E^{\varepsilon}_{i-j}} |\mathrm{grad}\ u|^2\, dx = I^{\varepsilon}_{ij} + II^{\varepsilon}_{ij} + III^{\varepsilon}_{ij} + IV^{\varepsilon}_{ij}.$$

We now estimate the lim sup, for i and j fixed and ε converging to zero, of each of the four terms of the right-hand side of (2.20).

In view of Lemma 2.3, $\limsup_{\varepsilon \to 0} I^{\varepsilon}_{ij}$ is small provided j is sufficiently large; indeed, since

$$\|\mathrm{grad}\ T_k(v)\|_{(L^2(\Omega))^N} \leq \|\mathrm{grad}\ v\|_{(L^2(\Omega))^N},\ \forall k > 0,\ \forall v \in H^1_0(\Omega),$$

the right-hand side of (2.15) is bounded by

$$\frac{1}{\alpha}\left[\ \|f\|_{H^{-1}(\Omega)} + \|A\|_{(L^{\infty}(\Omega))^{N\times N}} \|\mathrm{grad}\ u\|_{(L^2(\Omega))^N}\right] \|\mathrm{grad}(u-T_j(u)\|_{(L^2(\Omega))^N}$$

which tends to zero when j tends to infinity.

Since F^{ε}_{ij} is a subset of Ω

$$\limsup_{\varepsilon\to 0} II^{\varepsilon}_{ij} \leq 2\int_{\Omega} |\mathrm{grad}(T_j(u)-u)|^2\, dx,$$

which is small when j is large.

In view of Lemma 2.2, $\limsup_{\varepsilon\to 0} III^{\varepsilon}_{ij}$ is small provided i-j is sufficiently large.

Finally, splitting Ω into the union of the sets $\{x|\ |u(x)| \neq i-j\}$ and $\{x|\ |u(x)| = i-j\}$ (note that grad u = 0 a.e. on the second set by virtue of (2.5)), and using Lebesgue's dominated convergence theorem, it is easy to prove that

$$\limsup_{\varepsilon\to 0} IV_{ij}^{\varepsilon} \leq 2 \int_{\{x||u(x)|>i-j\}} |\text{grad } u|^2 \, dx,$$

which is small when i-j is large.

Hence $\limsup_{\varepsilon\to 0} \int_{\Omega} |\text{grad}(u^{\varepsilon}-u)|^2 \, dx$ is small if we choose j, next i-j, sufficiently large. Theorem 2.1. is proved. □

<u>END OF THE PROOF OF THEOREM 1.1</u>: Let h belong to $C_c^1(\mathbb{R})$ and φ belongs to $\mathcal{D}(\Omega)$. Multiplying (2.3) by $h(u^{\varepsilon})\varphi$ which belongs to $H_0^1(\Omega)$ and integrating by parts we obtain

$$\int_{\Omega} (A \text{ grad } u^{\varepsilon} + \phi^{\varepsilon}(u^{\varepsilon}))(h'(u^{\varepsilon})\varphi \text{ grad } u^{\varepsilon} + h(u^{\varepsilon}) \text{ grad } \varphi) \, dx = \tag{2.21}$$
$$= \langle f, h(u^{\varepsilon})\varphi \rangle$$

Since h and h' have compact support on $\mathbb{R}$, we have for ε sufficiently small

$$\phi^{\varepsilon}(t)h(t) = \phi(T_{1/\varepsilon}(t))h(t) = \phi(t)h(t),$$
$$\phi^{\varepsilon}(t)h'(t) = \phi(T_{1/\varepsilon}(t))h'(t) = \phi(t)h'(t), \tag{2.22}$$

and the right-hand sides in (2.22) are $(C^0(\mathbb{R}) \cap L^{\infty}(\mathbb{R}))^N$ functions. Because u^{ε} converges to u strongly in $H_0^1(\Omega)$ (see Theorem 2.1) it is easy to pass to the limit successively in each term of (2.21); this yields

$$\int_{\Omega} (A \text{ grad } u + \phi(u))(h'(u)\varphi \text{ grad } u + h(u) \text{ grad } \varphi) \, dx$$
$$= \langle f, h(u)\varphi \rangle, \ \forall h \in C_c^1(\mathbb{R}), \ \forall\varphi \in \mathcal{D}(\Omega)$$

which is equivalent to (1.8). Theorem 1.1 is proved. □

3. <u>PROOFS OF THEOREMS 1.2 AND 1.3</u>

<u>PROOF OF THEOREM 1.2</u>: Let u in $H_0^1(\Omega)$ be a solution of (1.8). Consider a function $H \in \mathcal{D}(\mathbb{R})$ such that

$$H(t) = 1 \text{ if } |t| \leqq 1,\ H(t) = 0 \text{ if } |t| \geqq 2,$$

$$|H(t)| \leqq 1,\ |H'(t)| \leqq 2,\ \forall t \in \mathbb{R},$$

and define for $\varepsilon > 0$ h_ε by:

$$h_\varepsilon(t) = \begin{cases} H(t + 1/\varepsilon) & \text{if } t + 1/\varepsilon \leqq 0, \\ 1 & \text{if } |t| \leqq 1/\varepsilon, \\ H(t - 1/\varepsilon) & \text{if } t - 1/\varepsilon \geqq 0. \end{cases} \tag{3.1}$$

The function h^ε belongs to $C_c^1(\mathbb{R})$ and can be used in (1.8); multiplying (1.8) by $\varphi \in \mathcal{D}(\Omega)$ and integrating by parts we obtain

$$\begin{aligned} &\int_\Omega A \operatorname{grad} u\, (h_\varepsilon'(u)\varphi \operatorname{grad} u + h_\varepsilon(u) \operatorname{grad} \varphi)\, dx \\ &+ \int_\Omega \phi(u) h_\varepsilon(u) \operatorname{grad} \varphi\, dx + \int_\Omega \varphi\phi(u) h_\varepsilon'(u) \operatorname{grad} u\, dx \\ &= \langle f, h_\varepsilon(u)\varphi\rangle. \end{aligned} \tag{3.2}$$

Since for any $t \in \mathbb{R}$,

$$|h_\varepsilon(t)| \leqq 1, \quad h_\varepsilon(t) \to 1 \quad \text{as } \varepsilon \to 0, \tag{3.3}$$

$$|h_\varepsilon'(t)| \leqq 2, \quad h_\varepsilon'(t) \to 0 \quad \text{as } \varepsilon \to 0, \tag{3.4}$$

Lebesgue's dominated convergence theorem implies that

$$h_\varepsilon(u) \to 1 \text{ in } H^1(\Omega) \text{ strongly.}$$

We can thus pass to the limit in the first and last terms of (3.2).

If we assume now that $\phi(u) \in (L^1_{loc}(\Omega))^N$ (hypothesis (1.11)), Lebesgue's dominated convergence and (3.3) allow to pass to the limit in the second term of (3.2), since in this case

$$\phi(u) h^\varepsilon(u) \to \phi(u) \text{ in } (L^1_{loc}(\Omega))^N \text{ strongly.}$$

Define now

$$\tilde{\psi}_\varepsilon(t) = \int_0^t \phi(s)h'_\varepsilon(s)ds;$$

then by virtue of (3.4) and of the definition (1.10) of $\tilde{\psi}$

$$|\tilde{\psi}_\varepsilon(t)| \leqq 2|\tilde{\psi}(t)| \quad \tilde{\psi}_\varepsilon(t) \to 0 \text{ as } \varepsilon \to 0, \forall t \in \mathbf{R}.$$

If we assume that $\tilde{\psi}(u) \in L^1_{loc}(\Omega)$ (hypothesis (1.12)), this implies (using once again Lebesgue's dominated convergence theorem) that

$$\tilde{\psi}_\varepsilon(u) \to 0 \text{ in } L^1_{loc}(\Omega) \text{ strongly.}$$

Thus the third term of (3.2), which is equal to $\langle -\mathrm{div}(\tilde{\psi}_\varepsilon(u)),\varphi\rangle_{\mathcal{D}'(\Omega),\mathcal{D}(\Omega)}$ (see Lemma 2.1), tends to zero, when (1.12) holds true.

We have proved that for any φ in $\mathcal{D}(\Omega)$

$$\int_\Omega A \text{ grad } u \text{ grad } \varphi \, dx + \int_\Omega \phi(u) \text{ grad } \varphi \, dx = \langle f,\varphi\rangle$$

which is equivalent to (1.1). Theorem 1.2 is proved. □

PROOF OF THEOREM 1.3: Let u be a solution of (1.7), (1.8) and s be a Lipschitz continuous, piecewise C^1 function from **R** to **R** such that $s(0) = 0$.

First step. We assume here that s is bounded. Then $s(u)$ belongs to $H^1_0(\Omega) \cap L^\infty(\Omega)$. In this case there exists a sequence φ_n such that

$$\begin{cases} \varphi_n \in \mathcal{D}(\Omega), & \|\varphi_n\|_{L^\infty(\Omega)} \leqq C, \\ \varphi_n \to s(u) \text{ in } H^1_0(\Omega) \text{ strongly as } n \to +\infty. \end{cases}$$

Then for any $h \in C^1_c(\mathbf{R})$

$$h(u)\varphi_n \to h(u)s(u) \text{ in } H^1_0(\Omega) \text{ strongly as } n \to \infty.$$

Using $\varphi_n \in \mathcal{D}(\Omega)$ as a test function in (1.8) and passing to the limit yields:

$$\langle -\mathrm{div}(A \text{ grad } u), h(u)s(u)\rangle$$

$$+ \int_\Omega \phi(u)h(u) \text{ grad } s(u)\, dx + \int_\Omega s(u)\phi(u)h'(u) \text{ grad } u\, dx \tag{3.5}$$

$$= \langle f, h(u)s(u)\rangle, \quad \forall h \in C_c^1(R).$$

Lemma 2.1, applied first to the $(L^\infty(R))^N$, piecewise continuous function $\phi(t) = \phi(t)h(t)s'(t)$, next to the $(C^0(R) \cap L^\infty(R))^N$ function $\theta(t) = s(t)\phi(t)h'(t)$, proves that the second and third terms in (3.5) are zero. Thus

$$\langle -\mathrm{div}(A \text{ grad } u), h(u)s(u)\rangle = \langle f, h(u)s(u)\rangle, \quad \forall h \in C_c^1(R). \tag{3.6}$$

Use now in (3.6) the functions h_ε defined in (3.1); from (3.3) and (3.4) it is easy to prove that for any Lipschitz continuous, piecewise C^1 function s such that $s(0) = 0$ *which is bounded*, one has:

$$h_\varepsilon(u)s(u) \to s(u) \text{ in } H_0^1(\Omega) \text{ strongly as } \varepsilon \to 0.$$

Thus passing to the limit in (3.6) proves that

$$\langle -\mathrm{div}(A \text{ grad } u), s(u)\rangle = \langle f, s(u)\rangle \tag{3.7}$$

for any Lipschitz continuous, piecewise C^1 function s such that $s(0) = 0$, $s \in L^\infty(R)$.

Second Step. Consider the general case where the function s is not assumed to be bounded, and let for $m > 0$ $s_m(t) = T_m(s(t))$ be the truncation of s at the level m (see (2.1)); we can use s_m in (3.7). On the other hand, it is easy to prove that

$$s_m(u) = T_m(s(u)) \to s(u) \text{ in } H_0^1(\Omega) \text{ strongly as } m \to \infty.$$

Passing to the limit in (3.7) gives

$$\langle -\mathrm{div}(A \text{ grad } u), s(u)\rangle = \langle f, s(u)\rangle$$

for any Lipschitz continuous, piecewise continuous s,

such that $s(0) = 0$,

which is equivalent to (1.16). Theorem 1.3 is proved. □

REFERENCES

[1] P. Benilan, L. Boccardo, T. Gallouet, R. Gariepy, M. Pierre and J.L. Vazquez. On the p-Laplacian on L^1 (to appear).

[2] A. Bensoussan, L. Boccardo and F. Murat. On a nonlinear partial differential equation having natural growth terms and unbounded solution. Ann. Inst. Henri Poincaré, Analyse non linéaire, 5 (4), (1988), 347-364.

[3] L. Boccardo, I. Diaz, D. Giachetti and F. Murat. (to appear.)

[4] L. Boccardo and D. Giachetti. Alcune osservazioni sulla regolarità delle soluzioni di problemi fortemente non lineari e applicazioni. Ric. Mat., 34 (2), (1985), 309-323.

[5] L. Boccardo and D. Giachetti. Existence results via regularity for some nonlinear elliptic problems. Comm. P.D.E. (to appear).

[6] L. Boccardo and F. Murat. Remarques sur l'homogénéisation de certains problèmes quasi-lineaires. Port. Math., 41 (1-4) (1982), 535-562.

[7] L. Boccardo, F. Murat and J.P. Puel. Existence de solutions non bornées pour certaines équations quasi-linéaires. Port. Math., 41 (1-4), (1982), 507-534.

[8] J. Carrillo. Unicité des solutions du type Kruskov pour des problemes elliptiques avec des termes de transport non linéaires. C.R. Acad. Sc. Paris, Serie I, 303 (5), (1986), 189-192.

[9] J. Carrillo and M. Chipot. On some nonlinear elliptic equations involving derivatives of the nonlinearity. Proc. Roy. Soc. Edinburgh, A, 100, (1985), 281-294.

[10] M. Chipot and G. Michaille. Uniqueness results and monotonicity properties for strongly nonlinear variational inequalities (to appear).

[11] D. Kinderlehrer and G. Stampacchia. An Introduction to Variational Inequalities and Their Applications. Academic Press, New York, 1980.

[12] R.J. Di Perna and P.L. Lions. On the Fokker-Planck-Boltzmann equation (to appear).

[13] R.J. Di Perna and P.L. Lions. On the Cauchy problem for Boltzmann equations: global existence and weak stability (to appear).

L. Boccardo
Dipartimento di Matematica
Università di Roma I
Piazza A. Moro 2
00185 Roma
Italy

J.I. Diaz
Departamento de Matematica Aplicada
Universidad Complutense de Madrid
28040 Madrid
Spain

D. Giachetti
Dipartimento di Matematica
Università dell' Aquila
Via Roma
67 100 L'Aquila
Italy

F. Murat
Laboratoire d'Analyse Numérique
Université Paris VI
Tour 55-65, 5ème étage,
4 place Jussieu
75252 Paris Cedex 05
France

L. BOCCARDO AND F. MURAT

Strongly nonlinear Gauchy problems with gradient-dependent lower order nonlinearity

In this article we prove the existence of solutions of nonlinear parabolic equations of the type

$$u' + A(u) + g(x,t,u,Du) = h \text{ in } Q = \Omega \times)0,T(; \; u(x,0) = \Omega, \qquad (1.1)$$

where A is a Leray-Lions operator from $L^p(0,T; W_0^{1,p}(\Omega))$ into its dual, $h \in L^{p'}(0,T; W^{-1,p'}(\Omega))$ and g is a nonlinear lower order term having growth of order q $(q < p)$ with respect to $|Du|$. With respect to $|u|$, we do not assume any growth restriction on g, but we assume the "sign-condition"

$$g(x,t,s,\xi)s \geqq 0. \qquad (1.2)$$

It will turn out that, for any solution u, g(x,t,u,Du) will be in $L^1(Q)$, but, for each $v \in L^p(0,T; W_0^{1,p}(\Omega))$, g(x,t,v,Dv) can be very odd. In the present article, the main features are the "sign-condition" and the non-smoothness of the right-hand side h.

When g does not depend on Du, existence results for this type of problems have been proved in [BB1], [BB2], [LM]. The case where g has growth of order p is an open problem for parabolic equations. In the elliptic case the problem is completely solved in [BBM], where an extensive set of references and comments can be found, as well as in [La].

In order to prove existence of solutions of (1.1), consider a sequence of approximating problems, with solutions u_ε. In the convergence proof, the essential tool is a new compactness theorem ([BM]) which asserts that the gradients of the solutions of nonlinear parabolic problems with Leray-Lions operators and right-hand side bounded in $L^1(Q)$ are compact in $L^q(Q)$, for any $q < p$.

2. STATEMENT OF THE RESULT

Let Ω be a bounded subset of $\mathbb{R}^N$ and let Q be $\Omega \times)0,T($, $0 < T < \infty$. Let $1 < p < \infty$ be fixed and let A be a nonlinear operator from $L^p(0,T; W_0^{1,p}(\Omega))$ into its dual $L^{p'}(0,T, W^{-1,p'}(\Omega))$ defined by

$$A(u) = -\mathrm{div}(a(x,t,u,Du)),$$

where $a(x,t,s,\xi)$ is a Carathéodory function such that

$$|a(x,t,s,\xi)| \leqq \beta[|s|^{p-1} + |\xi|^{p-1} + k(x,t)],\ k(x,t) \in L^{p'}(Q),\ \beta > 0,$$

$$[a(x,t,s,\xi) - a(x,t,s,\eta)]\cdot[\xi-\eta] > 0,\ \forall\xi \neq \eta, \tag{2.1}$$

$$a(x,t,s,\xi)\ \xi \geqq \alpha|\xi|^p, \quad \alpha > 0.$$

Let g be a Carathéodory function such that

$$g(x,t,s,\xi)s \geqq 0,$$

$$|g(x,t,s,\xi)| \leqq b(|s|)(|\xi|^q + c(x,t)),\ q < p, \tag{2.2}$$

where b is continuous and increasing with finite values in R^+, $c \in L^1(Q)$.

For a given right-hand side h

$$h \in L^{p'}(0,T;\ W^{-1,p'}(\Omega)), \tag{2.3}$$

consider the Cauchy problem

$$u' + A(u) + g(x,t,u,Du) = h,\ u(x,0) = 0,$$

$$u \in L^p(0,T;\ W_0^{1,p}(\Omega)) \cap C^0([0,T];\ W^{-1,r}(\Omega)),\ 1 \leqq r < \inf(p',\ \tfrac{N}{N-1}), \tag{2.4}$$

$$g(x,t,u,Du),\ ug(x,t,u,Du) \in L^1(Q).$$

Our objective is to prove the following:

THEOREM 2.1: Under the assumptions (2.1), (2.2), (2.3) there exists at least a solution of (2.4). □

Before giving the proof of this theorem let us emphasize that the main difficulty stems from the fact u is unbounded and from the growth of g with respect to $|Du|$, whose order q can be greater than p-1.

3. PROOF OF THEOREM 2.1

Let us define

$$g_\varepsilon(x,t,s,\xi) = \frac{g(x,t,s,\xi)}{1 + \varepsilon|g(x,t,s,\xi)|} \tag{3.1}$$

and let us consider the Cauchy problem

$$\begin{aligned} &u_\varepsilon' + A(u_\varepsilon) + g_\varepsilon(x,t,u_\varepsilon,Du_\varepsilon) = h, \\ &u_\varepsilon(x,0) = 0, \\ &u_\varepsilon \in L^p(0,T; W_0^{1,p}(\Omega)), \end{aligned} \tag{3.2}$$

which has a solution by the classical result of Lions [Li].

Multiplying (3.2) by u_ε and using (2.2) we get, denoting by $\langle\ ,\ \rangle$ the duality pairing between $L^p(0,T; W_0^{1,p}(\Omega))$ and its dual

$$\langle A(u_\varepsilon),u_\varepsilon\rangle \leq \langle h,u_\varepsilon\rangle,$$

hence

$$\|u_\varepsilon\|_{L^p(0,T;W_0^{1,p}(\Omega))} \leq C_1. \tag{3.3}$$

Thus we can extract a subsequence, still denoted by u_ε, such that

$$u_\varepsilon \rightharpoonup u \text{ in } L^p(0,T;W_0^{1,p}(\Omega)) \text{ weakly.} \tag{3.4}$$

We have also that

$$\int_Q u_\varepsilon g_\varepsilon(x,t,u_\varepsilon,Du_\varepsilon) \leq C_2. \tag{3.5}$$

Now we adapt a method of [W] to prove the equi-integrability of $g_\varepsilon(x,t,u_\varepsilon,Du_\varepsilon)$. Denoting

$$G_m = \{(x,t) \in Q: |u| > m\},$$

$$P_m = \{(x,t) \in Q: |u| \leq m\},$$

we have for any measurable subset $E \subset Q$

$$\int_E |g_\varepsilon(x,t,u_\varepsilon,Du_\varepsilon)| = \int_{E\cap G_m} |g_\varepsilon(x,t,u_\varepsilon,Du_\varepsilon)| + \int_{E\cap P_m} |g_\varepsilon(x,t,u_\varepsilon,Du_\varepsilon)|$$

$$\leq \int_{E\cap G_m} \frac{1}{m} u_\varepsilon g_\varepsilon(x,t,u_\varepsilon,Du_\varepsilon) + \int_{E\cap P_m} b(m)(|Du_\varepsilon|^q + c(x,t)) \qquad (3.6)$$

$$\leq \frac{1}{m} C_2 + b(m)\, C_1^{q/p}\, |E|^{1-q/p} + b(m) \int_E c(x,t).$$

So (3.6) implies the equi-integrability of $g_\varepsilon(x,t,u_\varepsilon,Du_\varepsilon)$, that is the weak-$L^1$ compactness. If we prove that, for some subsequence, $(u_\varepsilon,Du_\varepsilon)$ converges a.e. to (u,Du), Vitali's theorem yields

$$g_\varepsilon(x,t,u_\varepsilon,Du_\varepsilon) \to g(x,t,u,Du) \text{ strongly in } L^1(Q). \qquad (3.7)$$

We first claim that the sequence u_ε is relatively compact in $L^p(Q)$. Indeed, it is sufficient to use the compactness result of Proposition 1 of [BB1] because u_ε is bounded in $L^p(0,T;W_0^{1,p}(\Omega))$ and

$$u_\varepsilon' = (-A(u_\varepsilon) + h) - g_\varepsilon(x,t,u_\varepsilon,Du_\varepsilon),$$

where $(-A(u_\varepsilon) + h)$ is bounded in $L^{p'}(0,T;W^{-1,p'}(\Omega))$ and g_ε is sequentially weakly compact in $L^1(Q)$ (see also [S]).

The almost pointwise convergence of $Du_\varepsilon(x,t)$ follows from Theorem 4.1.below. Thus we can deduce (3.7). The almost pointwise convergence of u_ε and Du_ε also yields

$$a(x,t,u_\varepsilon,Du_\varepsilon) \longrightarrow a(x,t,u,Du) \text{ in } L^{p'}(Q) \text{ weakly}, \qquad (3.8)$$

which implies

$$A(u_\varepsilon) \rightharpoonup A(u) \text{ in } L^{p'}(0,T;W^{-1,p'}(\Omega)) \text{ weakly.} \tag{3.9}$$

Then (3.7) and (3.9) allow to pass to the limit in (3.2). Moreover since $g_\varepsilon(x,t,u_\varepsilon,Du_\varepsilon)u_\varepsilon \geq 0$ a.e., it follows from (3.5) that $u\, g(x,t,u,Du) \in L^1(Q)$, which implies by a proof similar to (3.6) that $g(x,t,u,Du) \in L^1(Q)$. Now $u' = - A(u) + h - g(x,t,u,Du) \in L^{p'}(0,T;W^{-1,p'}(\Omega)) + L^1(0,T;L^1(\Omega)) \subset L^1(0,T;W^{-1,r}(\Omega))$ for any $r < \inf(p',N/N-1)$. Thus u belongs to $C^0(0,T;W^{-1,r}(\Omega))$ and Theorem 2.1 is proved.

4. ALMOST POINTWISE CONVERGENCE OF THE GRADIENTS

In this section we state (and we prove in the linear case) the following compactness theorem, which is the most important new tool in the proof of Theorem 2.1.

THEOREM 4.1: Let u_ε be a bounded sequence in $L^p(0,T;W^{1,p}(\Omega))$ such that

$$u'_\varepsilon + A(u_\varepsilon) = h + y_\varepsilon, \tag{4.1}$$

$$h \in L^{p'}(0,T;W^{-1,p'}(\Omega)), \tag{4.2}$$

$$y_\varepsilon \text{ is bounded in } L^1(Q), \tag{4.3}$$

where A is an operator satisfying (2.1). Then Du_ε is strongly compact in $L^q(Q)$, for any $q < p$. □

This theorem is proved in [BM]. We give here a different short proof under the additional assumptions that p = 2 and A is linear, i.e.

$$A(v) = - \operatorname{div}(b(x,t)Dv), \tag{4.4}$$

for some $L^\infty(Q)$ matrix b.

PROOF OF THEOREM 4.1: (We assume here (4.4)). Since u_ε is bounded in $L^2(0,T;H_0^1)$ and u_ε' is bounded in $L^2(0,T;H^{-1}) + L^1(0,T;L^1)$, it follows from a result of [S] that u_ε is bompact in $L^2(0,T;L^2)$. Extracting a subsequence we have

$$u_\varepsilon \rightharpoonup u \text{ in } L^2(0,T;H_0^1) \text{ weak,} \tag{4.5}$$

$$u_\varepsilon \to u \quad \text{in } L^2(Q) \text{ and a.e.} \tag{4.6}$$

Since y_ε is bounded in $L^1(Q)$, we can also assume that, for some subsequence

$$y_\varepsilon \rightharpoonup y \text{ in } \mathfrak{m}(Q) \text{ weak star,} \tag{4.7}$$

where y is a measure on Q.

Let us define v_ε as the solution of the Cauchy problem

$$\begin{aligned} & v_\varepsilon \in L^2(0,T;\ H_0^1(\Omega)), \\ & -v_\varepsilon' + A^*(v) = -\mathrm{div}\Big(\frac{D(u_\varepsilon-u)}{1+|D(u_\varepsilon-u)|^2}\Big), \\ & v_\varepsilon(x,T) = 0, \end{aligned} \tag{4.8}$$

where the adjoint operator A^* is of course given by

$$A^*(v) = -\mathrm{div}(b^*(x,t)Dv).$$

We multiply (4.1) for ε and η by $v_\varepsilon Q$, $\in \mathcal{D}(Q)$, and (4.13) by $(u_\varepsilon-u_\eta)\phi$; we obtain

$$\begin{aligned} \int_Q (y_\varepsilon - y_\eta)v_\varepsilon\phi &= \langle(u_\varepsilon-u_\eta)',\ v_\varepsilon\phi\rangle + \int_Q b(x,t)D(u_\varepsilon-u_\eta)Dv_\varepsilon\,\phi + \\ &+ \int_Q b(x,t)\ D(u_\varepsilon-u_\eta)D\phi\ v_\varepsilon = \\ &= -\langle u_\varepsilon-u_\eta,\ v_\varepsilon\phi'\rangle - \langle(u_\varepsilon-u_\eta)\phi, v_\varepsilon'\rangle + \\ &+ \int_Q b^*(x,t)Dv_\varepsilon\ [(u_\varepsilon-u_\eta)\phi] - \int_Q b^*(x,t)Dv_\varepsilon D\phi\ (u_\varepsilon-u_\eta) + \end{aligned} \tag{4.9}$$

$$+\int_Q b(x,t)\, D(u_\varepsilon-u_\eta)D\phi\; v_\varepsilon =$$

$$= -\int_Q (u_\varepsilon-u_\eta)\; v_\varepsilon\phi' + \int_Q \frac{D(u_\varepsilon-u)\; D(u_\varepsilon-u_\eta)}{1+|D(u_\varepsilon-u)|^2}\phi + \int_Q \frac{D(u_\varepsilon-u)D\phi}{1+|D(u_\varepsilon-u)|^2}(u_\varepsilon-u_\eta) \quad (4.9)$$

$$-\int_Q b^*(x,t)Dv_\varepsilon D\varphi(u_\varepsilon-u_\eta) + \int_Q b(x,t)D(u_\varepsilon-u_\eta)\, D\phi\;\; v_\varepsilon$$

Note that the integrations by parts above are licit since $(u_\varepsilon-u_\eta)'$ as well as v_ε' belong to $L^2(0,T;H^{-1})$. For ε fixed it is easy to pass to the limit in (4.9), for $\eta \to 0$, using (4.5)-(4.7) and the local Hölder continuity of v_ε on Q (see [LSU]). This gives

$$\int_Q y_\varepsilon v_\varepsilon\phi - \int_Q v_\varepsilon\phi\; dy + \int_Q (u_\varepsilon-u)v_\varepsilon\phi' -$$

$$-\int_Q \frac{D(u_\varepsilon-u)\cdot D\phi}{1+|D(u_\varepsilon-u)|^2}(u_\varepsilon-u) + \int_Q b^*(x,t)Dv_\varepsilon\; D\phi(u_\varepsilon-u) +$$

$$+\int_Q b(x,t)D(u_\varepsilon-u)D\phi\; v_\varepsilon = \int_Q \frac{|D(u_\varepsilon-u)|^2}{1+|D(u\;-u)|^2}\phi. \quad (4.10)$$

Now let ε tend to zero. The left-hand side tends to zero in view of (4.5)-(4.7) and of the uniform convergence of v_ε to some v locally on Q (since v_ε is bounded in $C^{0,\,}_{loc}(Q)$ for some γ 0, see [LSU]). Thus

$$\int_Q \frac{|D(u_\varepsilon-u)\Delta^2}{1+|D(u_\varepsilon-u)|^2}\phi \to 0,$$

which implies that (for a subsequence)

$$Du_\varepsilon \to Du \quad \text{a.e.} \quad \text{in } Q. \quad (4.11)$$

Since Du_ε is bounded in $L^2(Q)$, the convergence of Du_ε to Du in $L^q(Q)$, for any $q < 2$, follows from (4.11) and Vitali's theorem.

ACKNOWLEDGEMENTS: This work was done while the first author was "professeur associé" at the University of (Old) Orléans and the second one visiting professor of C.N.R. at the University of Roma I. The paper was written while the first author was "professeur associé" at the University of Besançon.

REFERENCES

[BBM] A. Bensoussan, L. Boccardo and F. Murat , On a nonlinear P.D.E. having natural growth terms and unbounded solutions, Annales I.H.P. Anal. Nonlin. 5 (1988), 347-364.

[BG] L. Boccardo and T. Gallouet , Nonlinear elliptic and parabolic equations involving measure data, J. Funct. Anal. (to appear).

[BM] L. Boccardo and F. Murat, Compactness of the gradients of solutions to nonlinear heat equations (in preparation).

[BB1] H. Brezis and F.E. Browder, Strongly nonlinear parabolic initial boundary value problems, Proc. Nat. Acad. Sci. USA 76 (1979), 38-40.

[BB2] H. Brezis and F.E. Browder, Strongly nonlinear parabolic variational inequalities, Proc. Nat. Acad. Sci. USA 77 (1980), 713-715.

[La] R. Landes, Solvability of perturbated elliptic equations with critical growth exponent for the gradient (to appear).

[Li] J.L. Lions, Quelques méthodes de résolution des problèmes aux limites non linéaires, Dunod, Paris, 1969.

[LM] R. Landes and V. Mustonen, A strongly nonlinear parabolic initial boundary value problem, Ark. Mat. 25 (1987), 29-40.

[LSU] O. Ladyzhenskaya, V. Solonnikov and N. Uralceva, Linear and quasi-linear equations of parabolic type, Trans. of Math. Monographs, Vol. 23, A.M.S., Providence, 1968.

[S] J. Simon, Compact sets in the space $L^p(0,T;B)$, Annali Mat. Pura Appl. 146 (1987), 65-96.

[W] J. Webb, Boundary value problems for strongly nonlinear elliptic equations, J. London Math. Soc. 21 (1980), 123-132.

L. Boccardo
Università di Roma I
Piazza A. Moro 2
00185 Roma
Italy

F. Murat
Laboratoire d'Analyse Numérique
Université Paris VI
Tour 55-65, 5ème étage,
4, place Jussieu
75252 Paris Cedex 05
France

M. CHIPOT, A. FEGGOUS AND G. MICHAILLE

Monotonicity properties for variational inequalities associated with strongly nonlinear diagonal systems

1. INTRODUCTION AND NOTATION

Let Ω be a Lipschitz, bounded open set of R^n and let Γ be its boundary. For $p > 1$, $N \in \mathbb{N}^*$, let us denote by $L_p(\Omega)$ the space $(L_p(\Omega))^N$ equipped with the norm

$$|u|_p = \sum_{i=1}^{N} |u^i|_p, \ \forall u = (u^1, u^2, \ldots, u^N),$$

where $| \ |_p$ is the usual norm in $L_p(\Omega)$ and by $W^{1,p}(\Omega)$ the Sobolev space $(W^{1,p}(\Omega))^N$. The norm in $W^{1,p}(\Omega)$ is, for instance,

$$|u|_{1,p} = \sum_{i=1}^{N} |u^i|_{1,p}, \ \forall u = (u^1, u^2, \ldots, u^N),$$

where $| \ |_{1,p}$ is the usual norm in $W^{1,p}(\Omega)$ (See [11]).

If K is a closed convex set in $W^{1,p}(\Omega)$, denote by V the closed subspace of $W^{1,p}(\Omega)$ spanned by

$$K - K = \{k - k'/k, k' \in K\}.$$

Let V^* be the dual of V endowed with the $W^{1,p}(\Omega)$-topology, and denote by $\langle \ , \ \rangle$ the duality bracket between V^* and V.

We would like to prove monotonicity properties with respect to the data for the solution u of variational inequalities

$$u \in K, \quad \langle A(x,u,Du), v - u\rangle \geq \langle f, v - u\rangle, \ \forall v \in K. \tag{1}$$

Here A denotes a nonlinear operator from K into V^*, given by

$$\langle A(x,u,Du), v\rangle = \sum_{j=1}^{N} \langle A_j(x, u^j, \nabla u^j), v^j\rangle \tag{2}$$

with

$$\langle A_j(x,u^j,\nabla u^j),v^j\rangle =$$

$$\int_\Omega A_j^\beta(x,u^j,\nabla u^j)D_\beta v^j + \int_\Omega a_j(x,u^j)v^j + \int_\Gamma \gamma_j(x,u^j)v^j \qquad (3)$$

for every $u = (u^1,u^2,\ldots,u^N) \in K$ and $v = (v^1,v^2,\ldots,v^N) \in V$ (In the above formula we use the summation convention in β – but not in j – and for convenience we drop the measures of integration dx in the integrals over Ω and $d\sigma$ in the integral over Γ. We will do so throughout the paper. $Du = (D_\alpha u^j)$ is the Jacobian matrix with $D_\alpha u^j = \partial u^j/\partial x_\alpha$, $\forall\alpha = 1,\ldots,n$, $\forall j = 1,\ldots,N$, ∇ is the usual gradient of a function.)

Our results will hold for a large class of convex sets K, including in particular the convex set of the N-membranes problem (see [5]) and, for $N = 1$, most convex sets defined by pointwise constraints (see Section 3). The paper generalizes in the case of systems results obtained for equations in [1], [2], [3], [4], [6], [7], [9], [10], [11]. We will not address here the question of existence and refer the interested reader to [6], [8], [12], [13].

In order for (2), (3) to make sense we assume that for $j = 1,\ldots,N$, $\beta = 1,\ldots,n$

$$A_j^\beta(x,s,\xi),\quad a_j(x,s),\quad \gamma_j(x,s) \text{ are Carathéodory functions} \qquad (4)$$

(i.e. measurable in x and continuous in the other variables) such that

there exist a constant C, functions $C' \in L_{p'}(\Omega)$,

$C'' \in L_{p'}(\Gamma)$, $C, C', C'' \geqq 0$, $p' = p/p - 1$, such that

$$|A_j^\beta(x,s,\xi)| \leqq C(|s|^{p-1} + |\xi|^{p-1}) + C'(x),$$

$$|a_j(x,s)| \leqq C(|s|^{p-1}) + C'(x),$$

a.e. x in , $\forall s \in \mathbf{R}$, $\forall\xi \in \mathbf{R}^n$,

$$|\gamma_j(x,s)| \leqq C|s|^{p-1} + C''(x) \text{ a.e. } x \text{ in } \Gamma,\ \forall s \in \mathbf{R},$$

(| | denotes either the Euclidean norm or the absolute value). Moreover, we will need the following ellipticity assumption:

$$\exists\nu > 0 \text{ such that } A_j^\beta(x,s,\xi) - A_j^\beta(x,s,\xi')\cdot(\xi_\beta - \xi'_\beta) \geq \nu|\xi - \xi'|^p$$
$$\forall j = 1,\ldots,N,\ \forall s \in \mathbb{R},\ \forall \xi,\xi' \in \mathbb{R}^n, \tag{6}$$

(with a summation in β) and the fact that for any $j = 1,\ldots,N$

$$s \mapsto a_j(x,s) \text{ is nondecreasing a.e. } x \in \Omega, \tag{7}$$

$$s \mapsto \gamma_j(x,s) \text{ is nondecreasing a.e. } x \in \Gamma. \tag{8}$$

To complete our assumptions on the structure of A we will assume also that there exist a positive, nondecreasing, continuous function ω, a constant C and a function $g \in L_{p'}(\Omega)$ such that

$$|A_j^\beta(x,s,\xi) - A_j^\beta(x,t,\xi)| \leq C\omega(|s - t|)(|\xi|^{p-1} + g(x)),$$
$$\forall j = 1,\ldots,N,\quad \forall\beta = 1,\ldots,n,\ \forall s,t \in \mathbb{R},\ \forall \xi \in \mathbb{R}^n,\ \text{a.e. } x \in \Omega. \tag{9}$$

For ω, we will consider the two assumptions

$$\int_{0+} \frac{ds}{\omega(s)^{1/p-1}} = +\infty, \tag{10}$$

$$\int_{0+} \frac{ds}{\omega(s)^{p'}} = +\infty. \tag{11}$$

<u>REMARK 1</u>: Clearly (10) implies (11) and when $p \leq 2$, (10) holds for A_j^β's which are Hölder continuous in s with a Hölder exponent greater than or equal to p - 1, and similarly (11) holds for A_j^β's which are Hölder continuous in s with a Hölder exponent greater than or equal to 1/p', ω being nothing but the modulus of continuity of A_j^β in s. For $p > 2$, (10) does not hold unless the A_j^β's do not depend on s.

2. A GENERAL MONOTONICITY PROPERTY

Let us introduce the following property for a pair of convex sets. We will say that (K_1,K_2) satisfies the property (H) if and only if

$$\text{(H)} \qquad u_1 \in K_1 \text{ and } u_2 \in K_2$$

$$\Rightarrow (u_1^j + F(u_2^j - u_1^j))_j \in K_1 \text{ and } (u_2^j - F(u_2^j - u_1^j))_j \in K_2$$

for any nonnegative, nondecreasing Lipschitz continuous function F having a Lipschitz modulus less than 1 and such that $F(x) = 0$ for $x \leq 0$. (H) appears, with our choice of test function, to be the minimal property necessary to obtain the following result:

THEOREM 1: Let K_i $(i = 1,2)$ be two closed convex sets in $W^{1,p}(\Omega)$ satisfying (H). Let V_i be the space spanned by $K_i - K_i$, V_i^* its dual and $f_i \in V_i^*$ be such that

$$\langle f_1,v\rangle \geq \langle f_2,v\rangle, \quad \forall v \in V_1 \cap V_2,\ v \geq 0. \tag{12}$$

Assume that (2) to (9) hold for K_1 and K_2 and let u_i $(i = 1,2)$ be a solution of

$$u_i \in K_i,\ \langle A(x,u_i,Du_i),v - u_i\rangle \geq \langle f_i,v - u_i\rangle,\ \forall v \in K_i. \tag{13}$$

Then if

(i) (11) holds and for $j = 1,\dots,N$

$$s \to a_j(x,s) \text{ is increasing a.e. } x \in \Omega; \tag{14}$$

(ii) (10) holds and for $j = 1,\dots,N$

$$s \mapsto a_j(x,s) \text{ is increasing on a set of positive measure of } \Omega, \tag{15}$$

or

$$s \mapsto \gamma_j(x,s) \text{ is increasing on a set of positive measure of } \Gamma, \tag{16}$$

or

there exists a constant C such that $|v|_p \leq \sum_{j=1}^{N} \|\nabla v^i\|_p$, $\forall v \in V_1 \cap V_2$ (17)

we have

$$u_2(x) \leq u_1(x) \quad \text{a.e.} \quad x \in \Omega. \tag{18}$$

(In (12) 0 is the vector 0, $v(x) \leq w(x)$ means that each component $w(x)$ is greater than or equal to each component of $v(x)$).

<u>PROOF</u>: Set

$$F_\varepsilon(t) = \begin{cases} \dfrac{1}{I(\varepsilon)} \displaystyle\int_\varepsilon^t \frac{ds}{\omega(s)^{p'}} & \text{if } t \geq \varepsilon, \\ 0 & \text{if } t \leq \varepsilon, \end{cases}$$

where

$$I(\varepsilon) = \int_\varepsilon^{+\infty} \frac{ds}{\omega(s)^{p'}}$$

(taking ω large enough in (9) we can assume without loss of generality that $I(\varepsilon) < +\infty$). For δ small enough, by (H), we have clearly $v_1 = (u_1^j + \delta F_\varepsilon(u_2^j - u_1^j)) \in K_1$ and $v_2 = (u_2^j - \delta F_\varepsilon(u_2^j - u_1^j)) \in K_2$. Hence, substituting these two functions in (13), adding and using (12), we get

$$\langle A(x,u_1,Du_1) - A(x,u_2,Du_2),\ \delta F_\varepsilon(u_2 - u_1)\rangle \geq 0$$

(with the notation $\delta F_\varepsilon(u_2 - u_1) = (\delta F_\varepsilon(u_2^j - u_1^j))_j$). This also reads

$$\langle A(x,u_2,Du_2) - A(x,u_2,Du_1), F_\varepsilon(u_2 - u_1)\rangle \leq$$

$$\langle A(x,u_1,Du_1) - A(x,u_2,Du_1), F_\varepsilon(u_2 - u_1)\rangle.$$

Taking into account (6), (7), (8), (9) and using the Young inequality we deduce easily for some positive constant C

$$\sum_{j=1}^{N} \frac{\nu}{2I(\varepsilon)} \int_{[u_2^j-u_1^j\geq\varepsilon]} \frac{|\nabla(u_2^j-u_1^j|^p}{\omega^{p'}(|u_2^j-u_1^j|)} + \sum_{j=1}^{N} \int_\Omega (a_j(x,u_2^j)-a_j(x,u_1^j))F_\varepsilon(u_2^j-u_1^j)$$

$$+ \sum_{j=1}^{N} \int_\Gamma (\gamma_j(x,u_2^j) - \gamma_j(x,u_1^j))F_\varepsilon(u_2^j - u_1^j)$$

$$\leq \frac{C}{I(\varepsilon)} \sum_{j=1}^{N} \int_\Omega (|\nabla u_1^j|^{p-1} + g(x))^{p'}. \tag{19}$$

Let us first consider the case (i).

By (8), (19) we get

$$\sum_{j=1}^{N} \int_\Omega (a_j(x,u_2^j) - a_j(x,u_1^j))F_\varepsilon(u_2^j - u_1^j) \leq \frac{C}{I(\varepsilon)} \sum_{j=1}^{N} \int_\Omega (|\nabla u_1^j|^{p-1} + g(x))^{p'}$$

and letting $\varepsilon \to 0$ we obtain

$$\int_{[u_2^j-u_1^j\geq 0]} a_j(x,u_2^j) - a_j(x,u_1^j) \leq 0, \quad \forall j = 1,\ldots,N.$$

which, with (14), gives our result.

Assume now that (ii) holds.

Recalling (11) we get by letting $\varepsilon \to 0$

$$\int_{[u_2^j-u_1^j\geq 0]} a_j(x,u_2^j)-a_j(x,u_1^j)+ \int_{[u_2^j-u_1^j\geq 0]} \gamma_j(x,u_2^j)-\gamma_j(x,u_1^j) \leq 0, \forall j=1,\ldots,N,$$

and in the two first cases of (ii), each component of $u_2 - u_1$ is nonpositive on a part of positive measure of Ω or Γ. Again, by (19), we have

$$\sum_{j=1}^{N} \int_{[u_2^j-u_1^j\geq\varepsilon]} \frac{|\nabla(u_2^j-u_1^j)|^p}{\omega^{p'}(|u_2^j-u_1^j|)} \leq \frac{C}{\nu} \sum_{j=1}^{N} \int_\Omega (|\nabla u_1^j|^{p-1} + g(x))^{p'} \leq C',$$

where C' is a constant independent of ε. If we set

$$S_\varepsilon(t) = \begin{cases} \displaystyle\int_\varepsilon^t \frac{ds}{\omega(s)^{1/p-1}} & \text{if } t \geq \varepsilon, \\ 0 & \text{if } t \leq \varepsilon, \end{cases}$$

we obtain

$$\sum_{j=1}^{N} \int_\Omega |\nabla S_\varepsilon(u_2^j - u_1^j)|^p \leq C'.$$

But it is easy to see that $S_\varepsilon(u_2 - u_1) = (S_\varepsilon(u_2^j - u_1^j))_j \in V_1 \cap V_2$ (see (H)) and that the Poincaré inequality holds (in case (17) this is a part of the assumption, in cases (15) and (16) this results from the fact that $S_\varepsilon(u_2 - u_1)$ vanishes on a part of positive measure on Ω or Γ). Thus, we obtain

$$\int_\Omega |S_\varepsilon(u_2^j - u_1^j)|^p \leq C', \quad \forall j = 1,\dots,N,$$

and the conclusion of Theorem 1 follows by letting $\varepsilon \to 0$ due to (10). □

REMARK 2: In case (ii), if we only assume that $s \to a_j(x,s)$ is nondecreasing then uniqueness can fail (see [6], [13]). However, some results are preserved in this case (see [3], [6], [8], [13]). We do not know, when $\omega(t)$ is only assumed to tend to 0 as t goes to 0, if (ii) of Theorem 1 holds. Some progress in this direction is made in [4].

3. SOME APPLICATIONS

3.1. The N-membranes problem

We would like to show that Theorem 1 leads, in particular, to uniqueness of a solution to the so-called N-membranes problem. For this, let us consider (1) with the two convex sets

$$K_i = \{v = (v^1,\dots,v^N) \in W^{1,p}(\Omega) \mid v^j = \phi_i^j \text{ on } \Gamma_0^j,$$
$$v^1(x) \geq v^2(x) \geq \dots \geq v^N(x) \text{ a.e. } x \text{ in } \Omega\},$$

where, for i = 1,2, $\phi_i \in W^{1,p}(\Omega)$. Then it is easy to show (see [6]) that if $\phi_2^j(x) \leq \phi_1^j(x)$ a.e. $x \in \Gamma_0^j$, then (K_1,K_2) verifies (H). Thus we have the following corollary:

COROLLARY 1: Under the assumptions of Theorem 1, if u_i (i = 1,2) is a solution of (1) corresponding to K_i, or if $\phi_2^j(x) \leq \phi_1^j(x)$ a.e. $x \in \Gamma_0^j$, then we have

$$u_2(x) \leqq u_1(x) \text{ a.e. } x \text{ in } \Omega.$$

In particular, we get uniqueness for (1) corresponding to $K = K_1$ or K_2.

Let us now examine what kind of operator can be used here. If, for simplicity, we restrict ourselves to the case $p = 2$ and consider $A_j^\beta(x,s,\xi) = a_j^{\alpha,\beta}(x,s)\cdot\xi_\alpha$ and if we assume that for C, $\nu > 0$, and for every j

$$|a_j^{\alpha,\beta}(x,s) - a_j^{\alpha,\beta}(x,t)| \leq C\omega(|s-t|),\ a_j^{\alpha,\beta}(x,s)\xi_\alpha\xi_\beta \geqq \nu|\xi|^2,\ \forall\xi \in R^n,$$

then the assumptions of Theorem 1 are fulfilled. Taking, in particular, $\Gamma_0^j = \Gamma$, $\forall j = 1,\ldots,N$, i.e.

$$K = \{v = (v^1,\ldots,v^N) \in W^{1,2}(\Omega)\,|\,v = \phi \text{ on } \Gamma,$$

$$v^1(x) \geqq v^2(x) \geqq \ldots \geqq v^N(x) \text{ a.e. } x \text{ in } \Omega\},$$

we have uniqueness and monotonicity with respect to the data for this problem when the operator is quasi-linear. (For complementary results see [8].)

3.2. Monotonicity results for variational inequalities

Here $N = 1$. We are going to show that Theorem 1 leads to uniqueness of the solution of any equation or, more generally, any variational inequality associated with the standard convex sets K defined by pointwise constraints. Let us introduce some closed convex sets and for $i = 1,2$ let us consider the functions

$$\phi_i : \Gamma \to \bar{R}, \qquad \psi_i : \Gamma \to \bar{R},$$

$$\Phi_i : \Omega \to \bar{R}, \qquad \Psi_i : \Omega \to \bar{R}.$$

Set

$$K_i = \{v \in W^{1,p}(\Omega) \mid \phi_i(x) \leqq v(x) \leqq \psi_i(x) \quad \text{a.e. } x \in \Gamma,$$

$$\Phi_i(x) \leqq v(x) \leqq \Psi_i(x), \quad \nabla v(x) \in C(x) \quad \text{a.e. } x \in \Omega\},$$

where, for a.e. $x \in \Omega$, $C(x)$ is a closed convex set of $\mathbf{R}^n$ and the restriction of v to Γ is taken in the trace sense. It is easy to prove (see [6], [13]) that if $(\phi_2,\psi_2,\Phi_2,\Psi_2) \leqq (\phi_1,\psi_1,\Phi_1,\Psi_1)$ (i.e. if each component of the first vector is less than or equal, a.e. on Γ or a.e. on Ω, to each component of the second one), then the pair $(K_1;K_2)$ satisfies (H). So, as an obvious consequence of Theorem 1, we have:

<u>COROLLARY 2</u>: With the assumptions of Theorem 1, if u_i is a solution of (13) with the above convex sets K_i, and if $(f_2,\phi_2,\psi_2,\Phi_2,\Psi_2) \leqq (f_1,\phi_1,\psi_1,\Phi_1,\Psi_1)$, then $u_2(x) \leqq u_1(x)$ a.e. $x \in \Omega$. In particular, for

$$K = \{v \in W^{1,p}(\Omega) \mid \phi(x) \leqq v(x) \leqq \psi(x) \text{ a.e. } x \in \Gamma,$$

$$\Phi(x) \leqq v(x) \leqq \Psi(x), \ \nabla v(x) \in C(x) \text{ a.e. } x \in \Omega\} \tag{20}$$

the problem (1) has a unique solution.

The above result gives us uniqueness or, more generally, monotonicity properties for numerous problems. Let us list a few of them.

(a) Nonlinear elliptic boundary value problems.

Choose here $\Phi \equiv -\infty$, $\Psi \equiv +\infty$, $C(x) = \mathbf{R}^n$, $\forall x \in \Omega$. Select some function ϕ_0 in $W^{1,p}(\Omega)$ and choose $\phi = \psi = \phi_0$ on a subset Γ_0 of Γ, $\phi = -\infty$, $\psi = +\infty$ elsewhere. Then we have $K = \phi_0 + V$ where V is defined as the space $V = \{v \in W^{1,p}(\Omega) : v = 0 \text{ on } \Gamma_0\}$. So, for $f \in V^*$ defined by

$$\langle f,v\rangle = \int_\Omega f_1 \cdot v + \int_\Gamma f_2 \cdot v$$

(with f_1, f_2 in $L_{p'}(\Omega)$ or $L_{p'}(\Gamma)$, respectively, $A_1^\beta = A^\beta$, $a_1 = a, \gamma_1 = \gamma$) u is the solution of the nonlinear problem

$$- \frac{\partial A^{\beta}(x,u,\nabla u)}{\partial x^{\beta}} + a(x,u) = f_1 \text{ in } \Omega,$$

$$u = \phi_0 \text{ on } \Gamma_0,$$

$$A^{\beta}(x,u,\nabla u)\cdot n_{\beta} + \gamma(x,u) = f_2 \text{ on } \Gamma \setminus \Gamma_0$$

(n is the outward normal to Γ).

So, in case (i) or (ii), by the previous result, and if a solution is known to exist, then this solution is unique and depends monotonically on the data ϕ_0, f_1, f_2. If Γ_0 has a positive measure it is well known that the Poincaré inequality holds and we are in case (ii). In the particular case where $\Gamma_0 = \Gamma$ we have a nonlinear Dirichlet problem. In the case where $\Gamma_0 = \emptyset$, the problem is a problem of the Neumann type for which we have uniqueness in cases (i), (ii).

(b) Obstacle problems.

As above take here $\phi = \psi = \phi_0$ on Γ_0 where Γ_0 is a subset of Γ, $\phi = -\infty$, $\psi = +\infty$ elsewhere, $C(x) = \mathbb{R}^n$, $\forall x \in \Omega$. Then if E, F are two measurable sets in Ω take

$$\Phi = \Phi \text{ on } E, \ \Phi = -\infty \text{ on } \Omega \setminus E,$$

$$\Psi = \Psi \text{ on } F, \ \Psi = +\infty \text{ on } \Omega \setminus F.$$

Then K becomes

$$K = \{v \in W^{1,p}(\Omega) \mid v(x) = \phi_0 \text{ on } \Gamma_0,$$

$$\Phi(x) \leq v(x) \text{ a.e. } x \in E, \ v(x) \leq \Psi(x), \text{ a.e. } x \in F\}$$

and for such a convex set we get uniqueness of a solution as well as monotone dependence with respect to the data f, ϕ_0, Φ, Ψ. Note that when $E = \Omega$, $F = \emptyset$ we have the usual one-obstacle problem and when $E = F = \Omega$ the double-obstacle problem.

(c) Signorini's problems or thin obstacle problems.

Take $\Phi \equiv -\infty$, $\Psi \equiv +\infty$, $C(x) = R^n$, $\forall x \in \Omega$, and if E, F are two measurable sets in Γ take

$$\phi = \phi \text{ on } E,\ \phi = -\infty \text{ on } \Gamma\setminus E,$$

$$\psi = \psi \text{ on } F,\ \psi = +\infty \text{ on } \Gamma\setminus F.$$

Then K becomes

$$K = \{v \in W^{1,p}(\Omega) \mid \phi(x) \leq v(x) \text{ a.e. } x \in E,\ v(x) \leq \psi(x), \text{ a.e. } x \in F\},$$

and for such a convex set, and provided that we are in case (i) or (ii) we have uniqueness and monotonicity in f,ϕ,ψ for the solution.

(d) Problems with constraints on the derivatives.

Take, for instance, $\phi = \psi = \phi_0$ on Γ, $\Phi \equiv -\infty$, $\Psi \equiv +\infty$, then K becomes

$$K = \{v \in W^{1,p}(\Omega) \mid v = \phi_0 \text{ on } \Gamma,\ \nabla v(x) \in C(x) \text{ a.e. } x \in \Omega\}.$$

In the case $p = 2$, $\phi_0 \equiv 0$, $C(x) = B_1$, where B_1 is the unit ball of $\mathbf{R}^n$ we get convex set of the elastic-plastic torsion problem

$$K = \{v \in W^{1,2}(\Omega) \mid v = 0 \text{ on } \Gamma,\ \ |\nabla v(x)| \leq 1 \text{ a.e. } x \in \Omega\}.$$

Now if $A: R^n \to \mathbf{R}^q$ is a linear map and C is a closed convex set of $\mathbf{R}^q$, $A^{-1}C = \{\xi \mid A\xi \in C\}$ is a closed convex set of $\mathbf{R}^n$. So, if A(x) is a matrix defined on Ω, uniqueness holds also for

$$K = \{v \in W^{1,p}(\Omega) \mid v = \phi_0 \text{ on } \Gamma,\ A(x)\nabla v(x) \in C(x) \text{ a.e. } x \in \Omega\}.$$

REFERENCES

[1] M. Artola, Sur une classe de problèmes paraboliques quasilinéaires, Bolletino U.M.I. 6, 5-B (1986), 51-70.

[2] H. Brezis, D. Kinderlehrer and G. Stampacchia, Sur une nouvelle formulation du problème de l'écoulement à travers une digue, C.R. Acad. Sc. Paris 303, Série A (1987), 711-714.

[3] J. Carrillo and M. Chipot, On nonlinear elliptic equations involving derivative of the nonlinearity, Proceeding of the Royal Society of Edinburgh 100A (1985), 281-294.

[4] J. Carrillo, Uncité des solutions de type Kruskov pour des problèmes elliptiques avec des termes de transport non linéaires, C.R. Acad. Sc. Paris 303, Série I (1986), 189-192.

[5] M. Chipot and G. Vergara Caffarelli, The N-membranes problem, Appl. Math. Optimization 13 (1985), 231-249.

[6] M. Chipot and G. Michaille, Uniqueness results and monotonicity properties for strongly nonlinear elliptic variational inequalities, Preprint #347, Institute For Mathematics And Its Applications, University of Minnesota, 1987 (to appear).

[7] J. Douglas Jr., T. Dupont and J. Serrin, Uniqueness and comparison theorems for nonlinear elliptic equations in divergence form, Arch. Rat. Mech. Anal. 42 (1972), 157-168.

[8] A. Feggous, Thèse, Université de Metz, 1988.

[9] G. Gagneux, Une approche analytique nouvelle des modèles de la récupération secondée en ingéniérie pétrolière. J. Mécanique Théor. Appl. 5, N1 (1986), 3-20.

[10] G. Gagneux and F. Guerfi, Approximation de la fonction de Heaviside et résultats d'unicité pour une classe de problèmes quasi-linéaires stationnaires, Preprint, Université de Pau et des Pays de L'Adour.

[11] D. Gilbarg and N.S. Trudinger, Elliptic Partial Differential Equations of Second Order, Springer, 1985.

[12] J.L. Lions, Quelques methodes de résolution des problèmes aux limites non linéaires. Dunod-Gauthier-Villars, 1969.

[13] G. Michaille, Thèse, Université de Metz, 1988.

[14] N.S. Trudinger, On the comparison principle for quasilinear divergence structure equations, Arch. Rat. Mech. Anal. 57 (1974), 128-133.

M. Chipot, A. Feggous and G. Michaille
Departement de Mathematiques
Université de Metz
Ile du Saulcy, 57045 Metz Cedex
France

P. CLÉMENT AND G. SWEERS

On subsolutions to a semilinear elliptic problem

ABSTRACT: The relation between the existence of a subsolution for the problem $-\Delta u = f(u)$ with 0 – Dirichlet boundary value on a bounded domain and on a ball of R^N is considered. As a consequence a necessary condition for the existence of solutions when f changes sign is given.

1. INTRODUCTION AND STATEMENT OF RESULTS

In this paper we consider the following problem:

$$\begin{aligned} -\Delta u &= f(u) \text{ in } \Omega, \\ u &= 0 \quad \text{on } \partial\Omega, \end{aligned} \tag{1.1}$$

where Ω is a bounded domain of R^N, $N > 1$, and the function $f: R \to R$ is assumed to be only continuous.

We call a function $u \in C(\bar{\Omega})$ a "*subfunction on* Ω" if the following differential inequality

$$\int_\Omega [u(-\Delta\phi) - f(u)\phi]dx \leqq 0 \tag{1.2}$$

holds for every $\phi \in D^+(\Omega)$, where $D^+(\Omega)$ consists of all nonnegative functions in $C_0^\infty(\Omega)$.

We shall make use of the following result (see [8, Lemma A.4, p. 105]): let u and v be subfunctions on Ω, then $\max(u,v)$ is also a subfunction on Ω. If the function u also satisfies

$$u \leqq 0 \text{ on } \partial\Omega, \tag{1.3}$$

then we shall call u a "*subsolution on* Ω". If the inequality in (1.2) (resp. (1.2) and (1.3)) is reversed, then we call u a superfunction on Ω (resp. a supersolution on Ω). A *solution on* Ω is a function u which is both a sub- and a supersolution on Ω. We shall use the following notation:

for $x_0 \in \mathbb{R}^N$ and $R > 0$, $B(x_0,R) := \{x \in \mathbb{R}^N;\ |x-x_0| < R\}$ and $B_R = B(0,R)$.

THEOREM 1: Let u be a subsolution on Ω satisfying $\max_\Omega u > 0$. Then there exist $R > 0$ and a subsolution v on B_R with $\max_{B_R} v = \max_\Omega u$, satisfying:

(i) v is positive on B_R and $v = 0$ on ∂B_R,

(ii) v is radially symmetric.

REMARK: The function v can even be chosen such that $|x| \to v(x)$ is non-increasing on $[0,R]$. As a consequence, we have

THEOREM 2: If u is a subsolution on Ω with $\max_\Omega u > 0$, then

$$\int_s^{\max u} f(t)\,dt > 0 \text{ holds for all } s \in [0,\max u). \tag{1.4}$$

A first result in this direction was obtained by De Figueiredo in [6]. With additional regularity on f and for *positive* solutions, condition (1.4) has been proved to be necessary in Dancer and Schmitt [5]. See also for related results [2], [3], [9]. In the proof of Theorems 1 and 2, we avoid the use of a theorem of Gidas, Ni and Nirenberg [7], which requires the positivity of u and more regularity on f.

Concerning a partial converse of Theorem 1, we mention the following result [10]: Let v be a subsolution on B_1 satisfying $\max_{B_1} v > 0$. Suppose either $f(0) \geq 0$ or Ω satisfies a uniform interior sphere condition [1]. Then there exist $\lambda > 0$ and $u \in C(\bar{\Omega})$ satisfying

$$\int_\Omega u(-\Delta\phi)dx \leq \lambda \int_\Omega f(u)\phi\,dx, \text{ for every } \phi \in D^+(\Omega),\ \max u = \max v,$$

$$u \text{ is positive on } \Omega, \text{ and} \tag{1.5}$$

$$u = 0 \text{ on } \partial\Omega.$$

If $f(0) < 0$ and Ω does not satisfy a uniform sphere condition, it may happen that the conclusion fails.

EXAMPLE: Let $f(u) = -\cos u$. Then if the boundary of Ω is of class C^3, there exists a pair $(\lambda,u) \in \mathbb{R}^+ \times C^d(\bar{\Omega}) := \mathbb{R}^+ \times C^2(\Omega) \cap C^0(\bar{\Omega})$ satisfying

$$\begin{aligned} -\Delta u &= \lambda f(u) \quad \text{in } \Omega, \\ u &= 0 \quad \text{on } \partial\Omega, \end{aligned} \tag{1.6}$$

with u positive in Ω and $\max u \in (\pi, 3\pi/2)$ (see, for example, [3]). This is true in particular if Ω is the unit ball in $\mathbb{R}^N$. However, it is shown in [8] that if Ω is a hypercube there is no pair $(\lambda,u) \in \mathbb{R}^+ \times C^d(\bar{\Omega})$ satisfying (1.6), with u positive in Ω and $\max u \in (\pi, 3\pi/2)$. Since $3\pi/2$ is a supersolution of (1.6) for every $\lambda > 0$, there is no positive subsolution of (1.6) with maximum lying in $(\pi, 3\pi/2)$ for every $\lambda > 0$. For a proof of this statement and for more results in this direction, we refer the reader to [8] and [10].

2. PROOFS

PROOF OF THEOREM 1: Let u be a subsolution on Ω satisfying $\max_{\Omega} u > 0$. Without loss of generality we may assume, by using a translation, that the maximum of u is achieved at the origin. Since Ω is bounded, we may also assume, by using Tietze's theorem, that the function u is the restriction on $\bar{\Omega}$ of some continuous function on $\mathbb{R}^N$, nonpositive outside of Ω and zero outside of a ball large enough. We shall still denote the extended function by u. Define

$$u^*(r) = \max\{u(x);\ |x| = r\} \text{ for every } r \geq 0.$$

Observe that u^* is continuous since the following inequality holds:

$$|u^*(r_1) - u^*(r_2)| \leq \max_{|\theta|=1} |u(r_1\theta) - u(r_2\theta)| \tag{2.1}$$

and u is uniformly continuous on $\mathbb{R}^N$.

We also have

$$u^*(0) = u(0) = \max u = \max u^* > 0.$$

Denote by R the first zero of u^*. Define

$$v(x) = u^*(|x|) \text{ for } |x| \in [0,R].$$

Then v is continuous on $\bar{B}_R$, positive on B_R, $v = 0$ on ∂B_R, $\max_{B_R} v = \max_{\Omega} u$, and v is radially symmetric.

We shall prove that v is a subfunction on B_R, and therefore v satisfies all the required properties. By using partitions of unity and the compactness of the support of the test functions ϕ's, it is sufficient to prove that for every $x_0 \in B_R$ there is $r_0 > 0$ such that $B(x_0,r_0)$ lies in B_R and that v is a subfunction on $B(x_0,r_0)$. Let $x_0 \in B_R$ and $\alpha = v(x_0) > 0$. From the uniform continuity of u on $\mathbb{R}^N$, one finds $r_0 > 0$ such that

$$|u(x) - u(y)| < \frac{1}{3}\alpha \text{ for } |x-y| < r_0, \ x,y \in \mathbb{R}^N. \tag{2.2}$$

From (2.1), we also have

$$|u^*(s_1) - u^*(s_2)| < \frac{1}{3}\alpha \text{ for } |s_1-s_2| < r_0, s_1, s_2 \geqq 0. \tag{2.3}$$

From (2.3) we get

$$u^*(r) > \frac{2}{2}\alpha > 0 \text{ for } |r - |x_0|| < r_0. \tag{2.4}$$

One finds that $B(x_0,r_0)$ lies in B_R, since u^* is continuous and vanishes at R. Define

$$\Sigma = \{\theta \in \mathbb{R}^N; \ |\theta| = 1\} \text{ and}$$

$$\Sigma' = \{\theta \in \Sigma; \ |x_0|\theta \in \Omega \text{ and } d(|x_0|\theta,\partial\Omega) \geqq r_0\}. \tag{2.5}$$

Let $\theta \in \Sigma\setminus\Sigma'$ and $y \in B(|x_0|\theta, r_0)$, then it follows from (2.2) and the fact that $u \leqq 0$ outside of Ω, that $u(y) < \frac{2}{3}\alpha$.

Recalling (2.4) we obtain

$$u^*(r) = \max\{u(r\theta); \ \theta \in \Sigma\} = \max\{u(r\theta); \ \theta \in \Sigma'\} \text{ for } |r-|x_0|| < r_0.$$

We can even choose a countable dense subset Σ'' of Σ' such that

$$u^*(r) = \max\{u(r\theta);\ \theta \in \Sigma''\} \text{ for } |r-|x_0|| < r_0.$$

Denote by $\theta_0,\theta_1,\theta_2,\ldots,\theta_n,\ldots$ the elements of Σ''. It follows from the definition of Σ', (2.5), that $B(|x_0|\theta_n,r_0) \subset \Omega$ for all $n \in \mathbb{N}$. Moreover, u is a subfunction on $B(|x_0|\theta_n,r_0)$ $n \in \mathbb{N}$. For every $n \in \mathbb{N}$, there is a rotation R_n which maps $B(x_0,r_0)$ onto $B(|x_0|\theta_n,r_0)$ and such that

$$v(x) = \max\{u(R_n x);\ n \in \mathbb{N}\},\ x \in B(x_0,r_0).$$

From the rotation invariance in (1.2), it follows that the functions w_n defined by $w_n(x) = u(R_n x)$ are subfunctions on $B(x_0,r_0)$. Then

$$v_n = \max\{w_k;\ 0 \leqq k \leqq n\}$$

is an increasing sequence of subfunctions on $B(x_0,r_0)$ and by Dini's theorem $v = \sup_{n\geqq 0} v_n = \lim_{n\to\infty} v_n$ is a subfunction on $B(x_0,r_0)$. This completes the proof of Theorem 1. □

PROOF OF THEOREM 2: Let u be a subsolution of Ω with $\rho = \max u > 0$. Let $R > 0$ and let v be the subsolution on B_R from Theorem 1.

Suppose that there is $s \in [0,\rho)$ such that $\int_s^\rho f(t)\,dt \leqq 0$. We will obtain a contradiction. For $n = 1,2,\ldots,$ define

$$f_n(t) = \begin{cases} f(t) & \text{for } t \leqq \rho, \\ f(\rho)[1+n(\rho-t)] & \text{for } t > \rho\ . \end{cases}$$

Then $f_n \in C(\mathbb{R})$ and $f_n(\rho + 1/n) = 0$. Then v is a radially symmetric subsolution to the problem

$$(P_n) \qquad \begin{cases} -\Delta u = f_n(u) & \text{on } B_R, \\ u = 0 & \text{on } \partial B_R, \end{cases}$$

and $\rho + 1/n$ is a radially symmetric supersolution of (P_n) satisfying $v \leqq \rho + 1/n$. By using a result of [4], there exists a solution u_n of (P_n), satisfying $v \leqq u_n \leqq \rho + 1/n$. By using a slight modification of the argument of [4], namely by applying the Schauder fixed-point theorem on the space of continuous functions on $\bar{B}_R$ which are radially symmetric, one can assume that u_n is radially symmetric. Then we obtain, since $u_n \in C^2(\bar{B}_R)$,

$$\begin{aligned} -u_n''(r) - \frac{N-1}{r} u_n'(r) &= f_n(u_n(r)), \quad r \in [0,R], \\ u_n'(r) = 0, \qquad & \qquad u_n(R) = 0. \end{aligned} \tag{2.6}$$

By integration, we obtain

$$\int_{u_n(r)}^{u_n(0)} f_n(s)\,ds \geq (N-1) \int_0^r \frac{1}{s} u_n'^2(s)\,ds, \; r \in [0,R]. \tag{2.7}$$

Since the functions $f_n(u_n)$ are uniformly bounded on $[0,R]$, it follows from (2.6) that the functions u_n, u_n' and u_n'' are uniformly bounded on $[0,R]$. For every $n \in \mathbb{N}$, there exists $r_n \in [0,R]$ such that $u_n(r_n) = s$. There exist a function $\bar{u} \in C^1[0,R]$, $\bar{r} \in [0,R]$ and a subsequence which we still denote by (u_n,r_n) such that u_n converges to $\bar{u}$ in $C^1[0,R]$ and r_n to $\bar{r}$.

Since $\rho > s = \lim_{n\to\infty} u_n(r_n) = \bar{u}(r)$ and $\bar{u}(0) = \rho$, we have $\bar{r} > 0$. From (2.7), we obtain

$$\int_0^{\bar{r}} \frac{1}{t} \bar{u}'^2(t)\,dt = \lim_{n\to\infty} \int_0^{r_n} \frac{1}{t} u_n'^2(t)\,dt = 0.$$

Hence $\bar{u}(t) = \rho$ on $[0,\bar{r}]$, and $s = \bar{u}(\bar{r}) = \rho$, a contradiction. This completes the proof of Theorem 2. □

REFERENCES

[1] R.A. Adams, Sobolev Spaces, Academic Press, New York, 1975.

[2] Ph. Clément and G. Sweers, Existence et multiplicité des solutions d'un problème aux valeurs propres elliptique semilinéaire, C.R. Acad. Sci. Paris 302, Série I, 19 (1986), 681-683.

[3] Ph. Clément and G. Sweers, Existence and multiplicity results for a semilinear elliptic eigenvalue problem, Ann. Scuola Norm. Sup. Pisa, Cl. Sci. 13 (1987), 97-121.

[4] Ph. Clément and G. Sweers, Getting a solution between sub- and super-solutions without monotone iteration, Rend. dell'Instituto di Matematica dell'Università di Trieste, Vol. XIX (1987), 189-194.

[5] E.N. Dancer and K. Schmitt, On positive solutions of semilinear elliptic equations, Proc. A.M.S. 101 (1987), 445-452.

[6] D.G. De Figueiredo, On the uniqueness of positive solutions of the Dirichlet problem $-\Delta u = \lambda \sin u$, Nonlinear P.D.E. Appl. Collège de France Seminar, Vol. 7, Pitman, 1985, pp. 80-83.

[7] B. Gidas, W.M. Ni and L. Nirenberg, Symmetry and related properties via the maximum principle, Comm. Math. Phys. 68 (1979), 209-243.

[8] G. Sweers, Doctoral Thesis, Delft, 1988.

[9] G. Sweers, On the maximum of solutions for a semilinear elliptic problem, Proceedings of the Royal Soc. of Edinburgh, 108A (1988), 357-370.

[10] G. Sweers, Semilinear elliptic problems on domains with corners, to appear in Commun. Partial Differ. Eq.

Ph. Clement and G. Sweers
Faculty of Mathematics
Delft University of Technology
Julianalaan 132
2628 BL Delft
THE NETHERLANDS

I. FONSECA

Remarks on phase transitions

A typical problem of phase transitions consists of minimizing an energy functional of the type

$$E(u) := \int_{\Omega} W(u)\, dx$$

subject to appropriate boundary conditions, where $\Omega \subset \mathbb{R}^N$ is an open bounded smooth domain, u maps Ω into $\mathbb{R}^n$ and W supports more than one phase (i.e. $W \geqq 0$ and there exist $a \neq b$ such that $W(a) = 0 = W(b)$).

As an example, for materials with crystalline structure, W is periodic in many directions and has several potential wells (see Chipot and Kinderlehrer [5], Ericksen [7], Fonseca [8], Kinderlehrer [17]). In this case, W is the strain energy density and u is a deformation gradient.

The main feature of these functionals is that either they do not have a minimizer in the class of admissible functions (see Ball and James [3]) or, when interfaces are allowed to form without an increase in energy, we are, in general, faced with a striking nonuniqueness of solutions. It is then natural to search for a model that singles out the solutions that are more likely to be observed.

I. REGULARIZATION

In a first approach, we study the asymptotic behaviour of minimizers of the perturbed problem

(P_ε) Minimize

$$E_\varepsilon(u) := \int_{\Omega} W(u)\, dx + \varepsilon^2 \int_{\Omega} |\nabla u|^2 dx$$

on

$$\{u \in W^{1,1}(\Omega; \mathbb{R}^n) \mid \int_{\Omega} u(x)\, dx = m\}, \text{ where } m = \theta a + (1-\theta)b \text{ for some } \theta \in (0,1)$$

We suppose that Ω is an open bounded strongly Lipschitz domain of $\mathbb{R}^N$ with $meas(\Omega) = 1$ and that W satisfies the following properties:

(H1) W is a nonnegative locally Lipschitz function such that

$$W(u) = 0 \text{ if and only if } u \in \{a,b\}, \text{ where } a \neq b;$$

(H2) there exist $\alpha, \delta > 0$ such that

$$|u - a| < \delta \Rightarrow \alpha\, |u - a|^2 \leq W(u) \leq \frac{1}{\alpha}\, |u - a|^2$$

and

$$|u - b| < \delta \Rightarrow \alpha\, |u - b|^2 \leq W(u) \leq \frac{1}{\alpha}\, |u - b|^2;$$

(H3) there exist $C, R > 0$ such that

$$\text{if } |u| > R \text{ then } W(u) \geq C|u|.$$

Consider the functionals

$$J_\varepsilon(u) := \begin{cases} \frac{1}{\varepsilon}\int_\Omega W(u)\, dx + \varepsilon \int_\Omega |\nabla u|^2\, dx & \text{if } u \in W^{1,1}(\Omega;\mathbb{R}^n) \text{ and } \int_\Omega u(x)dx=m, \\ +\infty & \text{otherwise,} \end{cases}$$

and

$$J_0(u) := \begin{cases} K\, Per_\Omega(\{u=a\}) & \text{if } u \in BV(\Omega), u(x) \in \{a,b\} \text{ a.e. and } \int_\Omega u(x)dx = m, \\ +\infty & \text{otherwise,} \end{cases}$$

where K is the length of a geodesic curve, namely

$$K := 2 \inf \{\int_{-1}^{1} \sqrt{W(g(s))}\, |g'(s)|ds \,|\, g \text{ is piecewise } C^1,\ g(-1) = a, g(1)=b\}.$$

THEOREM I.1 ([10]):

(i) Any family (v_ε) such that $J_\varepsilon(v_\varepsilon) \leq$ const. $< +\infty$ for all $\varepsilon > 0$ is compact in $L^1(\Omega; R^n)$;

(ii) if $v_\varepsilon \to v_0$ in $L^1(\Omega; R^n)$ then $\liminf J_\varepsilon(v_\varepsilon) \geq J_0(v_0)$;

(iii) for any $v_0 \in L^1(\Omega; R^n)$ there exists a family (v_ε) such that $v_\varepsilon \to v_0$ in $L^1(\Omega; R^n)$ and $\lim J_\varepsilon(v_\varepsilon) = J_0(v_0)$.

Properties (ii) and (iii) state that J_0 is the $\Gamma(L^1(\Omega))$-limit of J_ε.

It turns out that "physically preferred" solutions of E_0 are those that minimize the area of the interface $\{x \in \Omega | x \in \partial A\}$ where $A := \{u = a\}$.

THEOREM I.2 ([10]): Every sequence of solutions (u_ε) of (P_ε) admits a subsequence converging in $L^1(\Omega; R^n)$ to a solution u_0 of (P_0) with minimal interfacial area, i.e. u_0 is a solution of the geometric variational problem:

$$\text{Minimize } \mathrm{Per}_\varepsilon(\{u = a\})$$

on

$$\{u \in BV(\Omega) | W(u) = 0 \text{ a.e. and } \int_\Omega u(x)dx = m\}.$$

The case n = 1 was studied by Carr, Gurtin and Slemrod [4], Gurtin [14], Kohn and Sternberg [18], Modica [19], Owen [20, [21], Owen and Sternberg [22], Sternberg [26]. More recently, the case $n \geq 2$ was analysed by Fonseca and Tartar [10].

In the context of nonlinear elasticity, u is a deformation gradient and there are necessary compatibility conditions to be satisfied. In order to accommodate these, we take a different approach where the interfaces are directly penalized (see Fonseca [9], Gurtin [12], [13], Parry [23], Pitteri [24]).

II. DIRECT PENALIZATION OF THE INTERFACE

Here, we adopt the point of view that, in the dead loading case, equilibria correspond to local minima of an energy functional of the form

$$E(u) := \int_{\Omega} W(\nabla u)\, dx + \sum_{I \in F} \int_I \Gamma\, dS - \int_{\Omega} f \cdot u\, dx - \int_{\partial\Omega_2} t \cdot u\, dS,$$

where W is the strain energy density, Γ is a surface energy density accounting for jumps in the deformation gradient, F is the family of phase boundaries, f represents the body forces and t is the surface traction on a portion $\partial\Omega_2$ of the boundary $\partial\Omega$ of the strongly Lipschitz bounded domain $\Omega \subset \mathbb{R}^3$.

The constitutive hypotheses for Γ are motivated by the previous work of Herring [15] and Parry [23], who based their analysis on molecular considerations of surface energies arising from interatomic interaction of finite range in solid crystals.

In what follows, an "*admissible deformation*" is a mapping $u \in W^{1,\infty}(\Omega; \mathbb{R}^3)$ such that there exist a finite partition of Ω, $\{\Omega_i\}_{i=1,\dots,P}$ and a set E of area zero verifying

$$u \in C^2(\bar{\Omega}_i; \mathbb{R}^3) \text{ and } \det \nabla u > 0 \text{ on } \bar{\Omega}_i,$$

$$I_{ij} := \partial\Omega_i \cap \partial\Omega_j \cap \partial\Omega\ ,\ 1 \leq i < j \leq P, \text{ is a } C^2 \text{ surface at every } x \in I_{ij} \setminus E,$$

and

$$\text{area } (I_{ij} \cap I_{k,l}) = 0 \text{ if } (i,j) \neq (k,l).$$

We call the surfaces I_{ij}, $1 \leq i < j \leq P$, the "*interfaces of u*" (see Fonseca [8], Gurtin [11]). In most cases of interest, E reduces to a finite union of piecewise C^2 curves.

As a kinematical restriction, if n is the direction of the normal to the interface I and if F_+ and F_- are the gradients of the deformations that form the adjoint phases of an admissible u then

$$F_+ = F_- + a \otimes n$$

and we set

$$\Gamma = \Gamma(F_-; a, n).$$

The considerationsof Herring [15] and Parry [23] suggest that Γ presents a a certain lack of differentiability with respect to some crystallographically simple directions $n \in N \subset \partial B(0.1)$. Accordingly we assume that

(H1) $W \geqq 0$ and $W \in C^1(M_+^{3\times 3})$, where $M_+^{3\times 3}$ is the set of 3×3 real matrices with positive determinant;

(H2) $\Gamma \geqq 0, \Gamma \in W^{1,\infty}_{loc}(X) \cap C^2(X \setminus X^*)$ and $\Gamma(\cdot;a,n) \in C^2$, where

$X := \{(F;a,n) \in M_+^{3\times 3} \times \mathbb{R}^3 \times \mathbb{R}^3 | F + a \otimes n \in M_+^{3\times 3}\}$ and

$X^* := \{(F;a,n) \in X | n \in N\}$;

(H3) $\Gamma(F;a,n) = \Gamma(F + a \otimes n;-a,n)$ for all $(F;a,n) \in X$;

(H4) $\Gamma(F;\lambda a,n) = \Gamma(F;a,\lambda n)$ for all $a \in \mathbb{R}^3$, $\lambda \in \mathbb{R}$ and $(F;a,n) \in X$;

(H5) $\Gamma(F;0,n) = \Gamma(F;a,0)$ for all $(F,a,n) \in M_+^{3\times 3} \times \mathbb{R}^3 \times \mathbb{R}^3$.

It turns out that this lack of differentiability permits the selection of the class N of directions n of the interfaces where the gradients F_+ and F_- of a piecewise local minimizer may not satisfy the Maxwell rule (see James [16]). Precisely

THEOREM II.1 ([9]): Let u be a piecewise affine deformation such that $E(u) \leqq E(u + \varepsilon)$ and $E(u) \leqq E(u(\cdot + \phi(\cdot))$ for all $\phi \in C^\infty(\Omega;\mathbb{R}^3)$ with $\phi = 0$ on $\partial\Omega$ and $||\phi||_{1,\infty}$ small. If n is the normal to an interface I of u and if $n \notin N$, then:

(i) (the traction is continuous across the interface)

$$[\frac{\partial W}{\partial F}]\, n = 0 \text{ on } I;$$

(ii) (Maxwell rule)

$$W(F_+) - W(F_-) - \frac{\partial W}{\partial F}(F_-).(F_+ - F_-) = 0 \text{ on } I.$$

This confirms Parry's [23] conjecture for the arrangement of thin shear bands in unloaded crystals and it is in agreement with photographs by Richman [25] where the strain energy densities of F_+ and F_- are, in general, lying in potential wells of W of different depth. Therefore, the inclusion of surface energy terms may help explain the structure of the interfaces observed and the mechanism of selection of the "preferred" solution.

Moreover, we show that there is "necking" near the phase boundary of a two-phase piecewise affine metastable deformation.

THEOREM II.2 ([9]): Let u be a pairwise affine deformation such that $E(u) \leq E(v)$ wheenver $v(x + \varphi(x)) = u(x) + \phi(x)$ for $\varphi, \phi \in C^\infty(\Omega;\mathbb{R}^3)$ with $||\varphi||_{1,\infty}$, $||\phi||_{1,\infty}$ small and $v = u$ on $\partial\Omega_1$. If $I = \{x \in \Omega | x \cdot n = k\}$ is the interface with normal $n \notin N$ and if closure (I) $\cap$ closure $(\partial\Omega \setminus \partial\Omega_1) = \emptyset$ then either $\Gamma = 0$ or

$$\limsup_{\varepsilon \to 0} \frac{A(\varepsilon) - A(0)}{\varepsilon} \leqq 0 \leqq \liminf_{\varepsilon \to 0^+} \frac{A(\varepsilon) - A(0)}{\varepsilon},$$

where $A(\varepsilon) := \{x \in \Omega | x \cdot n = k + \varepsilon\}$.

It is well known that, when the interfacial energy is not taken into account, if u is a metastable deformation, then W is quasi-convex and rank one convex at $\nabla u(x_0)$ whenever u is smooth near x_0 (see Ball [1], Ball and Murat [3]). In [9] we obtain analogs of the quasi-convexity condition and rank one condition for Γ. In the case of elastic crystals, these properties, together with the frame indifference and the symmetry invariance inherited by the molecular structure, permit us to conclude that at metastable states

$$\frac{\partial \Gamma}{\partial F} F_-^T$$

presents characteristics similar to those of the Cauchy stress tensor.

We recall that (see Ericksen [6]), when the interfacial energy contribution is disregarded, the Cauchy stress tensor

$$\frac{1}{\det F} \frac{\partial W}{\partial F} F^T$$

is a "pressure", i.e. there exists $\alpha \in \mathbb{R}$ such that

$$\frac{1}{\det F}\frac{\partial W}{\partial F}F^T = \alpha \mathbf{1}.$$

In addition to (H1)-(H5), assume that:

(H6) (*frame indifference*) $W(RF) = W(F)$ for all $F \in M_+^{3\times 3}$ and for all proper orthogonal matrix R;

(H7) (*frame indifference*) $\Gamma(RF;Ra,n) = \Gamma(F;a,n)$ for all $(F;a,n) \in X$ and for all proper orthogonal matrix R;

(H8) (*invariance under the change of lattice basis*) $\Gamma(FM;a,M^T n) = \Gamma(F;a,n)$ for all $M \in G$ and for all $(F;a,n) \in X$, where $G := \{M \in M_+^{3\times 3} | M_{ij} \in \mathbb{Z}$ and $\det M = 1\}$.

Moreover, considerations regarding variants related by symmetry lead us to

(H9) $\Gamma(F;Fa,b) = 0$ for every a, $b \in \mathbb{Z}^3$ such that $a\cdot b = 0$.

<u>THEOREM II.3</u> [9]): Let u be an admissible deformation such that $E(u) \leq E(u+\varphi)$ for all $\varphi \in C^\infty(\Omega)$ with $\varphi = 0$ on $\partial\Omega$ and $\|\varphi\|_{1,p}$ small, $1 < p < \infty$, and $E(u) \leq E(v)$ whenever v is piecewise C^2, v has the same discontinuity surfaces of u and $\|u-v\|_{1,\infty}$ is small. If $n \notin N$ then

$$\frac{\partial \Gamma}{\partial F}F_-^T$$

is a symmetric matrix that reduces to a "pressure", up to a rank one matrix, i.e. there exists

$$\lambda \leq -\frac{\Gamma}{\|F_-^{-T}n\|^2}$$

such that

$$\frac{\partial \Gamma}{\partial F}F_-^T = -\lambda\|F_-^{-T}n\|^2\,\mathbb{1} + \lambda(F_-^{-T}n)\otimes(F_-^{-T}n).$$

REFERENCES

[1] Ball, J.M., Convexity conditions and existence theorems in nonlinear elasticity, Arch. Rat. Mech. Anal. 63 (1977), 337-403.

[2] Ball, J.M. and James, R.D., Fine phase mixtures as minimizers of energy, Arch. Rat. Mech. Anal. 100 (1987), 13-52.

[3] Ball, J.M. and Murat, F., $W^{1,p}$-quasiconvexity and variational problems for multiple integrals, J. Funct. Anal. 58 (1984), 225-253.

[4] Carr, J., Gurtin, M.E. and Slemrod, M., Structured phase transitions on a finite interval, Arch. Rat. Mech. Anal. 86 (1984), 317-351.

[5] Chipot, M. and Kinderlehrer, D., Equilibrium configurations of crystals, 103 (1988), 237-277.

[6] Ericksen, J.L., Loading devices and stability of equilibrium, in Nonlinear Elasticity, Academic Press, New York 1973, pp. 161-173.

[7] Ericksen, J.L., Twinning of crystals, in Metastability and Incompletely Posed Problems, Antman, S., Ericksen, J.L., Kinderlehrer, D. Müller, I., eds., Springer, 1987, pp. 77-94.

[8] Fonseca, I., Variational methods for elastic crystals, Arch. Rat. Mech. Anal. 97 (1987), 189-220.

[9] Fonseca, I., Interfacial energy and the Maxwell rule, (to appear).

[10] Fonseca, I. and Tartar, L., The gradient theory of phase transitions for systems with two potential wells, (to appear).

[11] Gurtin, M.E. Two phase deformations of elastic solids, Arch. Rat. Mech. Anal. 84 (1983), 1-29.

[12] Gurtin, M.E., On a theory of phase transitions with interfacial energy, Arch. Rat. Mech. Anal. 87 (1984), 187-212.

[13] Gurtin, M.E., On phase transitions with bulk, interfacial, and boundary energy, Arch. Rat. Mech. Anal. 96 (1986), 243-264.

[14] Gurtin, M.E. Some results and conjectures in the gradient theory of phase transitions, in Metastability and Incompletely Posed Problems, Antman, S., Ericksen, J.L., Kinderlehrer, D., Müller, I., eds., Springer, 1987, pp. 135-146.

[15] Herring, C., Some theorems on the free energies of crystal surfaces, Phys. Rev. 82 (1951), 87-93.

[16] James, R.D., Finite deformations by mechanical twinning, Arch. Rat. Mech. Anal. 77 (1981), 143-176.

[17] Kinderlehrer, D., Twinning of crystals II, in Metastability and Incompletely Posed Problems, Antman, S., Ericksen, J.L., Kinderlehrer, D., Müller, I., eds., Springer, 1987, pp. 185-211.

[18] Kohn, R. and Sternberg, P., Local minimizers and singular perturbations, Proc. Royal Soc. Edinburgh (to appear).

[19] Modica, L., Gradient theory of phase transitions and minimal interface criterion, Arch. Rat. Mech. Anal. 98 (1987), 123-142.

[20] Owen, N.C., Existence and stability of necking deformations for nonlinear elastic rods, Arch. Rat. Mech. Anal. 98 (1987), 357-383.

[21] Owen, N.C., Nonconvex variational problems with general singular perturbations, Trans. A.M.S. (to appear).

[22] Owen, N.C. and Sternberg, P., Nonconvex variational problems with antisotropic perturbations (to appear).

[23] Parry, G.P., On shear bands in unloaded crystals (to appear).

[24] Pitteri, M., On 1- and 3-dimensional models in "non-convex" elasticity (to appear).

[25] Richman, R.H., The diversity of twinning in body-centred cubic structures, in Deformation Twinning, R.E. Reed-Hill, J.P. Hirth and H.C. Rogers, eds., Gordon and Breach Science Publishers Inc., New York, London, 1964.

[26] Sternberg, P., The effect of a singular perturbation on nonconvex variational problems, Arch. Rat. Mech. Anal. (to appear).

I. Fonseca
Department of Mathematics
University Carnegie-Mellon
Pittsburgh, P.A. 15213-3890
U.S.A.

A. HENROT AND M. PIERRE

About existence of a free boundary in electromagnetic shaping

1. INTRODUCTION

We discuss here the existence of a free boundary in a problem arising in electromagnetic shaping. It concerns the flow of a liquid metal falling under gravity in an electromagnetic field with high frequency. We are interested in the shape of a horizontal section of this flow.

This papers deals with the particular situation where *the superficial tension of the liquid is zero* for which explicit computations can be made. The generic case (positive superficial tension) is studied in [5]. We first describe here the corresponding model in the general case.

We follow the same two-dimensional model as in [4] and as introduced in [1], [2]. If $\vec{B}$ denotes the total magnetic induction vector and $\vec{j}_0$ the current density vector, then equilibrium is characterized by the following set of equations:

$$
(S_0)\quad
\begin{cases}
\nabla \wedge \vec{B} = \mu_0 \vec{j}_0 & \text{in } \Omega, \\
\nabla \cdot \vec{B} = 0 & \text{in } \Omega, \\
\vec{B} \cdot \vec{n} = 0 & \text{on } \Gamma = \partial\Omega, \\
\|\vec{B}\|^2 / 2\mu_0 + \sigma C = \text{constant on } \Gamma, \\
\vec{B} \text{ tend to 0 at infinity,}
\end{cases}
$$

where Ω is the exterior of the two-dimensional section of the region occupied by the liquid, μ_0 is the magnetic permeability of vacuum, $\vec{n}$ is the unit normal vector at the boundary Γ, $\|\cdot\|$ is the Euclidian norm, C is the curvature of Γ and $\sigma \geq 0$ is the superficial tension of the liquid.

Thus the problem consists of finding Γ and $\vec{B}$ satisfying (S_0) when $\vec{j}_0$ is given. We are particularly interested in constructing the free boundary Γ.

This problem has been solved explicitly in some particular situations and numerical approaches have also been considered (see [6], [3] and references in [4]). The "inverse" problem which consists of finding $\vec{j}_0$ when the curve Γ is given is treated in [4]. The method employed in the latter suggested to us an approach to the direct problem above. Indeed, the curve Γ was then represented (at least in the simplest cases) as the image of the unit circle Γ_0 under a univalent holomorphic mapping of the exterior Ω_0 of the unit disk onto the exterior Ω of the domain occupied by the liquid. We then proved that, assuming a solution exists, the function

$$v(x,y) = \ln|\phi'(x + iy)|, \tag{1.1}$$

where ϕ denotes the conformal mapping mentioned above, can be obtained as the solution of the system

$$(S_1)\quad \begin{cases} \Delta v = 0 & \text{on } \Omega_0 = \text{exterior of the unit disk,} \\ -\tau \dfrac{\partial v}{\partial r} = g^2 e^{-v} - Pe^{v} + \tau \text{ on } \Gamma_0, & \\ v & \text{is bounded at infinity,} \end{cases}$$

where $\tau = 2\sigma\mu_0$, P is the constant appearing in (S_0), $\partial v/\partial r$ is the radial derivative and

$$g = g(\theta) = \varepsilon\sqrt{(P-\tau C)}|\phi'(e^{i\theta})|, \ \varepsilon = \pm 1 \quad \text{(see [4]).}$$

Note that the boundary condition in (S_1) is nothing but the corresponding one in (S_0).

Obviously, g directly depends on Γ and is a priori unknown when looking for Γ. On the other hand, if g were known, we could determine v as the solution of (S_1) and then obtain ϕ (and therefore the free boundary Γ) by integrating (1.1).

It turns out that much information can be obtained a priori on g when the magnetic field is created by linear vertical conductors, that is, when $\vec{j}_0 = (0,0,j_0)$ with

$$j_0 = \sum_{k=1}^{N} \beta_k \delta_{Z_k}, \; \beta_k \in \mathbb{R}, \; Z_k \in \mathbb{C}, \tag{1.2}$$

δ_{Z_k} = Dirac mass at Z_k. Indeed, one can prove the following:

PROPOSITION A: Let j_0 be as in (1.2). Assume that (S_0) has a solution $(\Omega,\vec{B})$ with

$$Z_k \in \Omega, \; k = 1,\ldots,N, \tag{1.3}$$

there exist a closed Jordan curve Γ of class C^2 and

$$\phi \in C^2(\bar{\Omega}_0), \text{ holomorphic on } \Omega_0, \text{ one-to-one on } \bar{\Omega}_0 \text{ such that} \tag{1.4}$$

$$\Omega = \phi(\Omega_0), \; \Omega \text{ is unbounded}, \; \Gamma = \partial\Omega = \phi(\Gamma_0).$$

Then, $v = \ln|\phi'|$ is a solution of (S_1) with

$$g(\theta) = \sum_{k=1}^{N} \alpha_k[e^{i\theta}/(z_k - e^{i\theta}) + e^{-i\theta}/(\bar{z}_k - e^{-i\theta}) + 1], \tag{1.5}$$

where

$$z_k = \phi^{-1}(Z_k), \; \alpha_k = -\beta_k\mu_0/2\pi, \; k = 1,\ldots,N. \tag{1.6}$$

This proposition shows that g is completely determined from the knowledge of β_k and z_k. If the z_k are not known a priori, since they depend on ϕ, some information on their position can however be obtained in some cases using, for instance, symmetry considerations.

In any case, the idea of finding the free boundary Γ in (S_0) is to solve (S_1) with g given as in (1.5). Then ϕ is deduced from v — when possible — by integrating $v = \ln|\phi'|$. Then we analyse the possibility of making the (z_k,α_k) fit with prescribed (Z_k,β_k) and the prescribed surface of the liquid.

The above proposition will not be established here. Moreover, we will not consider the general situation, whose analysis would be too long for this communication. We will rather develop the particular case of zero superficial tension σ, which is simpler and relies on direct and explicit computations. This turns out to be very significant for applications to liquids with small superficial tension. The case $\sigma \neq 0$ will be treated in [5] as well as the proof of Proposition A.

2. THE CASE OF ZERO SUPERFICIAL TENSION

As above, we denote

$$\Omega_0 = \{z \in \mathbb{C};\ |z| > 1\},\ \Gamma_0 = \{z \in \mathbb{C};\ |z| = 1\}.$$

The points z_k are given in Ω_0 and we set (see (1.5)):

$$g(\theta) = \sum_{k=1}^{N} \alpha_k[e^{i\theta}/(z_k-e^{i\theta}) + e^{-i\theta}/(\bar{z}_k-e^{-i\theta}) + 1], \tag{2.1}$$

where α_k, $k = 1,\dots,N$, are real numbers.

Here $\sigma = 0$ and (S_1) becomes

$$\begin{aligned} &\Delta v = 0 \text{ on } \Omega_0, \\ &v = \ln|g(\theta)P^{-\frac{1}{2}}| \text{ on } \Gamma_0, \\ &v \quad \text{is bounded at infinity.} \end{aligned} \tag{2.2}$$

Our program is as follows:

(A) We will explicitly solve (2.2) using elementary computations.

(B) We will determine a necessary and sufficient condition on g (actually on the data z_k, α_k) ensuring that

$$v = \ln|\phi'| \text{ on } \Omega_0, \tag{2.3}$$

where ϕ is holomorphic on Ω_0. We will see that this condition is automatically satisfied if g is π-periodic.

(C) When ϕ exists, it is locally injective on Ω_0 and holomorphic on a neighbourhood of Γ_0. We then set

$$\Gamma = \phi(\Gamma_0),\ \Omega = \phi(\Omega_0). \tag{2.4}$$

In general, ϕ is not one-to-one on $\bar{\Omega}_0$. But, if this is the case, we will obtain a solution to our problem. Note that ϕ is always one-to-one when $\tau > 0$, as proved in [5].

<u>PROPOSITION 2.1</u>: Let v be the solution of (2.2). Assume there exists ϕ holomorphic on Ω_0 and injective on $\bar{\Omega}_0$ such that (2.3) holds. Then $\Gamma = \phi(\Gamma_0)$ is a solution to (S_0) with $\sigma = 0$ and

$$j_0 = -2\pi\ \mu_0^{-1} \sum_k \alpha_k\ \delta_{\phi(z_k)}\ . \tag{2.5}$$

Moreover, the field $\vec{B}$ is given by

$$\vec{B} = (\psi_Y, -\psi_X),\ \psi(Z) = \text{Real part } (f(\phi^{-1}(Z))), \tag{2.6}$$

where f is defined on Ω_0 by $f'(z) = -G(z)/z$ and

$$G(z) = \sum_{k=1}^{N} \alpha_k[z/(z_k-z) + 1/(z\bar{z}_k - 1) + 1]. \tag{2.7}$$

(D) We will see that ϕ (and therefore Γ) can be explicitly determined. Now it remains to go back to the original problem where $Z_k = \phi(z_k)$ are given and the surface of the liquid (i.e. its quantity here) is prescribed. We will give the results in the case of four conductors.

<u>Step A</u>: Let G be given by (2.7). Remark that $G(e^{i\theta}) = g(\theta)$. We write

$$G(z) = Az^r \frac{\prod_{j=1}^{p} (z-X_j)(z-1/\bar{X}_j) \prod_{j=1}^{m} (z-e^{i\theta_j})}{\prod_{k=1}^{N} (z_k-z)(z\bar{z}_k-1)}, \tag{2.8}$$

where

$$\begin{aligned} &X_j,\ j = 1,\dots,p \text{ are the zeros of G of magnitude greater than 1,} \\ &e^{i\theta_j},\ j = 1,\dots,m \text{ are the zeros of G of magnitude 1.} \end{aligned} \tag{2.9}$$

Note that the zeros of G of magnitude less than 1 (but 0) are $1/\bar{X}_j$, $j = 1,\ldots,p$, since $G(1/\bar{z}) = \overline{G(z)}$. For the same reason, m is even and

$$r = N-p-m/2 \geq 1, \tag{2.10}$$

the latter coming from $G(0) = 0$.

We solve (2.2) using Fourier expansion of

$$h(\theta) = \ln|g(\theta)P^{-\frac{1}{2}}| = \sum_{n\in\mathbb{Z}} h_n e^{in\theta}. \tag{2.11}$$

PROPOSITION 2.2: The Fourier coefficients of h are given by

$$h_n = \begin{cases} -\sum_{j=1}^{m} e^{-in\theta_j}/2n - \sum_{j=1}^{p} 1/n\, X_j^n + \sum_{k=1}^{N} 1/n\, z_k^n & \text{for } n \geq 1, \\ \ln(|A|P^{-\frac{1}{2}}) + \sum_{j=1}^{p} \ln|X_j| - 2\sum_{k=1}^{N} \ln|z_k| & \text{for } n = 0, \\ \bar{h}_{-n} & \text{for } n \leq -1. \end{cases} \tag{2.12}$$

PROOF: We use (2.8) and the identities

- if $|X| > 1$, $(2\pi)^{-1} \int_0^{2\pi} \ln|e^{i\theta}-X|e^{-in\theta}\, d\theta = \begin{cases} -1/2n\, X^n & \text{if } n > 0, \\ \ln|X| & \text{if } n = 0; \end{cases}$
- if $|X| < 1$, $(2\pi)^{-1} \int_0^{2\pi} \ln|e^{i\theta}-X|e^{-in\theta}\, d\theta = \begin{cases} -\bar{X}^n/2n & \text{if } n > 0, \\ 0 & \text{if } n = 0; \end{cases}$
- $(2\pi)^{-1} \int_0^{2} \ln|e^{i\theta}-e^{i\alpha}|e^{-in\theta}\, d\theta = \begin{cases} -e^{-in\alpha}/2n & \text{if } n > 0, \\ 0 & \text{if } n = 0. \end{cases}$

Now we set

$$V(z) = h_0 + 2\sum_{n=1}^{\infty} h_{-n} z^{-n}, \tag{2.13}$$

which is convergent for $|z| > 1$. Since $h_{-n} = \bar{h}_n$, we check that

$$\mathcal{Re}(V(e^{i\theta})) = h(\theta).$$

Therefore, the solution of (2.2) is given by

$$v(r,\theta) = \mathcal{Re}(V(re^{i\theta})) = \sum_{n\in\mathbb{Z}} h_n\, r^{-|n|} e^{in\theta}. \tag{2.14}$$

Step B: We now want to determine ϕ holomorphic on Ω_0 such that

$$v(r,\theta) = \ln|\phi'(re^{i\theta})|,\ r > 1,\ \theta \in R, \tag{2.15}$$

or, equivalently (up to multiplying by a constant of magnitude 1)

$$\phi'(z) = e^{V(z)} \text{ on } \Omega_0. \tag{2.16}$$

This obviously happens if and only if the coefficient of $1/z$ in the Laurent expansion of $e^{V(z)}$ vanishes. But it is equal to $2e^{h_0}h_{-1}$. Therefore

PROPOSITION 2.3: The solution of (2.2) satisfies (2.15), (2.16) if and only if

$$\sum_{k=1}^{N} 1/z_k = \sum_{j=1}^{p} 1/X_j + \sum_{j=1}^{m} e^{-i\theta_j/2}. \tag{2.17}$$

This is the case if $\{(z_k,\alpha_k),\ k = 1,\ldots,N\} = \{(R_q z_k,\ \varepsilon\alpha_k);\ k = 1,\ldots,N\}$ where $R_q = e^{i2\pi/q}$, q is an integer greater than 1 and $\varepsilon = \pm 1$.

PROOF: Relation (2.17) comes from (2.12) with $n = -1$. If the set $\{z_k,\ k = 1,\ldots,N\}$ is invariant by a rotation of angle $2\pi/q$, then $\sum_{k=1}^{N} 1/z_k = 0$. If the above property holds for the whole set $\{(z_k,\alpha_k)\}$, then one also has $G(e^{-i2\pi/q}z) = \varepsilon G(z)$, so that the zeros of G are also invariant under the same rotation and the right-hand side of (2.17) is zero. □

REMARK: For $N = 1$ (case of only one mass), (2.17) does not hold.

Assuming (2.17), we are able to compute ϕ' and ϕ explicitly.

PROPOSITION 2.4: Assume (2.17) holds. Then

$$\phi'(z) = \frac{|A_1|}{\sqrt{(P)}} z^{2r} \frac{\prod_{j=1}^{m} (z-e^{i\theta_j}) \prod_{j=1}^{p} (z-1/\bar{X}_j)^2}{\prod_{k=1}^{N} (z-1/\bar{z}_k)^2}, \tag{2.18}$$

where

$$A_1 = A \prod_{j=1}^{p} X_j / \prod_{k=1}^{N} z_k^2$$

and

$$\phi(z) = \frac{|A_1|}{\sqrt{(P)}} \{z + \sum_{k=1}^{N} \frac{a_k}{z-1/\bar{z}_k} + \sum_{k=1}^{N} b_k \ln(1-1/z\bar{z}_k)\}, \tag{2.19}$$

where a_k, b_k are complex constants such that

$$(z-1/\bar{z}_k)^2\phi'(z) = \frac{|A_1|}{\sqrt{(P)}} \{-a_k + (z-1/\bar{z}_k)b_k + (z-1/\bar{z}_k)^2 H_k(z)\} \tag{2.20}$$

with H_k holomorphic in a neighbourhood of $1/\bar{z}_k$.

<u>REMARK</u>: Note that $z \to \sum_{k=1}^{N} b_k \ln(1-1/z\bar{z}_k)$ is well defined on $|z| > 1/\min_k |z_k|$.

<u>PROOF OF PROPOSITION 2.4</u>: From (2.13) and (2.12), we obtain

$$V(z) = \ln|A_1|P^{-\frac{1}{2}} + \sum_{j=1}^{m} \ln(1-e^{i\theta_j}/z) + 2 \sum_{j=1}^{p} \ln(1-1/z\bar{X}_j)$$
$$- 2 \sum_{k=1}^{N} \ln(1-1/z\bar{z}_k).$$

Relation (2.18) follows using (2.10). To obtain (2.19), we integrate the corresponding decomposition of $\phi'(z)$ and we use $\Sigma b_k = 0$ coming from (2.17). □

<u>Step C.</u> <u>PROOF OF PROPOSITION 2.1</u>: Here Γ is a closed Jordan curve. It separates the plane into two components: the bounded one is the region occupied by the liquid; the other one is the set Ω in (S_0). Moreover, we have

$$f(z) = \sum_{k=1}^{N} \alpha_k[\ln(z-z_k)-\ln(1-z\bar{z}_k)], \tag{2.21}$$

so that $\psi(Z) = Re(f(\phi^{-1}(Z)))$ is harmonic on $\Omega\setminus\{\phi(z_k), k = 1,\ldots,N\}$ and satisfies (2.5). Moreover, one easily checks that

$$\psi = 0 \text{ on } \Gamma, \quad \lim_{|Z|\to\infty} |\nabla\psi(Z)| = 0, \quad \psi_X^2 + \psi_Y^2 = P \text{ on } \Gamma.$$

Therefore $\vec{B} = (\psi_Y,-\psi_X)$ is a solution of (S_0) with $\sigma = 0$.

Step D: We suppose that we are now given $\{(Z_k,\beta_k), k = 1,\ldots,N\}$ as well as the surface S of the liquid. We would like to decide whether there is a solution of (S_0) satisfying (1.3), (1.4) (i.e. the liquid occupies a simply connected region Ω_i bounded by a Jordan curve Γ and surrounded by the Z_k). According to the analysis, we need to find $\{z_k, k = 1,\ldots,N\}$ in Ω_0 such that (2.17) holds and the function ϕ defined by (2.19) satisfies

$$Z_k = \phi(z_k), \; k = 1,\ldots,N, \tag{2.22}$$

$$S_\phi = \text{Surface of } \Omega_i = \frac{1}{2}\int_{\Gamma_0} \overline{\phi(z)}\phi'(z)\,dz = S. \tag{2.23}$$

These three conditions are rather implicit. However, if $\{(Z_k,\beta_k); k = 1,\ldots,N\}$ has some invariance property of the type mentioned in Proposition (2.3), we expect $\{(z_k,\alpha_k)\}$ to have the same. Hence, the idea is to start from such a family (z_k,α_k) for which (2.17) holds thanks to the invariance property. Then we look at the $\{\phi(z_k)\}$ for the corresponding ϕ satisfying (2.23) and check if $\{Z_k\}$ can be reached.

The computations can be easily done on a computer. Next, to give a more precise idea, we give explicit results in the case of four masses at the vertices of a square, namely

$$\{Z_k\} = \{1,i1,-1,-i1\}, \; \{\beta_k\} = \{\lambda,-\lambda,\lambda,-\lambda\}. \tag{2.24}$$

This analysis can be compared to the one made in [6] in this specific case.

Here we choose

$$\{z_k\} = \{a,ia,-a,-ia\}, \ \{\alpha_k\} = \{K,-K,K,-K\}, \tag{2.25}$$

with $K = -\lambda\mu_0/2\pi$. Obviously, (2.17) holds and we have

$$G(z) = 4Ka^2(a^4-1)z^2(z^4+1)/(a^4-z^4)(a^4z^4-1), \tag{2.26}$$

$$\phi'(z) = 4Ka^2(a^4-1)P^{-\frac{1}{2}}\, z^4(z^4+1)/(a^4z^4-1)^2, \tag{2.27}$$

$$\phi(z) = \phi_0 \{z - \frac{a^4+1}{4}\frac{z}{a^4z^4-1} + \frac{a^4+5}{16a}[\ln \frac{1-1/az}{1+1/az} + i \ln \frac{1-i/az}{1+i/az}]\}, \tag{2.28}$$

with $\phi_0 = 4K(a^4-1)a^{-6}P^{-\frac{1}{2}}$ and ln is the usual complex logarithm.

It turns out that ϕ is one-to-one from $\bar{\Omega}_0$ onto $\bar{\Omega} = \phi(\bar{\Omega}_0)$ if and only if

$$a \geq 7^{1/4}. \tag{2.29}$$

Moreover, we have

$$\ell = \phi(a) = KP^{-\frac{1}{2}}\{\frac{4a^4-5}{a^5} + \frac{(a^4-1)(a^4+5)}{4a^7}[\ln \frac{a^2-1}{a^2+1} + 2 \operatorname{Arc\,tan} a^{-2}]\}. \tag{2.30}$$

$$S_\phi = 4\pi K^2P^{-1}(a^4-1)^2a^{-12}\{\frac{11a^8-40a^4+25}{(a^4-1)^2} + \frac{(a^4+5)^2}{4a^2}[\ln \frac{a^2-1}{a^2+1} + 2 \operatorname{Arc\,tan} a^{-2}]\}. \tag{2.31}$$

One easily verifies that $a \to \theta(a) = S_\phi(a)/\ell^2(a)$ is decreasing from $[7^{1/4},\infty)$ onto $(0,c]$ with $c \simeq 0.581$.

<u>CONCLUSION</u>: Let (Z_k,β_k) be given as in (2.24) and let S be the surface of the liquid. Then there will exist a free boundary Γ if

$$\ell^2 \geq S/c. \tag{2.32}$$

In that case, $\Gamma = \Phi(\Gamma_0)$ where Φ is given by (2.28) with

$$a = \theta^{-1}(S/\ell^2) \tag{2.33}$$

and $KP^{-\frac{1}{2}}$ is determined by (2.31) where $S_\phi = S$.

REFERENCES

[1] Brancher, J.P. and Sero-Guillaume, O., Sur l'équilibre des liquides magnétiques. Application à la magnétostatique, J.M.T.A. 2, No. 2 (1983), 265-283.

[2] Brancher, J.P., Etay, J. and Sero-Guillaume, O., Formage d'une lame métallique liquide: calculs et expériences, J.M.T.A. 2, No. 6 (1983), 976-989.

[3] Brancher, J.P., De Framond, R. and Sero-Guillaume O., Shaping of liquid metal cylinders, 4th Beer Sheva Seminar on MHD Flows and Turbulence, Israël, 1984.

[4] Henrot, A. and Pierre, M., Un problème inverse en formage des métaux liquides, Rairo, M^2AN (to appear).

[5] Henrot, A. and Pierre, M., Existence of equilibria in electromagnetic shaping (to appear).

[6] Shercliff, J.A., Magnetic shaping molten metal columns, Proc. R. Soc. London A 375 (1981), 455-473.

A. Henrot and M. Pierre
Département de Mathématiques
Université de Nancy I
B.P. 239 54506 Vandoeuvre-Les-Nancy
France

N. LABANI AND C. PICARD

Homogenization of a nonlinear Dirichlet problem in a periodically perforated domain

ABSTRACT: Consider the following nonlinear Dirichlet problem in $\Omega\setminus T_\varepsilon$:

$$-\mathrm{div}(|\nabla u_\varepsilon|^{p-2}\nabla u_\varepsilon) = f \text{ in } \Omega\setminus T_\varepsilon,$$
$$u_\varepsilon \in W_0^{1,p}(\Omega\setminus T_\varepsilon), \qquad (1_\varepsilon)$$

where $p \geq 2$ and T_ε is the union of small inclusions contained in an open bounded subset Ω of R^N $(N \geq 2)$ which are ε-periodically distributed and of size $r_\varepsilon < \varepsilon$. In this paper, the limit problems of (1_ε) as ε goes to zero are determined according to the values of r_ε; some results about correctors and speed of convergence of u_ε to the solution of the limit problem are given.

1. INTRODUCTION

Let Ω be an open bounded subset of R^N $(N \geq 2)$ which contains small inclusions $T_{\varepsilon j}$, $j \in J \subset \mathbb{N}$, periodically distributed in Ω with periodic ε such that $T_{\varepsilon j}$ are deduced on $r_\varepsilon T$ by translation, where T is a closed smooth subset of R^N and $r_\varepsilon < \varepsilon\,(\mathrm{diam}\,T)^{-1}$; let T_ε be the union of all $T_{\varepsilon j}$ contained in Ω.

Given $N \geq p \geq 2$ and $f \in L^{p'}(\Omega)$ $(1/p + 1/p' = 1)$, we consider the following nonlinear Dirichlet problem in the periodically perforated domain $\Omega_\varepsilon = \Omega\setminus T_\varepsilon$:

$$-\Delta_p u_\varepsilon = f \text{ in } \Omega_\varepsilon,$$
$$u_\varepsilon = 0 \text{ on } \partial\Omega_\varepsilon, \qquad (1_\varepsilon)$$

where Δ_p is the p-Laplacian operator, that is $\Delta_p u = \mathrm{div}(|\nabla u|^{p-2}\nabla u)$.

Problem (1_ε) is the Euler equation of the minimization problem

$$(D_\varepsilon) \quad \mathrm{Min}\,\{\int_\Omega |\nabla v|^p\,dx - p\int_\Omega fv\,dx;\ v \in W_0^{1,p}(\Omega),\ v = 0 \text{ on } T_\varepsilon\}.$$

PROPOSITION 1.1: There exists a unique $u_\varepsilon \in W_0^{1,p}(\Omega)$, $u_\varepsilon = 0$ on T_ε solution of problem (D_ε). Moreover, the family $(u_\varepsilon)_{\varepsilon>0}$ is bounded in $W_0^{1,p}(\Omega)$.

CONSEQUENCE: There exists a subsequence of (u_ε), still denoted (u_ε), which converges to u in $w\text{-}W_0^{1,p}(\Omega)$ as ε goes to zero.

The question is to characterize u as a solution of a variational problem (the limit problem or (1_ε) of (D_ε)) according to the behaviour of r_ε and to summarize the convergence of u_ε to u, in particular, some speeds of convergence and correctors.

In the case p = 2, the limit problem has been studied a few years ago (see, [4],[10],[11]); the method used consists of inserting suitable test functions in the variational formulation of (1_ε) and to pass to the limit.

For $p \geq 2$, the limit problem has been determined in [2] in the case T = B (the unit ball of $\mathbb{R}^N$) and it was deduced from a general compactness theorem for a class of convex functionals defined on $L^p(\Omega)$ of the type

$$F_\varepsilon(v) = \int_\Omega j(x,\nabla v(x))\, dx + G_\varepsilon(v),$$

where G_ε belongs to a class of constraint functionals of obstacle type (see [1], [2], [5], [6], [7]).

In this article, the epi-limit of the sequence of functionals

$$F_\varepsilon(v) = \int_\Omega |\nabla v|^p\, dx - p\int_\Omega fv\, dx + I\{v \in W_0^{1,p}(\Omega),\ v = 0 \text{ on } T_\varepsilon\}(v)$$

(where $IK(v) = 0$ if $v \in K$ and $IK(v) = +\infty$ if $v \notin K$) is determined by a direct proof making use of suitable test functions. Thus, from properties of the epi-convergence notion (see [1], [5], [7]), we get the limit problem of (D_ε). This direct method, and also a better study of the properties of those test functions, allow us to obtain more information about the convergence of u_ε to u.

2. LIMIT PROBLEMS AND CONVERGENCE OF (u_ε) TO u

In this section, we assume that $r_\varepsilon \ll \varepsilon$.

THEOREM 2.1: Let

$$a = \begin{cases} \lim_{\varepsilon \to 0} \varepsilon^{-N} r_\varepsilon^{N-p} \in [0, +\infty] & \text{if } N > p, \\ \lim_{\varepsilon \to 0} \varepsilon^{-N} (-\mathrm{Ln}\, r_\varepsilon)^{-(N-1)} & \text{if } N = p. \end{cases}$$

Let

$$C = \begin{cases} \inf \{ \int_{R^N} |\nabla v|^p \, dx;\ v \in W^{1,p}(R^N);\ v = 1 \text{ on } T\} & \text{if } N > p, \\ S_N, \text{ the area of the unit sphere of } \mathbb{R}^N & \text{if } N = p. \end{cases}$$

(1) If $a = 0$, then u_ε converges to u in s-$W_0^{1,p}(\Omega)$ and u is the solution of

$$-\Delta_p u = f \text{ in } \Omega,$$
$$u = 0 \text{ on } \partial\Omega.$$

(2) If $a \in]0, +\infty[$, then u_ε converges to u in w-$W_0^{1,p}(\Omega)$ and u is the solution of

$$-\Delta_p u + aC|u|^{p-2}u = f \text{ in } \Omega$$
$$u = 0 \text{ on } \partial\Omega$$

Moreover, $\int_\Omega |\nabla u_\varepsilon|^p dx \to \int_\Omega |\nabla u|^p \, dx + a\, C \int_\Omega |u|^p \, dx$

(3) If $a = +\infty$ and $r_\varepsilon \ll \varepsilon$, then u_ε converges to 0 in s-$W_0^{1,p}(\Omega)$.

REMARK 2.2:

(a) The case $a \in]0, +\infty[$ means $r_\varepsilon \approx k\varepsilon^{N/N-p}$ if $N > p$ and $r_\varepsilon \approx \exp(-k\varepsilon^{-N/(N-1)})$ if $N > p$.

(b) If $N > p$ and $T = B$, the unit ball of $\mathbb{R}^N$, then $C = S_N((N-p)/(p-1))^{p-1}$.

Theorem 2.1 easily follows, using properties of epi-converge (see [1], for instance), from:

THEOREM 2.3: The functionals

$$F_\varepsilon(v) = \int_\Omega |\nabla v|^p \, dx + I\{v \in W_0^{1,p}(\Omega),\ v = 0 \text{ on } T_\varepsilon\}(v)$$

epi-converge in $L^p(\Omega)$, as ε goes to zero, to the functional

$$F(v) = \int_\Omega |\nabla v|^p \, dx + a\, C \int_\Omega |v|^p \, dx + I\{v \in W_0^{1,p}(\Omega)\}(v)$$

which means

(i) $\forall v \in L^p(\Omega)$, $\exists v_\varepsilon \to v$ in $L^p(\Omega)$: $\limsup F_\varepsilon(v_\varepsilon) \leqq F(v)$;

(ii) $v_\varepsilon \to v$ in $L^p(\Omega) \Rightarrow \liminf F_\varepsilon(v_\varepsilon) \geqq F(v)$.

Sketch of the proof. First step. Test functions. In order to construct the functions v_ε of the previous assertion (i), let us introduce the test functions w_ε. Let $P_{\varepsilon j}$ be a sequence of cubes with side ε, ε-periodically distributed such that the ball $B_{\varepsilon j}$ of radius $\varepsilon/2$ contained in $P_{\varepsilon j}$ contains $T_{\varepsilon j}$ in its interior. We define w_ε on each ball $B_{\varepsilon j}$ as the p-capacitary potential of $T_{\varepsilon j}$ into $B_{\varepsilon j}$. □

PROPOSITION 2.4: Let w be defined as follows:

$$\Delta_p w_\varepsilon = 0 \text{ in } B_{\varepsilon j} \setminus T_{\varepsilon j},$$

$$w_\varepsilon = 1 \text{ on } T_{\varepsilon j},$$

$$w_\varepsilon = 0 \text{ on } P_{\varepsilon j} \setminus B_{\varepsilon j}.$$

Then

(1) $\int_\Omega |\nabla w_\varepsilon|^p \, dx \to a\, C \text{ meas } \Omega$.

(2) If $0 \leqq a < +\infty$, $w_\varepsilon \to 0$ in $s\text{-}L^p(\Omega)$, $w_\varepsilon \to 0$ in $w\text{-}W^{1,p}(\Omega)$ and $|\nabla w_\varepsilon|^p \, dx \to aC\, dx$ in $\sigma(M(\Omega), C(\Omega))$.

(3) If $0 \leqq a < +\infty$ and $0 < q < p$, then $w_\varepsilon \to 0$ in $s\text{-}W^{1,q}(\Omega)$.

PROOF: Parts (1) and (2) can be proved using similar methods as those used for the case p = 2 (see [2] and [4]). Part (3) can be deduced from the following lemma:

LEMMA 2.5: Assume $T \subset B$. For every $R > 1$, let w_R be the solution of

$$\Delta_p w_R = 0 \text{ in } B(R)\setminus T,$$

(2) $$w_R = 0 \text{ on } \partial B(R),$$

$$w_R = 1 \text{ on } T,$$

where B(R) is the ball of radius R.

Then there exists $\delta > 0$ and $K > 0$ such that, for every R large enough,

$$\int_\Omega |\nabla w_R|^{p-1}\, dx \leq KR^{(N/p)-\delta}.$$

PROOF OF LEMMA 2.5: Let us first recall that $\int_{B(R)\setminus T} |\nabla w_R|^p\, dx \to C$ as $R \to +\infty$. Let $1 < R^* < R$. Denote w instead of w_R. Write

$$\int_{B(R)\setminus T} |\nabla w|^{p-1}\, dx = \int_{B(R)\setminus B(R^*)} |\nabla w|^{p-1}\, dx + \int_{B(R^*)\setminus T} |\nabla w|^{p-1}\, dx.$$

From the Hölder inequality we get

$$\int_{B(R^*)\setminus T} |\nabla w|^{p-1}\, dx \leq R^{*N/p}\Big(\int_{B(R^*)\setminus T} |\nabla w|^p\, dx\Big)^{1-1/p}$$

$$\leq R^{*N/p}(C+1)^{1-1/p}, \text{ for every R large enough}$$

and

$$\int_{B(R)\setminus B(R^*)} |\nabla w|^{p-1}\, dx \leq R^{N/p}\Big(\int_{B(R)\setminus B(R^*)} |\nabla w|^p\, dx\Big)^{1-1/p}.$$

It remains to estimate $\int_{B(R)\setminus B(R^*)} |\nabla w|^p\, dx$. Using the Green formula we have

$$\int_{B(R)\setminus B(R^*)} |\nabla w|^p \, dx = - \int_{\partial B(R^*)} |\nabla w|^{p-2} w \, \partial w/\partial \nu \, d\sigma,$$

where ν is the exterior normal to $B(R^*)$. But the map $r \to - \int_{\partial B(r)} |\nabla w|^{p-2} w \, \partial w/\partial \nu \, d\sigma$ is a decreasing function on $]1,R[$. Hence

$$\begin{aligned}
\int_{B(R)\setminus B(R^*)} |\nabla w|^p \, dx &\leq - (R^*-1)^{-1} \int_1^{R^*} \left(\int_{\partial B(r)} |\nabla w|^{p-2} w \, \partial w/\partial \nu \, d\sigma\right) dr \\
&\leq (R^*-1)^{-1} \int_1^{R^*} \left(\int_{\partial B(r)} |\nabla w|^{p-1} |w| \, d\sigma\right) dr \\
&\leq (R^*-1)^{-1} \int_{B(R^*)\setminus B(1)} |\nabla w|^{p-1} |w| \, dx \\
&\leq (R^*-1)^{-1} (C+1)^{1-1/p} \left(\int_{B(R^*)\setminus B(1)} |w|^p \, dx\right)^{1/p}.
\end{aligned}$$

Let us denote by K any constant which does not depend on R. From the maximum principle (see [8]), we have that $0 \leq w \leq \underline{w}_R$ in $B(R)\setminus B(1)$ where $\underline{w}_R$ is the solution of (2) with $T = B(1)$. Let us investigate the case $N > p$. Since

$$\begin{aligned}
\underline{w}_R(\rho) &= (\rho^{-(N-p)/(p-1)} - R^{-(N-p)/(p-1)})(1 - R^{(N-p)/(p-1)})^{-1} \\
&\leq K \rho^{-(N-p)/(p-1)} \quad \text{for } 1 \leq \rho \leq R,
\end{aligned}$$

we obtain

$$\int_{B(R^*)\setminus B(1)} |w|^p \, dx \leq K \int_1^{R^*} \rho^{-p(N-p)/(p-1)+N-1} \, d\rho.$$

It is easy to deduce that

$$\int_{B(R^*)\setminus B(1)} |w|^p \, dx \leq K R^{*\alpha p} \text{ with } 0 \leq \alpha < 1.$$

Finally, we obtain

$$\int_{B(R)\setminus T} |\nabla w|^{p-1} \, dx \leq K(R^{N/p} R^{*(\alpha-1)(1-1/p)} + R^{*N/p}).$$

Choosing $R^* = R^\beta$ with $0 < \beta < 1$, the conclusion of Lemma 2.5 follows. □

PROOF OF PROPOSITION 2.4, PART (3): Since

$$\int_\Omega |\nabla w_\varepsilon|^{p-1}\, dx \sim \varepsilon^{-N} \text{ meas } \Omega \int_{B(\varepsilon/2)} |\nabla w|^{\,p-1}\, dx$$

$$= \varepsilon^{-N}\, r_\varepsilon^{N-p+1} \text{ meas } \Omega \int_{B(R)} |\nabla w_R|^{p-1}\, dx,$$

where $R = \varepsilon/2r_\varepsilon$, we deduce from Lemma 2.5 that, for ε small enough,

$$\int_\Omega |\nabla w_\varepsilon|^{p-1}\, dx \leqq Kr_\varepsilon \int_{B(R)} |\nabla w_R|^{p-1}\, dx \leqq K_1 r_\varepsilon R^{(N/p)-\delta} \leqq K_2 (r_\varepsilon/\varepsilon)^\delta .$$

Hence $\int_\Omega |\nabla w_\varepsilon|^{p-1}\, dx \to 0$ as $\varepsilon \to 0$. Since $\int_\Omega |\nabla w_\varepsilon|^p\, dx$ is bounded, it follows from the Hölder inequality that for all $0 < q < p$, $\int_\Omega |\nabla w_\varepsilon|^q\, dx \to 0$ as $\varepsilon \to 0$. □

Second step. Proof of (i). Suppose $0 \leqq a < +\infty$. Let $v \in D(\Omega)$. Consider $v_\varepsilon = v - w_\varepsilon v$ with $|\nabla w_\varepsilon| \to 0$ a.e. in Ω. We have $v_\varepsilon \to v$ in $L^p(\Omega)$ since $w_\varepsilon \to 0$ in $L^p(\Omega)$, $v_\varepsilon \in W_0^{1,p}(\Omega)$, $v_\varepsilon = 0$ on T_ε and

$$F_\varepsilon(v_\varepsilon) = \int_\Omega |\nabla v_\varepsilon|^p\, dx = \int_\Omega |\nabla v - w_\varepsilon \nabla v - \nabla v\, w_\varepsilon|^p\, dx.$$

Since $w_\varepsilon \to 0$ in $L^p(\Omega)$, it follows that

$$\limsup F_\varepsilon(v_\varepsilon) \leqq \limsup \int_\Omega (|\nabla v| + |v|\, |\nabla w_\varepsilon|)^p\, dx.$$

But, since $|\nabla w_\varepsilon|$ is bounded in $L^p(\Omega)$ and $|\nabla w_\varepsilon| \to 0$ a.e. in Ω, it follows from Theorem 1 of [3] that

$$\lim(\int_\Omega (|\nabla v| + |v|\, |\nabla w_\varepsilon|)^p\, dx - \int_\Omega |v|^p\, |\nabla w_\varepsilon|^p\, dx = \int_\Omega |\nabla v|^p\, dx.$$

Thus

$$\limsup F_\varepsilon(v_\varepsilon) \leqq \int_\Omega |\nabla v|^p\, dx + \limsup \int_\Omega |v|^p\, |\nabla w_\varepsilon|^p\, dx,$$

and from Proposition 2.4

$$\limsup F_\varepsilon(v_\varepsilon) \leqq \int_\Omega |\nabla v|^p \, dx + a\, C \int_\Omega |v|^p \, dx. \quad \square$$

Third step. Proof of (ii). It is rather technical and we refer to [9] for it.

Let us now state the main result about convergence of u_ε to u and correctors.

<u>THEOREM 2.6</u> Suppose $a \in \,]0, +\infty[$. Then:

(1) u_ε converges to u in $s\text{-}W_0^{1,q}(\Omega)$ for all $q < p$;

(2) if u belongs to $C_0^1(\Omega)$, then $u_\varepsilon - u + w_\varepsilon u \to 0$ in $s\text{-}W_0^{1,p}(\Omega)$.

<u>PROOF:</u> First let us prove that for every $\phi \in C_0^1(\Omega)$

$$\limsup \int_\Omega |\nabla u_\varepsilon - \nabla(\phi - w_\varepsilon \phi)|^p \, dx \leqq \|u-\phi\|_{1,p} \, C(\|\phi\|_{1,p}).$$

Let $\phi_\varepsilon = \phi - w_\varepsilon \phi$. There exists $\alpha > 0$ (see [8], p. 264, for instance) such that

$$\alpha \int_\Omega |\nabla u_\varepsilon - \nabla\phi_\varepsilon|^p \, dx \leqq \int_\Omega (|\nabla u_\varepsilon|^{p-2} \nabla u_\varepsilon - |\nabla\phi_\varepsilon|^{p-2} \nabla\phi_\varepsilon)(\nabla u_\varepsilon - \nabla\phi_\varepsilon) \, dx$$

$$\leqq \int_\Omega |\nabla u_\varepsilon|^p \, dx + \int_\Omega |\nabla\phi_\varepsilon|^p \, dx - \int_\Omega |\nabla u_\varepsilon|^{p-2} \nabla u_\varepsilon \nabla\phi_\varepsilon dx$$

$$- \int_\Omega |\nabla\phi_\varepsilon|^{p-2} \nabla\phi_\varepsilon \nabla u_\varepsilon \, dx.$$

But

$$\int_\Omega |\nabla u_\varepsilon|^p \, dx \to \int_\Omega |\nabla u|^p \, dx + a\, C \int_\Omega |u|^p \, dx, \text{ by Theorem 2.1,}$$

$$\limsup \int_\Omega |\nabla\phi_\varepsilon|^p \, dx \leqq \int_\Omega |\nabla\phi|^p \, dx + a\, C \int_\Omega |\phi|^p \, dx, \text{ by the second}$$

step of the proof of Theorem 2.1,

$$\int_\Omega |\nabla u_\varepsilon|^{p-2} \nabla u_\varepsilon \nabla\phi_\varepsilon dx = \int_\Omega f\phi_\varepsilon\, dx, \text{ variational formulation of } (1_\varepsilon),$$

$$\liminf \int_\Omega |\nabla\phi_\varepsilon|^{p-2} \nabla\phi_\varepsilon \nabla u_\varepsilon\, dx \geq \int_\Omega |\nabla\phi|^{p-2} \nabla\phi \nabla u\, dx + a\, C \int_\Omega |\phi|^{p-2}\phi u\, dx,$$

from the third step of the proof of Theorem 2.1.

Then

$$\limsup \alpha\int_\Omega |\nabla u_\varepsilon - \nabla\phi_\varepsilon|^p\, dx \leq \int_\Omega |\nabla u|^p\, dx + aC\int_\Omega |u|^p\, dx +$$

$$+ \int_\Omega |\nabla\phi|^p\, dx + aC\int_\Omega |\phi|^p\, dx$$

$$- \int_\Omega f\phi\, dx - \int_\Omega |\phi|^{p-2} \nabla\phi \nabla u\, dx - aC\int_\Omega |\phi|^{p-2}\phi u\, dx$$

$$\leq \int_\Omega f(u-\phi)\, dx + \int_\Omega |\nabla\phi|^{p-2} \nabla\phi(\nabla\phi-\nabla u)dx + aC\int_\Omega |\phi|^{p-2}\phi(\phi-u)\, dx$$

$$\leq \quad C\, \|u-\phi\|_{1,p}(\|f\|_{p'} + (||\phi||_{1,p})^{p-1}).$$

Hence, if $u \in C_0^1(\Omega)$, we deduce $\lim \int_\Omega |\nabla u_\varepsilon - \nabla(u-w_\varepsilon u)|^p\, dx = 0$, that is, $u_\varepsilon - u + w_\varepsilon u \to 0$ in $s\text{-}W_0^{1,p}(\Omega)$. Now let $q < p$. We have

$$\|u_\varepsilon - u\|_{1,q} \leq \|u_\varepsilon - \phi + w_\varepsilon\phi\|_{1,q} + \|u-\phi\|_{1,q} + \|w_\varepsilon\phi\|_{1,q}.$$

Using the previous estimate, it follows

$$\limsup \|u_\varepsilon - u\|_{1,q} \leq \|u-\phi\|_{1,p}\, C(||\phi||_{1,p}) + \limsup \|w_\varepsilon\phi\|_{1,q},$$

and since $w_\varepsilon \to 0$ in $s\text{-}W_0^{1,q}(\Omega)$,

$$\limsup \|u_\varepsilon - u\|_{1,q} \leq \|u-\phi\|_{1,p}\, C(||\phi||_{1,p}).$$

Making $\phi \to u$ in $s\text{-}W_0^{1,p}(\Omega)$, we get $u_\varepsilon \to u$ in $s\text{-}W_0^{1,q}(\Omega)$. □

3. STABILITY OF THE LIMIT PROBLEM OF (D_ε) WITH RESPECT TO SOME PERTURBATIONS

THEOREM 3.1: Let $1 \leqq s < p$ and let H be a convex continuous functional on $W^{1,s}(\Omega)$. Then the solution u_ε of

$$\text{Min } \{\int_\Omega |\nabla v|^p \, dx + H(v) - p \int_\Omega fv \, dx; \; v \in W_0^{1,p}(\Omega), \; v = 0 \text{ on } T_\varepsilon\}$$

converges in $w\text{-}W_0^{1,p}(\Omega)$ to the solution of

$$\text{Min } \{\int_\Omega |\nabla v|^p \, dx + H(v) + aC \int_\Omega |v|^p \, dx - p \int_\Omega fv \, dx; \; v \in W_0^{1,p}(\Omega) \; ,$$

where a and C are the same as in Theorem 2.1.

PROOF: The Proposition 2.4, part (3) is again used.

EXAMPLE: Let $h: R^N \to R^+$ be a convex continuous differentiable function such that $h(\nabla u) \leqq M|\nabla u|^s$ with $s < p$. Then the limit problem of

$$-\Delta_p u_\varepsilon - \text{div } (\partial h(\nabla u_\varepsilon) = f \text{ in } \Omega_\varepsilon,$$

$$u_\varepsilon = 0 \text{ on } \partial\Omega_\varepsilon,$$

is in the case $0 \leqq a < +\infty$

$$-\Delta_p u - \text{div } (\partial h(\nabla u) + aC|u|^{p-2}u = f \text{ in } \Omega,$$

$$u = 0 \text{ on } \partial\Omega.$$

Hence the "strange" term $aC|u|^{p-2}u$ does not depend on the perturbation $-\text{div}(\partial h(\nabla u)$.

4. THE CASE $r_\varepsilon = k\varepsilon$

We suppose here that the inclusions $T_{\varepsilon j}$ have a size of the same order as the periodic ε of the structure and $Y =]0,1[^N \supset T$. The case $p = 2$ has been studied in [11].

THEOREM 4.1: Suppose $r_\varepsilon = k\varepsilon$ with $k < 1/2$. Let u_ε be the solution of

$$-\Delta_p u_\varepsilon = f \text{ in } \Omega_\varepsilon,$$

$$u_\varepsilon = 0 \text{ on } \partial\Omega_\varepsilon.$$

Let w be the solution of

$$-\Delta_p w = 1 \text{ in } Y\setminus T$$

$$w = 0 \text{ on } T,$$

$$w \text{ is } Y\text{-periodic},$$

and let $w^* = \int_Y w \, dx$.

Then

(1) $u_\varepsilon \to 0$ in $s\text{-}W^{1,p}(\Omega)$;

(2) $\varepsilon^{-1/(p-1)} u_\varepsilon \to 0$ in $w\text{-}W^{1,p}(\Omega)$;

and

$$\int_\Omega |\nabla u_\varepsilon|^p \, dx \approx \varepsilon^{p(p-1)} \, w^* \int_\Omega |f|^{p/p-1} \, dx;$$

(3) $\varepsilon^{-p/(p-1)} u_\varepsilon \to z$ in $w\text{-}L^p(\Omega)$ and $|z|^{p-2} z = w^{*p-1} f$.

REFERENCES

[1] H. Attouch, Variational convergence for functions and operators, Applicable Mathematics Series, Pitman, London, 1984.

[2] H. Attouch and C. Picard, Variational inequalities with varying obstacles, J. Functional Analysis 50 (1983), 329-386.

[3] H. Brezis and E. Lieb, A relation between pointwise convergence of functions and convergence of functionals, Proc. A.M.S. 88 (1983), 486-490.

[4] D. Cioranescu and F. Murat, Un terme étrange venu d'aileurs, Collège de France Seminar, Research Notes in Mathematics, Pitman, London, 1982, No. 60, pp. 98-138, No. 70, pp. 154-178.

[5] G. Dal Maso and P. Longo, Γ-limits of ogstacles, Ann. Mat. Pura Appl. 128 (1981), 1-50.

[6] E. De Giorgi, G. Dal Maso and P. Longo, Γ-limiti di obstacoli, Rend. Acad. Naz. Lincei 68 (1980), 481-487.

[7] E. De Giorgi and T. Franzoni, Su un tipo di convergenza variazionale, Rend. Acad. Naz. Lincei 58 (1975), 842-850.

[8] I. Diaz, Nonlinear partial differential equations and free boundaries, Research Notes in Mathematics, Pitman, London, 1985, No. 106.

[9] N. Labani, Homogénéisation du problème de Dirichlet non linéaire dans un ouvert finement perforé, Thèse de 3° cycle, Université de Paris-Sud, Centre d'Orsay, No. 4129, 1987.

[10] C. Picard, Analyse limite d'équations variationnelles dans un domaine contenant une grille, Mathematical Modelling and Numerical Analysis 21 (1987) 293-326.

[11] L. Tartar, Cours Peccot au Collège de France, mars 1977 - partiellement rédigé dans F. Murat, H-convergence, Séminaire d'analyse fonctionnelle et numérique de l'Université d'Alger, 1977-78.

N. Labani
Laboratoire d'Analyse Numérique
Université de Paris-Sud
Bat. 425
F-91405-Orsay,
France

C. Picard
U.F.R. de Mathématiques et
Informatique
33 rue Saint Leu
80 039 Amiens
France

H. Labini
Département de Mathématiques
Faculté des Sciences
Route de Gray
25030 Besancon
France

J.P. PUEL

Some results on quasi-linear elliptic equations

1. INTRODUCTION

Our goal consists of studying quasilinear elliptic equations of the form

$$\begin{aligned} &A(u) + F(u,\nabla u) = 0 \text{ in } \Omega \text{ open bounded subset of } \mathbb{R}^N, \\ &u/\Gamma = 0, \text{ where } \Gamma = \partial\Omega, \end{aligned} \tag{1.1}$$

where A is an operator of the Leray-Lions type from $W_0^{1,p}(\Omega)$ into $W^{-1-p'}(\Omega)$ ($1 < p < +\infty$) and F is an operator associated with a Carathéodory function which has a growth of order p in the gradient variable.

In this paper, we prove a compactness theorem which is essential for studying (1.1), then give an existence result which improves those given in [2], [3], [4]. Next, we show some order properties related to the operator A and we apply these results to the existence, in a suitable context, of a maximal and a minimal solution for (1.1).

This work has been done in collaboration with L. Boccardo and F. Murat and is announced in [5].

We denote by Ω a bounded open domain in $\mathbb{R}^N$ and by Γ its boundary. Let p be a real number such that $1 < p < +\infty$ and p' its conjugate ($1/p+1/p' = 1$). We consider an operator of the Leray-Lions type ([6], [7]) of the form

$$A(v) = - \sum_{i=1}^{N} \frac{\partial}{\partial x_i} A_i(v,\nabla v),$$

where A_i are operators associated with Carathéodory functions, again denoted by A_i, from $\Omega \times \mathbb{R} \times \mathbb{R}^N$ into $\mathbb{R}$ satisfying

$$\begin{aligned} &\exists \beta_i \in \mathbb{R}, \exists h_i \in L^{p'}(\Omega), \forall s \in \mathbb{R}, \forall \xi \in \mathbb{R}^N, \\ &|A_i(x,s,\xi)| \leq \beta_i[h_i(x) + |s|^{p-1} + |\xi|^{p-1}], \text{ a.e. in } \Omega, \end{aligned} \tag{1.2}$$

$$\forall s \in R, \forall \xi, \xi^* \in R^N, \xi \neq \xi^*, \sum_{i=1}^{N} [A_i(x,s,\xi) - A_i(x,s,\xi^*)](\xi_i - \xi_i^*) > 0,$$

$$\text{a.e. in } \Omega, \tag{1.3}$$

$$\exists \rho > 0, \forall s \in R, \forall \xi \in R^N, \sum_{i=1}^{N} A_i(x,s,\xi)\xi_i \geq \rho|\xi|, \text{ a.e. in } \Omega. \tag{1.4}$$

Under these hypotheses, A is a bounded continuous pseudomonotone operator acting from $W_0^{1,p}(\Omega)$ into $W^{-1,p'}(\Omega)$.

Let now F be a Carathéodory function from $\Omega \times R \times R^N$ into R such that there exist an increasing function b from R^+ to R^+ and $k \in L^1(\Omega)$ with

$$|F(x,s,\xi)| \leq b(|s|)[k(x) + |\xi|^p], \text{ a.e. in } \Omega. \tag{1.5}$$

The associated Nemitskii operator will be denoted by $v \to F(v,\nabla v)$. We can give sense to (1.1) by considering solutions u of (1.1) with $u \in W_0^{1,p}(\Omega) \cap L^\infty(\Omega)$.

2. COMPACTNESS RESULT

We present here a compactness theorem which is essential for the study of (1.1), in particular for the existence results. This theorem gathers the main compactness arguments developed in [3] and [4]. The corresponding result for the parabolic situation has been given in [9].

THEOREM 2.1: Let (u_n) be a sequence of functions such that:

(i) $u_n \in W_0^{1,p}(\Omega) \cap L^\infty(\Omega)$, $A(u_n) \in L^1(\Omega)$;

(ii) $\exists M > 0, \forall n, \|u_n\|_{L^\infty(\Omega)} \leq M$;

(iii) $\exists C > 0, \exists h \in L^1(\Omega)$ such that $|A(u_n)(x)| \leq C[h(x) + |\nabla u_n(x)|^p]$

$$\text{a.e. in } \Omega. \tag{2.1}$$

Then (u_n) is relatively compact for the strong topology of $W_0^{1,p}(\Omega)$.

In order to make the exposition clear, we will present a sketch of the proof of this result in a simpler situation.

THEOREM 2.2: Let (u_n) be a sequence of functions such that:

(i) $u_n \in H_0^1(\Omega) \cap L^\infty(\Omega)$, $\Delta u_n \in L^1(\Omega)$;

(ii) $\exists M > 0$, $\forall n$, $\|u_n\|_{L^\infty(\Omega)} \leqq M$,

(iii) $\exists C > 0$, $\exists h \in L^1(\Omega)$ such that $|\Delta(u_n)(x)| \leqq C[h(x) + |\nabla u_n(x)|^2]$

$$\text{a.e. in } \Omega. \tag{2.2}$$

Then (u_n) is relatively compact for the strong topology of $H_0^1(\Omega)$.

PROOF: (a) We first show that (u_n) is bounded in $H_0^1(\Omega)$. Let $v_n = e^{\lambda u_n^2} u_n$, where $\lambda \in \mathbb{R}$ will be chosen later on equal to $C^2/2$. As $u_n \in H_0^1(\Omega) \cap L^\infty(\Omega)$, $v_n \in H_0^1(\Omega) \cap L^\infty(\Omega)$ and

$$\frac{\partial v_n}{\partial x_i} = e^{\lambda u_n^2} \frac{\partial u_n}{\partial x_i} + 2\lambda u_n^2 e^{\lambda u_n^2} \frac{\partial u_n}{\partial x_i} .$$

Denoting by $<\cdot,\cdot>$ the duality between $H^{-1}(\Omega)$ and $H_0^1(\Omega)$, we have

$$<-\Delta u_n, v_n> = \int_\Omega e^{\lambda u_n^2} |\nabla u_n|^2 \, dx + 2\lambda \int_\Omega u_n^2 e^{\lambda u_n^2} |\nabla u_n|^2 \, dx.$$

On the other hand, using (iii)

$$<-\Delta u_n, v_n> \leqq \int_\Omega C[h + |\nabla u_n|^2] e^{\lambda u_n^2} |u_n| dx.$$

Using (ii) and the Cauchy-Schwartz inequality we obtain

$$\int_\Omega e^{\lambda u_n^2} |\nabla u_n|^2 dx + 2\lambda \int_\Omega u_n^2 e^{\lambda u_n^2} |\nabla u_n|^2 dx \leqq C \int_\Omega e^{\lambda u_n^2} |u_n| \, |\nabla u_n|^2 dx + C(\lambda)$$

$$\leqq \frac{1}{2} \int e^{\lambda u_n^2} |\nabla u_n|^2 dx + \frac{C^2}{2} \int_\Omega e^{\lambda u_n^2} u_n^2 |\nabla u_n|^2 \, dx + C(\lambda),$$

where $C(\lambda)$ is a constant depending on λ. Choosing $\lambda = C^2/2$, we get

$$\int_\Omega e^{\lambda u_n^2} |\nabla u_n|^2 \, dx \leqq C',$$

where C' is a constant independent of n. Thus, (u_n) is bounded in $H_0^1(\Omega)$.

(b) Now we know that (u_n) is relatively compact for the weak topology of $H_0^1(\Omega)$. Let us extract a subsequence, again denoted by (u_n), such that

$$u_n \to u \text{ in } H_0^1(\Omega) \text{ weakly.}$$

We will show that in fact

$$u_n \to u \text{ in } H_0^1(\Omega) \text{ strongly.}$$

Because of (ii) we know that $u \in H_0^1(\Omega) \cap L^\infty(\Omega)$. Let $w_n = e^{\mu(u_n-u)^2}(u_n - u)$, where μ will be chosen equal to $2C^2$. Then

$$w_n \in H_0^1(\Omega) \cap L^\infty(\Omega),$$

$$w_n \to 0 \text{ in } H_0^1(\Omega) \text{ weakly,}$$

$$w_n \to 0 \text{ in } L^\infty(\Omega) \text{ weak}^*.$$

On the one hand, we have

$$\langle -\Delta(u_n-u), w_n\rangle = \int_\Omega e^{\mu(u_n-u)^2}|\nabla(u_n-u)|^2dx + 2\mu\int_\Omega e^{\mu(u_n-u)^2}(u_n-u)^2|\nabla(u_n - u)|^2\,dx.$$

On the other hand, using (iii) we get

$$\langle -\Delta(u_n-u),w_n\rangle \leqq \langle \Delta u,w_n\rangle + \int_\Omega C[h + |\nabla u_n|^2]e^{\mu(u_n-u)^2}|u_n - u|dx$$

$$\leqq \langle \Delta u,w_n\rangle + \int_\Omega C[h + 2|\nabla(u_n-u)|^2 + 2|\nabla u|^2]e^{\mu(u_n-u)^2}|u_n - u|dx$$

$$\leqq 2\int_\Omega Ce^{\mu(u_n-u)^2}|u_n-u||\nabla(u_n-u)|^2dx + \langle \Delta u,w_n\rangle + 2C\int_\Omega |\nabla u|^2|w_n|dx + C\int_\Omega h|w_n|dx$$

$$\leqq 2\int_\Omega Ce^{\mu(u_n-u)^2}|u_n - u||\nabla(u_n - u)|^2dx + o(1),$$

as $w_n \to 0$ in $H_0^1(\Omega)$ weakly and in $L^\infty(\Omega)$ weak*. Now we proceed as in step (a) and this shows that

$$\int_\Omega e^{\mu(u_n-u)^2} |\nabla(u_n - u)|^2 \, dx = o(1)$$

and finishes the proof of Theorem 2.2. The arguments for proving Theorem 2.1 are analogous but more technical. □

3. EXISTENCE RESULT

The previous compactness theorem enables us to prove the existence of a solution for problem (1.1) whenever we can find an approximation procedure which preserves an L^∞ bound and the inequality (2.1). This can be done in various contexts and in particular if we assume the existence of a subsolution smaller than a supersolution.

DEFINITION 3.1: A subsolution ϕ for problem (1.1) is an element of $W^{1,p}(\Omega) \cap L^\infty(\Omega)$ such that

$$\begin{aligned} &A(\phi) + F(\phi,\nabla\phi) \leqq 0 \text{ in } W^{-1,p'}(\Omega) + L^1(\Omega), \\ &\phi/\Gamma \leqq 0. \end{aligned} \tag{3.1}$$

A supersolution ψ for problem (1.1) is an element of $W^{1,p}(\Omega) \cap L^\infty(\Omega)$ such that

$$\begin{aligned} &A(\psi) + F(\psi,\nabla\psi) \geqq 0 \text{ in } W^{-1,p'}(\Omega) + L^1(\Omega), \\ &\psi/\Gamma \geqq 0. \end{aligned} \tag{3.2}$$

We then obtain:

THEOREM 3.1: Let us assume the existence of a subsolution ϕ and a supersolution ψ for problem (1.1) such that $\phi \leqq \psi$ a.e. in Ω. Then there exists a solution u of (1.1) such that $\phi \leqq u \leqq \psi$ a.e. in Ω.

COMMENT: The result of Theorem 3.1 was given in [3], Theorem 2 for the case of $\phi, \psi \in W^{1,\infty}(\Omega)$. In that situation, the proof is essentially based on Theorem 2.1 after a standard truncation argument which can be found, for example, in [2] and [3]. Here, we only assume that ϕ and ψ belong to $W^{1,p}(\Omega) \cap L^\infty(\Omega)$ and the difference is sometimes important as will be seen in Section 4. The proof of Theorem 3.1 is not a straightforward extension of the previous existence result. In fact, it follows, step by step, the argument of [3, Theorem 1], by considering the associated variational inequality on the convex set

$$K = \{v \in W_0^{1,p}(\Omega), \quad \phi - 1 \leqq v \leqq \psi + 1\},$$

after having made the standard truncation at ϕ and μ which is described in [3]. Then one remarks, as in [8], that the solution of this variational inequality is in fact the solution of (1.1).

4. ORDER PROPERTIES OF SOLUTIONS OF (1.1)

We begin with an inequality of the Kato type for the Leray-Lions operators. This result is well known for second-order linear elliptic operators, but it seems new in our context even though it is related to a result of [1].

THEOREM 4.1: Let us assume that in addition to (1.2), (1.3), (1.4) the functions A_i satisfy a Lipschitz condition in the s-variable

$$\begin{aligned} &\exists C_i > 0,\ \exists k_i \in L^{p'}(\Omega),\ \forall s,\ s^* \in \mathbb{R},\ \forall \xi \in \mathbb{R}^N, \\ &|A_i(x,s,\xi) - A_i(x,s^*,\xi)| \leqq C_i|s-s^*|[k_i(x) + |\xi|^{p-1}] \text{ a.e. in } \Omega. \end{aligned} \tag{4.1}$$

Let u_α, $\alpha = 1,2$, satisfy

$$A(u_\alpha) \leqq f_\alpha,$$

where $f_\alpha \in L^1(\Omega)$. If $u = \mathrm{Sup}(u_1,u_2)$ and

$$h_\theta = \begin{cases} f_1 \text{ if } u_1 > u_2, \\ f_2 \text{ if } u_2 > u_1, \\ (1-\theta)f_1 + \theta f_2 \text{ if } u_1 = u_2 \text{ where } \theta \in [0,1]. \end{cases}$$

Then

$$A(u) \leqq h_\theta, \ \forall\theta \in [0,1].$$

Of course, if $A(v_\alpha) \geqq f_\alpha$ and $v = \inf(v_1,v_2)$, we obtain $A(v) \geqq h_\theta$.

REMARK: If the operator A satisfies an hypothesis of strong monotonicity, i.e.

$$\exists\rho > 0, \ \forall s \in \mathbb{R}, \forall\xi,\xi^* \in \mathbb{R}^N,$$

$$\sum_{i=1}^{N} [A_i(x,s,\xi) - A_i(x,s,\xi^*)](\xi_i-\xi_i^*) > \rho|\xi - \xi^*|^p \text{ a.e. in } \Omega \tag{4.2}$$

the conclusion of Theorem 4.1 remains true if instead of (4.1) we assume that A satisfies the (weaker) Hölder condition of order $1/p'$

$$\begin{aligned} &\exists C_i > 0, \ \exists k_i \in L^{p'}(\Omega), \ \forall s, \ s^* \in \mathbb{R}, \ \forall\xi \in \mathbb{R}^N, \\ &|A_i(x,s,\xi) - A_i(x,s^*,\xi)| \leqq C_i|s - s^*|^{1/p'}[k_i(x) + |\xi|^{p-1}] \text{ a.e. in } \Omega. \end{aligned} \tag{4.3}$$

PROOF: For $\varepsilon > 0$ and $\theta \in [0,1]$, let β_ε^θ be an increasing regular function from $\mathbb{R}$ to $\mathbb{R}$ such that

$$\beta_\varepsilon^\theta(r) = \begin{cases} 0 \text{ if } r \leqq -\varepsilon, \\ \theta \text{ if } r = 0, \\ 1 \text{ if } r \geqq \varepsilon, \end{cases}$$

and

$$|\beta_\varepsilon'^\theta(r)| \leqq \frac{2}{\varepsilon}.$$

If $\chi \in \mathcal{D}^+(\Omega)$, we set

$$I_\varepsilon^\theta = \int_\Omega f_1[1 - \beta_\varepsilon^\theta(u_2-u_1)]\chi \, dx + \int_\Omega f_2 \, \beta_\varepsilon^\theta(u_2-u_1)\chi \, dx.$$

When $\varepsilon \to 0$, $I_\varepsilon^\theta \to I^\theta$, where

$$I^{\theta} = \int_{\Omega} f_1[1 - \beta^{\theta}(u_2-u_1)]\chi \, dx + \int_{\Omega} f_2 \quad {}^{\theta}(u_2-u_1)\chi \, dx,$$

and

$$\beta^{\theta}(r) = \begin{cases} 0 \text{ if } r < 0, \\ \theta \text{ if } r = 0, \\ 1 \text{ if } r > 0. \end{cases}$$

Then

$$I^{\theta} = \int_{\Omega} h_{\theta}\chi \, dx.$$

On the other hand,

$$I_{\varepsilon}^{\theta} = \int_{\Omega} [1 - \beta_{\varepsilon}^{\theta}(u_2 - u_1)]f_1\chi \, dx + \int_{\Omega} \beta_{\varepsilon}^{\theta}(u_2 - u_1)f_2\chi \, dx$$

$$\geq \langle A(u_1), [1 - \beta_{\varepsilon}^{\theta}(u_2 - u_1)]\chi\rangle + \langle A(u_2), \beta_{\varepsilon}^{\theta}(u_2 - u_1)\chi\rangle$$

$$= \int_{\Omega}\sum_{i=1}^{N} A_i(u_1,\nabla u_1)\frac{\partial\chi}{\partial x_i} \, dx + \int_{\Omega}\sum_{i=1}^{N} [A_i(u_2,\nabla u_2)-A_i(u_1,\nabla u_1)]\frac{\partial}{\partial x_i}[\beta_{\varepsilon}^{\theta}(u_2-u_1)\chi]dx$$

$$= \int_{\Omega}\sum_{i=1}^{N} A_i(u_1,\nabla u_1)\frac{\partial\chi}{\partial x_i} \, dx + \int_{\Omega}\sum_{i=1}^{N} [A_i(u_2,\nabla u_2)-A_i(u_1,\nabla u_1)]\beta_{\varepsilon}^{\theta}(u_2 - u_1)\frac{\partial\chi}{\partial x_i} \, dx$$

$$+ \int_{\Omega}\sum_{i=1}^{N} [A_i(u_2,\nabla u_2)-A_i(u_2,\nabla u_1)]\beta_{\varepsilon}'^{\theta}(u_2 - u_1) \frac{\partial}{\partial x_i} (u_2 - u_1)\chi \, dx$$

$$+ \int_{\Omega}\sum_{i=1}^{N} [A_i(u_2,\nabla u_1)-A_i(u_1,\nabla u_1)]\beta_{\varepsilon}'^{\theta}(u_2 - u_1) \frac{\partial}{\partial x_i} (u_2 - u_1)\chi \, dx.$$

We know that

$$\int_{\Omega}\sum_{i=1}^{N} [A_i(u_2,\nabla u_2) - A_i(u_2,\nabla u_1)]\beta_{\varepsilon}'^{\theta}(u_2 - u_1) \frac{\partial}{\partial x_i} (u_2 - u_1)\chi \, dx > 0.$$

Therefore

$$I_\varepsilon^\theta \geq \int_\Omega \sum_{i=1}^N A_i(u_1,\nabla u_1) \frac{\partial \chi}{\partial x_i}\, dx$$

$$+ \int_\Omega \sum_{i=1}^N [A_i(u_2,\nabla u_2) - A_i(u_1,\nabla u_1)]\beta_\varepsilon^\theta(u_2 - u_1) \frac{\partial \chi}{\partial x_i}\, dx + J_\varepsilon^\theta, \tag{4.4}$$

where

$$J_\varepsilon^\theta = \int_\Omega \sum_{i=1}^N [A_i(u_2,\nabla u_1) - A_i(u_1,\nabla u_1)]\beta_\varepsilon'^\theta(u_2 - u_1) \frac{\partial}{\partial x_i}(u_2 - u_1)\chi\, dx.$$

As $\beta_\varepsilon'^\theta(r) = 0$ for $|r| \geqq \varepsilon$ and $|\beta_\varepsilon'^\theta(r)| \leqq 2/\varepsilon$ for $|r| \leqq \varepsilon$, we obtain, using (4.1)

$$|J_\varepsilon^\theta| \leqq \int_{\{0<|u_2-u_1|\leqq\varepsilon\}} \sum_{i=1}^N C_i[k_i+|\nabla u_1|^{p-1}]|u_2-u_1| \frac{2}{\varepsilon} \left|\frac{\partial}{\partial x_i}(u_2-u_1)\right|\chi\, dx$$

$$\leqq \int_{\{0<|u_2-u_1|\leqq\varepsilon\}} \tilde{k}|\nabla(u_2 - u_1)|\, dx,$$

where $\tilde{k} \in L^{p'}(\Omega)$ (and $\nabla(u_2 - u_1) \in L^p(\Omega)$).

Therefore $J_\varepsilon^\theta \to 0$ if $\varepsilon \to 0$, and we can pass to the limit in (4.4) as $\varepsilon \to 0$ and we obtain

$$I^\theta \geqq \int_\Omega \sum_{i=1}^N A_i(u_1,\nabla u_1) \frac{\partial \chi}{\partial x_i}\, dx + \int_\Omega \sum_{i=1}^N [A_i(u_2,\nabla u_2)-A_i(u_1,\nabla u_1)]\beta^\theta(u_2 - u_1) \frac{\partial \chi}{\partial x_i}\, dx$$

$$= \int_{\{u_1>u_2\}} \sum_{i=1}^N A_i(u_1,\nabla u_1)\frac{\partial \chi}{\partial x_i}dx + \int_{\{u_2>u_1\}} \sum_{i=1}^N A_i(u_2,\nabla u_2) \frac{\partial \chi}{\partial x_i}\, dx$$

$$+ \int_{\{u_1=u_2\}} \sum_{i=1}^N [(1-\theta)A_i(u_1,\nabla u_1) + \theta A_i(u_2,\nabla u_2)] \frac{\partial \chi}{\partial x_i}\, dx.$$

On the set $\{u_1 = u_2\}$, $\nabla u_1 = \nabla u_2 = \nabla u$ a.e., and so

$$I^\theta \geqq \int_\Omega \sum_{i=1}^N A_i(u,\nabla u) \frac{\partial \chi}{\partial x_i}\, dx = \langle A(u),\chi\rangle.$$

This shows that

$$\forall \chi \in \mathcal{D}^+(\Omega),\ \langle A(u),\chi\rangle \leqq \langle h_\theta,\chi\rangle,$$

which is the result of Theorem 4.1. □

Now we can prove some order properties on solutions of problem (1.1) which make the result of Theorem 3.1 more precise.

THEOREM 4.2: Let us assume that in addition to (1.2), (1.3), (1.4), A satisfies the Lipschitz condition (4.1). If there exist a subsolution ϕ and a supersolution ψ for problem (1.1) such that $\phi \leqq \psi$ a.e. in Ω, then there exist a minimal solution u_{min} and a maximal solution u_{max} of (1.1) such that $\phi \leqq u_{min} \leqq u_{max} \leqq \psi$ a.e. in Ω (i.e. u_{min} and u_{max} are solutions and if u is a solution of (1.1) with $\phi \leqq u \leqq \psi$ a.e. in Ω, then $u_{min} \leqq u \leqq u_{max}$ a.e. in Ω).

REMARKS:

(1) The terminology of minimal and maximal solutions is somewhat improper. In fact, u_{min} and u_{max} are the smallest and the greatest solutions respectively.

(2) The result does not imply nonuniqueness for (1.1), as u_{min} may be equal to u_{max}.

PROOF: Let us define the set

$$S = \{v \in W_0^{1,p}(\Omega),\ \phi \leqq v \leqq \psi \text{ a.e. in } \Omega,\ v \text{ is a solution of (1.1)}\}.$$

We claim that S is inductive. Let (v_n) be an increasing sequence of elements of S, i.e.

$$\forall n,\ v_n \in S,\ v_n \leqq v_{n+1}.$$

Then there exists $v \in L^\infty(\Omega)$, $\phi \leqq v \leqq \psi$ a.e. in Ω, such that

$$v_n \to v \text{ in } L^q(\Omega),\ \forall q,\ 1 \leqq q < +\infty.$$

As $A(v_n) + F(v_n, \nabla v_n) = 0$, $\forall n$, and (v_n) is bounded in $L^\infty(\Omega)$, we can apply the compactness result of Theorem 2.1 which shows that

$$v_n \to v \text{ in } W_0^{1,p}(\Omega) \text{ strongly.}$$

Then, $v \in W_0^{1,p}(\Omega)$ and it is easy to show that

$$A(v) + F(v,\nabla v) = 0.$$

Therefore $v \in S$ and v is the least upper bound of (v_n); so S is inductive. From Zorn's lemma, S possesses a maximal element u_{max} (here maximal in the correct sense). Suppose u_{max} is not the greatest element of S. Then there exists $v \in S$ such that $v \not\leq u_{max}$. Let us set $w = \text{Sup}(v,u_{max})$. As

$$A(v) + F(v,\nabla v) = 0 \text{ and } A(u_{max}) + F(u_{max}, \nabla u_{max}) = 0,$$

applying Theorem 4.1 we obtain

$$A(w) + F(w,\nabla w) \leqq 0,$$

$$w \in W_0^{1,p}(\Omega) \cap L^\infty(\Omega).$$

So w is a subsolution for (1.1) and moreover $w \leqq \psi$ a.e. in Ω. Using Theorem 3.1, we know that there exists a solution u of (1.1) such that $w \leqq u \leqq \psi$ a.e. in Ω. Notice that here we need the result of Theorem 3.1 in its full generality because we do not know if w belongs to $W^{1,\infty}(\Omega)$. So $u \geqq u_{max}$ a.e. in Ω and $u \neq u_{max}$ (because $w \neq u_{max}$), which gives a contradiction to the fact that u_{max} is a maximal element of S. Thus, u_{max} is the greatest element of S, or the "maximal solution" of (1.1) as asserted in Theorem 4.2. In the same way, one can prove the existence of a minimal solution of (1.1) and the proof of Theorem 4.2 is complete. □

<u>COMMENT</u>: The result of Theorem 4.2 is well known for semilinear equations satisfying the maximum principle (see, for example, [10]). It was shown in [8] that it is also true for a restrictive class of quasi linear equations. In fact, it turns out to be very general and it only requires the fact that

the supremum of two solutions is a subsolution (application of Theorem 4.1). To prove this last result, we need a Lipschitz (or Hölder) condition in the u-variable (condition (4.1) or (4.3)). We do not know if this restriction is technical or if it is a "necessary" condition. The argument used in [8] was related to the uniqueness for the associated parabolic problem, and we have good reasons to think that such an argument could possibly work in the present context. It would be interesting to look for the possible relations between these conditions.

REFERENCES

[1] L. Barthélemy and P. Bénilan, Sous potentiels d'un opérateur non linéaire (to appear).

[2] L. Boccardo, F. Murat and J.P. Puel, Résultats d'existence pour certains problèmes elliptiques quasilinéaires. Ann. Scuola Norm. Sup. Pisa Cl. Sci. (4) 11 (1984), 213-235.

[3] L. Boccardo, F. Murat and J.P. Puel, Existence of bounded solutions for nonlinear elliptic unilateral problems. Ann. di Mat. Pura ed Appl. (to appear).

[4] L. Boccardo, F. Murat and J.P. Puel, Existence de solutions faibles pour des équations elliptiques quasilinéaires à croissance quadratique. Nonlinear Partial Differential Equations and Their Applications, Collège de France Seminar, Vol IV, Eds. H. Brezis and J.L. Lions, pp. 19-73 Pitman Research Notes in Mathematics, Vol. 84, 1983.

[5] L. Boccardo, F. Murat and J.P. Puel, Quelques propriétés des opérateurs elliptiques quasilinéaires, C.R.A.S. Paris t.307, Série I, pp.749-752, 1988.

[6] J.L. Lions, Quelques Méthodes de Résolution des Problèmes aux Limites non Linéaires, Paris, Dunod, 1969.

[7] J. Leray and J.L. Lions, Quelques résultats de Visik sur les problèmes elliptiques non linéaires par les méthodes de Minty-Browder. Bull. Soc. Math. France 93 (1965), 97-107.

[8] J.P. Puel, Existence, comportement à l'infini et stabilité dans certains problèmes quasilinéaires elliptiques et paraboliques d'ordre 2. Ann. Scuola Norm. Sup. Pisa 3 (1976), 89-119.

[9] J.P. Puel, A compactness theorem in quasilinear parabolic problems and application to an existence result. Nonlinear Parabolic Equations: Qualitative Properties of Solutions, eds., L. Boccardo and A. Tesei, pp. 189-199. Pitman Research Notes in Mathematics, Vol. 149, (1987).

[10] D. H. Sattinger, Monotone methods in nonlinear elliptic and parabolic boundary value problems. Indiana Univ. Math. Journal 21 (1972), 979-1000.

J.P. Puel
Département de Mathématiques
Université d'Orléans
B.P. 6759
45067 Orléans Cedex 2
France

and

Laboratorie d'Analyse Numérique
T.55-65, 5ème etage
Université Pierre et Marie Curie
4 Place Jussieu,
75252 Paris Cedex 05
France

J.F. RODRIGUES

Some remarks on the quasi-linear noncoercive elliptic obstacle problem

1. INTRODUCTION

In this paper we consider some "*noncoercive*" second-order, elliptic, quasi-linear obstacle problems in a bounded domain Ω of R^n $(n \geqq 2)$, namely, of the following type:

$$u \in K_\psi: \quad \langle Qu, v - u\rangle \geqq 0, \ \forall v \in K_\psi, \tag{1}$$

where, for an obstacle $\psi: \Omega \to [-\infty, +\infty[$,

$$K_\psi \equiv \{v \in H_0^1(\Omega): v \geqq \psi \text{ a.e. in } \Omega\} \neq \emptyset \tag{2}$$

and, for given Carathéodory functions A_i, $B: \Omega \times R \times R^n \to R$,

$$\langle Qu, v\rangle = \int_\Omega A_i(x,u,Du)D_i v\, dx + \int_\Omega B(x,u,Du)v\, dx, \ \forall u,v \in H_0^1(\Omega). \tag{3}$$

Here we use the sum convention on $i = 1,\dots,n$ and $Du = (D_1u,\dots,D_nu)$ denotes the gradient of u. The structure of the quasi-linear elliptic operator $Q: H^1(\Omega) \cap L^\sigma(\Omega) \to H^{-1}(\Omega)$ is defined by

$$[A_i(x,u,p) - A_i(x,u,q)](p_i - q_i) \geqq \alpha|p - q|^2, \tag{4}$$

$$|A_i(x,u,p) - A_i(x,v,p)| \leqq \omega(|u-v|)[g(x) + |p|], \tag{5}$$

$$|A_i(x,u,p)| \leqq f_i(x) + e_i(x)|u|^{\rho_i} + c_i|p| \quad (i = 1,\dots,n), \tag{6}$$

$$[B(x,u,p) - B(x,v,p)](u - v) \geqq 0, \tag{7}$$

$$|B(x,u,p) - B(x,u,q)| \leqq b(x)|p - q|, \tag{8}$$

$$|B(x,u,p)| \leqq f_0(x) + c_0(x)|u|^{\rho_0} + b(x)|p|, \tag{9}$$

for a.e. $x \in \Omega$, $u = u(x)$, $v = v(x)$, $p = Du(x)$, $q = Dv(x)$, where $\alpha > 0$ and $c_i \geq 0$ are constants, $f_i, g \in L^2(\Omega)$, $f_0 \in L^r(\Omega)$ ($r \geq 2n/(n+2)$ if $n \geq 3$ or $r > 1$ if $n = 2$), $b \in L^s(\Omega)$ ($s = n \geq 3$ or $s > n = 2$), the exponents $\rho_0 \in [0, (n+2)/(n-2)]$ and $\rho_i \in [0, n/(n-2)]$ ($0 \leq \rho_0, \rho_i < +\infty$ if $n = 2$) determine the assumptions on $c_0 \in L^{r_0}(\Omega)$ and $e_i \in L^{r_i}(\Omega)$ by $r_0 = 2n/[2n - (\rho_0 + 1)(n-2)]$ ($\forall r_0 > 1$ if $n = 2$) and $r_i = 2n/[n - \rho_i(n-2)]$ ($\forall r_i > 2$ if $n = 2$), so that $A_i(u,Du) = A_i(x,u(x), Du(x)) \in L^2(\Omega)$ and $B(u,Du) \in L^{\sigma'}(\Omega)$, $\sigma' = \sigma/(\sigma - 1)$ with $\sigma = 2n/(n-2)$ if $n \geq 3$, or $\sigma < \infty$ and $\sigma' > 1$ if $n = 2$. Here we have used the Sobolev imbedding $H_0^1(\Omega) \subset L^\sigma(\Omega)$.

In Section 2, we recall from [CM] a comparison principle for the case $b = 0$, under suitable assumptions on $\omega(t) \to 0$ as $t \to 0$ (ω is a modulus of continuity)

$$Qu = -D_i[A_i(x,u,Du)] + B(x,u). \tag{10}$$

including a special interesting case

$$Qu = -D_i[\alpha_{ij}(x,u)D_ju + \beta_i(x,u)] + B(x,u). \tag{11}$$

Using a technique of [T], we extend the comparison principle for the obstacle problem (1) to the novel case $\omega \equiv 0$

$$Qu = -D_i[A_i(x,Du)] + B(x,u,Du). \tag{12}$$

In these cases we obtain a L^∞-stability estimate for the solutions of (1) with respect to the L^∞-variation of the obstacles.

From that order principle, in Section 3, it is given a simple condition for the existence of a solution to the noncoercive obstacle problem (1) under the mild assumption $\psi \in L^\sigma(\Omega)$ in (2), $\sigma = 2s/(s-2)$, namely, the existence of a supersolution to (1). As a simple consequence we extend to this class of operators the property that the supremum of two supersolutions is still a supersolution.

Finally, in Section 4, we prove an *a priori* estimate in $H_0^1(\Omega)$ for the solution of (1) for the special case $e_i \equiv 0$, still "*noncoercive*", which yields a direct existence result without the uniqueness conditions (5) and

(8). We emphasize that no classical strong coerciveness assumptions are used here in order to control the nontrivial convective term in B. In particular, we do not need any further assumptions on b, or a L^∞-estimate for the solutions as in [RT] or in [BMP].

2. AN EXTENDED COMPARISON PRINCIPLE

We recall that a function $w \in H^1(\Omega) \cap L^\sigma(\Omega)$ is said to be a "*supersolution*" to the obstacle problem (1), if $w \geq \psi$ in Ω, $w \geq 0$ on $\partial\Omega$ and $Qw \geq 0$ in Ω, that is,

$$\langle Qw,v\rangle \geq 0, \quad \forall v \in H_0^1(\Omega): v \geq 0 \text{ in } \Omega. \tag{13}$$

THEOREM 1: Let w and u be, respectively, a supersolution and a solution to (1), under the assumptions (4)-(9). Then we have

$$w \geq u \text{ a.e. in } \Omega,$$

in each one of the following situations:

(i) $\omega \equiv 0$, i.e. in case (12);

(ii) $b \equiv 0$, i.e. in case (10) and $\int_{0+} dt/\omega(t) = +\infty$;

(iii) $b \equiv 0$, $\int_{0+} dt/\omega^2(t) = +\infty$ and B is strictly increasing in u, a.e. $x \in \Omega$;

(iv) $b \equiv 0$, $\int_{0+} dt/\omega^2(t) = +\infty$ and in case (11) the elliptic coefficients $a_{ij}(x,u)$ are Lipschitz continuous in $x \in \Omega$, uniformly in u, and there exist constants $\alpha_i \in \mathbb{R}$, at least one $\neq 0$, such that, $u \mapsto \alpha_i\ \beta_i(x,u)$ is monotone (increasing or decreasing).

PROOF: The cases (ii), (iii) and (iv) are a slightly more general restatement of the comparison results of [CM], which are given in terms of two solutions (see Corollary 1), as we have observed in [CR] for the parabolic case. So we limit ourselves to prove here the novel case (i).

Following [T], suppose for contradiction that $M = \sup z = \sup(u - w) > 0$ and define

$$z_k = (z - k)^+ = (z - k) \vee 0, \text{ for } 0 < k < M \leq \infty.$$

Remarking that $v = u - z_k \in K_\psi$, with this choice in (1) and with $v = z_k$ in (13) we obtain

$$\langle Qu - Qw, z_k \rangle \leq 0.$$

Setting $\Omega_k = \{x \in \Omega: z(x) > k, |Dz(x)| > 0\}$, we get

$$\alpha \int_\Omega |Dz_k|^2 \leq \int_\Omega [A_i(x,Du) - A_i(x,Dw)] D_i z_k$$

$$\leq \int_\Omega [B(w,Dw) - B(w,Du)] z_k + \int_{\{u>w+k\}} [B(w,Du) - B(u,Du)] z_k$$

$$\leq \int_\Omega b|D(w - u)| z_k = \int_\Omega b|Dz_k| z_k.$$

Using the Sobolev imbedding and recalling $b \in L^s(\Omega)$ ($s > n = 2$ or $s = n \geq 3$), it follows, by the Hölder inequality,

$$\alpha \|Dz_k\|^2_{L^2(\Omega)} \leq \int_\Omega b|Dz_k| z_k \leq \|b\|_{L^s(\Omega_k)} \|Dz_k\|_{L^2(\Omega)} \|z_k\|_{L^\sigma(\Omega)}$$

$$\leq C_\sigma \|b\|_{L^s(\Omega_k)} \|Dz_k\|^2_{L^2(\Omega)},$$

where $C_\sigma > 0$ and $\sigma = 2s/(s - 2)$. Therefore, we find

$$1 \leq \frac{C_\sigma}{\alpha} \|b\|_{L^s(\Omega_k)},$$

and this is impossible unless $M = 0$, since $\text{meas}(\Omega_k) \to 0$ as $k \to M$ (remember $\text{meas}\{x: z = M, |Dz| > 0\} = 0$ if $0 < M < +\infty$. □

<u>REMARK 1</u>: This proof actually shows that one also has the following

comparison principle: if $Qw \geqq Qu$ in Ω and $w \geqq u$ on $\partial\Omega$, then $w \geqq u$ in Ω. In this form, case (i) can be found in [T], under slightly different conditions, as well as case (ii) with $\omega(t) = ct(c > 0$, i.e. the Lipschitz case (see also [A])). Cases (iii) and (iv), covering the situations of the Hölder continuous nonlinearities with exponent $1/2 \leqq \lambda \leqq 1$, were treated in [CC] and extended in [CM].

REMARK 2: Case (i) can be generalized to the situation $\omega \neq 0$, provided (5) is replaced by

$$\int_{\{u>w\}} [A_i(x,u,Dw) - A_i(x,w,Dw)]D_i(u - w)dx \geqq 0. \tag{5'}$$

For instance, in the following more general situation

$$A_i(x,u,Dw) = a_i(x,Dw) + e_i(x)u, \tag{14}$$

or if Q is the linear operator

$$Qu = -D_i[a_{ij}(x)D_ju + e_i(x)u] + b_i(x)D_iu + c(x)u,$$

condition (5') is verified when

$$-D_ie_i \geqq 0 \text{ in } \mathcal{D}'(\Omega). \tag{15}$$

In the case of a linear operator the monotony condition (7) corresponds to the sign condition $c \geqq 0$ a.e. in Ω, which can actually be extended to $-D_ie_i + c \geqq 0$ in $\mathcal{D}'(\Omega)$ (see [R], for instance).

COROLLARY 1: Under the assumptions of Theorem 1, if $\hat{\psi} \geqq \psi$, then the corresponding solutions to (1) are such that $\hat{u} \geqq u$. In particular, there exists at most one solution to (1).

PROOF: It is immediate, since if $\hat{u}$ is a solution with $\hat{\psi}$ it is also a supersolution for the problem with ψ. The uniqueness follows then trivially. □

PROPOSITION 1: Under the assumption of Theorem 1 (and Remark 2) we have the estimate

$$\|u - \hat{u}\|_{L^\infty(\Omega)} \leq \|\psi - \hat{\psi}\|_{L^\infty(\Omega)}. \tag{16}$$

PROOF: Let $\ell \equiv \|\psi - \hat{\psi}\|_{L^\infty(\Omega)} > 0$ and observe that (7) implies

$$B(x,u + \ell,Du) \geq B(x,u,Du).$$

Therefore, in the first case (i), $u + \ell$ is a supersolution for the problem solved by $\hat{u}$ and $\hat{\psi}$, provided u solves (1) with ψ. Hence $\hat{u} \leq u + \ell$ and, reversing the role of u and $\hat{u}$, we easily conclude (16).

This argument does not apply in the other three cases. However, essentially the same proof of Theorem 1 gives the estimate (16) by considering as test functions $v = u + \delta\xi F_\varepsilon(\hat{u} - u - \ell) \in K_\psi$ in (1) for u and $v = \hat{u} - \delta\xi F_\varepsilon(\hat{u}-u-\ell)\in K_\psi$ in (1) for $\hat{u}$, where $\delta > 0$ is small enough, $\xi \in C^1(\bar{\Omega})$, $\xi \geq 0$ in Ω and F_ε is a nonnegative Lipschitz function ($F_\varepsilon(t) = 0$ if $t \leq \varepsilon$ and $F_\varepsilon(t) = \int_\varepsilon^t d\tau/\omega^2(\tau)]/[\int_\varepsilon^{+\infty} d\tau/\omega^2(\tau)]$ if $t > \varepsilon > 0$). Following the same arguments of [CM] it is easy to conclude that $\hat{u} \leq u + \ell$. The estimate $u \leq \hat{u} + \ell$ follows analogously. □

REMARK 3: The estimate (16) extends well-known results for the coercive obstacle problem (see [KS], for instance) and for noncoercive linear operators as in Remark 2 (see [R], for instance).

3. EXISTENCE VIA THE COMPARISON PRINCIPLE

In this section we use the comparison principle of Theorem 1 in order to give a result of existence (and uniqueness) for the solution of the problem (1).

THEOREM 2: Under the assumptions of Theorem 1, suppose the obstacle ψ in (2) is such that $\psi \in L^\sigma(\Omega)$ ($\sigma = 2s/(s - 2)$). Then the existence of a supersolution w, is a sufficient (and necessary) condition for the existence of a solution u to the obstacle problem (1).

PROOF: Any solution u to (1) is also a supersolution and, by the Sobolev imbedding, $u \in L^{\sigma}(\Omega)$, so that the existence condition is obviously necessary. On the other hand, if u exists, by Theorem 1, it must verify

$$\psi \leq u \leq w \quad \text{a.e. in } \Omega,$$

so that it is *a priori* bounded in $L^{\sigma}(\Omega)$, namely ($\|u\|_{\sigma} = \|u\|_{L^{\sigma}(\Omega)}$)

$$\|u\|_{\sigma} \leq \|u^{+}\|_{\sigma} + \|u^{-}\|_{\sigma} \leq \|w^{+}\|_{\sigma} + \|\psi^{-}\|_{\sigma}.$$

Taking any fixed $v_0 \in K_{\psi}$ in (1), this *a priori* boundness for the noncoercive terms of Q implies, by the structural assumptions (4), (6) and (9), that u is also *a priori* bounded in $H_0^1(\Omega)$. This fact follows easily from

$$\begin{aligned}\alpha\|D(u - v_0)\|_2^2 &\leq \int_{\Omega} [A_i(u,Du) - A_i(u,Dv_0)]D_i(u - v_0) \\ &\leq -\int_{\Omega} A_i(u,Dv_0)D_i(u - v_0) - \int_{\Omega} B(u,Du)(u - v_0) \\ &\leq \quad C_1 \|D(u - v_0)\|_2 + C_2,\end{aligned}$$

where $C_1 = \|\tilde{f}\|_2 + \|e_i\|_{r_i} \|u\|_{\sigma} + c\|Dv_0\|_2 + \|b\|_s \|u - v_0\|_{\sigma}$ and $C_2 = \|f_0\|_r + \|c_0\|_{r_0} \|u\|_{\sigma} + \|b\|_s \|u - v_0\|_{\sigma} \|Dv_0\|_2$.

Since Q is an operator of the calculus of variations (in the sense of Leray-Lions) we can apply a well-known existence result for elliptic variational inequalities in the bounded convex set $K_R = K_{\psi} \cap \{v \in H_0^1(\Omega): \|Dv\|_2 \leq R\}$ for large $R > 0$ (see Theorem 8.1 of [L], p. 245). By the above *a priori* estimate, if R is sufficiently large, it is well known that the solution u_R of

$$u_R \in K_R: \langle Qu_R, v - u_R \rangle \geq 0, \ \forall v \in K_R, \tag{17}$$

provided $\|Du_R\|_2 < R$, also solves (1), that is, $u = u_R$ is the solution of (1). □

<u>REMARK 4</u>: The restriction $\psi \in L^{\sigma}(\Omega)$ in Theorem 2 is related to the special structural conditions (6) and (9). In the special case where $\rho_0 = \rho_i = 1$

and $r_0 = r_i = +\infty$, we can actually take $\psi \in L^2(\Omega)$ and we do not need to require the supersolution w to be in $L^\sigma(\Omega)$ also, but only in $H^1(\Omega)$. Note this last condition is always fulfilled if Ω is smooth (say $\partial\Omega \in C^{0,1}$) since then $H^1(\Omega) \subset L^\sigma(\Omega)$. Notice that if $\psi \in L^\infty(\Omega)$ more general existence results can be obtained directly in $K_\psi \cap L^\infty(\Omega)$ under different assumptions on A_i and B, allowing a quadratic growth in p in the coefficient B (see [RT]).

REMARK 5: The comparison result of Corollary 1 can easily be extended to the two-obstacles problem (1) with K_ψ replaced by $K_\psi^\phi \equiv \{v \in H_0^1(\Omega): \phi \geq v \geq \psi$ a.e. in $\Omega\} \neq \emptyset$ (use $u - z_k \in K_\psi^\phi$ and $\hat{u} + z_k \in K_{\hat{\psi}}^{\hat{\phi}}$ and argue as in the proof of Theorem 1(i) for $z_k = (u - \hat{u} - k)^+$ to conclude $u \leq \hat{u}$; for the other cases, see [CM]). Then, for $\psi, \phi \in L^\sigma(\Omega)$ ($\sigma = 2$ if $\rho_0 = \rho_i = 1$ and $r_0 = r_i = +\infty$), it is clear that this proof of Theorem 2 gives the existence of a unique solution to the problem (1) for K_ψ^ϕ.

REMARK 6: Similarly to the preceding remark the existence of a supersolution w^* and of a subsolution w_* to the Dirichlet problem (i.e. if $\psi \equiv -\infty$)

$$u \in H_0^1(\Omega): Qu = 0 \text{ in } \Omega, \tag{18}$$

is also a sufficient (and necessary) condition for the (unique) solvability of this noncoercive problem. The use of super- and subsolutions have been used to give existence results for coercive problems with quadratic non-linearities (see [BMP] and the references there). □

COROLLARY 2: Let the assumptions of Theorem 1 hold. If w and v are two supersolutions to (1), then $v \wedge w = \inf(v,w)$ is also a supersolution to (1).

PROOF: By Theorem 2, we have the existence of a unique z solving

$$z \in K_{v \wedge w}: \langle Qz, \zeta - z\rangle \geq 0, \quad \forall \zeta \in K_{v\wedge w}.$$

Since v and w are also supersolutions to this problem, it follows, by Theorem 1, that $v \geq z$ and $w \geq z$. Then $v \wedge w \geq z$ and, consequently, $z = v \wedge w \geq \psi$ is also a supersolution to (1). □

REMARK 7: The preceding argument shows that under the assumptions (i)-(iv) of Theorem 1, the infimum of two supersolutions to the Dirichlet problem (18) is still a supersolution. Analogously, the supremum of two subsolutions to (18) is also a subsolution.

4. EXISTENCE VIA THE A PRIORI ESTIMATE

In this section we solve (1) under the following assumptions:

$$\psi^+ = \psi \vee 0 \in H_0^1(\Omega), \tag{19}$$

$$e_i \equiv 0, \quad i = 1,\dots,n. \tag{20}$$

For instance, (19) holds for any obstacle $\psi \in H^1(\Omega)$, such that $\psi \leqq 0$ on $\partial\Omega$ in the sense of traces, provided $\partial\Omega$ is a Lipschitzian boundary. However, (19) is more general because it does not require any information on $\psi^- = -(\psi \wedge 0)$ nor on $\partial\Omega$.

THEOREM 3: Under the structural conditions (4), (6), (7), (9) and (20) on Q, with an obstacle verifying (19), the variational inequality (1) has at least one solution.

PROOF: As in the proof of Theorem 2, it is enough to show the *a priori* estimate

$$\|Du\|_2 < R$$

for R sufficiently large (recall (17)). This will be done by decomposing $u = u^+ - u^-$ and estimating separately each nonnegative component.

First, to estimate u^+ we use the auxiliary function $z = u^+ - \psi^+ \in H_0^1(\Omega)$, $z \geqq 0$, which is decomposed following a technique of [BM] (see also [R], p. 120), by

$$z = \sum_{j=1}^{J} z_j, \; z_j = [(z-k_j) \vee 0] \wedge (k_{j-1}-k_j), \; j = 1,\dots, J, \tag{21}$$

where the constants $k_0 > k_1 > \dots > k_j > \dots > k_J = 0$ ($k_0 = +\infty$, by convention), are defined by the condition

$$\| b \|_{s,\Omega_j} = \left(\int_{\Omega_j} |b|^s\right)^{1/s} = \phi(k_j, k_{j-1}) \leq \alpha/2C_\sigma, \tag{22}$$

where $C_\sigma > 0$ is the Sobolev constant ($\| v \|_\sigma \leq C_\sigma \| Dv\|_2$, $v \in H_0^1(\Omega)$) and $\Omega_j = \{x \in \Omega: |Dz| > 0, k_j < z < k_{j-1}\}$, where $Dz_j = \chi_{\Omega_j} Dz$ for $j = 1,\ldots,J$. Note that $J \leq \|b\|_s^s (2C_\sigma/\alpha)^s + 1$, because we can choose k_m such that $\phi(k_m,k_{m-1}) = \alpha/2C_\sigma$ (if $\|b\|_{s,\Omega_m} \leq \alpha/2C_\sigma$ set $k_m = 0$ and $J = m$), $m = 1,\ldots,J$, due to the fact that $\phi(k_m,k_{m-1}) \to 0$ if $k_m \to k_{m-1}$, and

$$(J - 1)(\alpha/2C_\sigma)^s = \sum_{m=1}^{J-1} \|b\|_{s,\Omega_m}^s \leq \|b\|_s^s,$$

which implies that $J < +\infty$.

Now choosing $v = u - z_m \in K_\psi$ in (1) we find

$$\begin{aligned}\alpha \|Dz_m\|_2^2 &\leq \int_{\Omega_m} [A_i(u,Dz_m + D\psi^+) - A_i(u,D\psi^+)]D_iz_m \\ &\leq -\int_{\Omega_m} A_i(u,D\psi^+)D_iz_m - \int_{\{u>k_m+\psi^+\}} B(u,Du)z_m \\ &\leq C_1 \|Dz_m\|_2 - \int_{\{u>k_m+\psi^+\}} B(\psi^+,Du)z_m,\end{aligned} \tag{23}$$

where we have used conditions (4), (6), (20), (7) and set $C_1 = \|\bar{f}\|_2 + c\|D\psi^+\|_2$. The last term in (23) can be estimated using the property $z_mDz = z_m \sum_{j=1}^{m} Dz_j$ and, recalling (9) and (22),

$$\begin{aligned}-\int_\Omega B(\psi^+,Du)z_m &\leq \int_\Omega [f_0 + c_0|\psi^+|^{\rho_0} + b|D\psi^+| + b|Dz|]z_m \\ &\leq C_2 \|z_m\|_\sigma + \sum_{j=1}^{m} \int_{\Omega_j} z_m b[Dz_j] \\ &\leq C_2C_\sigma \|Dz_m\|_2 + \sum_{j=1}^{m} \|z_m\|_\sigma \|b\|_{s,\Omega_j} \|Dz_j\|_2 \\ &\leq C_3 \|Dz_m\|_2 + \frac{\alpha}{2} \|Dz_m\|_2 \sum_{j=1}^{m} \|Dz_j\|_2,\end{aligned} \tag{24}$$

where $C_2 = \| f_0 \|_r + \| c_0 \|_{r_0} \|\psi^+\|_\sigma + \|b\|_s \|D\psi^+\|_2$ and $m = 1,\ldots,J$.

Then, by induction on m, from (23) and (24), we find

$$\frac{\alpha}{2} \|Dz_m\|_2 \;\pi\; 2^{m-1}(C_1 + C_3), \; m = 1,\ldots,J,$$

and, recalling (21), we finally conclude

$$\|Du^+\|_2 \leqq \|D\psi^+\|_2 + \|Dz\|_2 \leqq \|D\psi^+\|_2 + (2^J - 1)(C_1 + C_3)(2/\alpha),$$

where the constants C_1, C_3 and J are independent of u.

In order to estimate u^-, we use a similar decomposition as in (21) by defining $u_j^- = [(u^- - \ell_j) \vee 0] \wedge (\ell_{j-1} - \ell_j)$, $j = 1,\ldots,L$, where the constants $\ell_0 = +\infty > \ell_1 > \ldots > \ell_L = 0$ are now defined through $\Omega_j^- = \{x \in \Omega: |Du^-| > 0, \ell_j < u^- < \ell_{j-1}\}$ and such that (22) holds (with k_j replaced by ℓ_j).

Choosing $v = u + u_m^- \in K_\psi$ in (1) and noting that $Du_m^- = - Du\, \chi_{\Omega_j^-}$, we have analogously to (23) and (24),

$$\begin{aligned}
\alpha\| Du_m^- \|_2^2 &\leqq \int_{\Omega_j^-} [A_i(u,Du) - A_i(u,0)]D_i u \\
&\leqq - \int_\Omega A_i(u,0)D_i u_m^- + \int_{\{u<-\ell_j\}} B(u,Du)u_m^- \\
&\leqq \|\bar{f}\|_2 \|Du_m^-\|_2 + \int_{\{u<-\ell_j\}} B(0,Du)u_m^- \\
&\leqq C_4 \|Du_m^-\|_2 + \frac{\alpha}{2} \|Du_m^-\|_2 \sum_{j=1}^{m} \|Du_j^-\|_2 ,
\end{aligned} \tag{25}$$

where $C_4 = \|\bar{f}\|_2 + C_\sigma \|f_0\|_r$ and $m = 1,\ldots,L$. Again by induction on m, we conclude, as before, that

$$\|Du^-\|_2 \leqq \sum_{m=1}^{L} \|Du_m^-\|_2 \leqq (2^L - 1)C_4(2/\alpha)$$

which completes the proof of the theorem. □

REMARK 8: In the case of the linear operator given in Remark 2, this *a priori* estimate requires $c - D_i e_i \geqq 0$ (see [R], p. 121, in line 12, where J should be $(2^J - 1)$). Note that following this argument, in the special case of a quasi-linear operator of the form (14), with the assumption (15), we can still prove the $H_0^1(\Omega)$ *a priori* estimate, by remarking that the additional term appearing in (23) can be easily controlled by

$$\int_{\Omega_m} e_i u D_i z = \int_{\Omega_m} e_i (z_m + k_m - \psi^+) D_i z_m = -\int_\Omega e_i \psi^+ D_i z_m + \int_\Omega e_i D_i (k_m z_m + \tfrac{1}{2} z_m^2)$$

$$\leqq -\int_\Omega e_i \psi^+ D_i z_m \leqq \|Dz_m\|_2 \, \|\psi^+\|_\sigma \, \|e_i\|_{r_i},$$

and, for the estimate of u^-, we have on the right-hand side of (25)

$$-\int_{\Omega_m^-} e_i u \, D_i u_m^- = \int_\Omega e_i D_i (\ell_m u_m^- + \tfrac{1}{2}(u_m^-)^2) \geqq 0.$$

REMARK 9: With suitable adaptations, most results of this paper can be extended to $W^{1,p}(\Omega)$ $(1 < p < \infty)$, eventually with different boundary conditions (see [CM]) and different convex sets as in [B]. As was observed in [B] (Oss. 4.5) for the linear case, we can also extend the previous results to certain quasi-linear degenerate operators, for example, with a A_2 *weight* as in [CF].

REMARK 10: In particular, we can consider $\psi \equiv -\infty$ in Theorem 3, which means that the Dirichlet problem

$$u \in H_0^1(\Omega): \; -D_i[A_i(x,u,Du) + E_i(x)u] + B_i(x,u,Du) = 0 \text{ in } \Omega,$$

under assumptions (4), (6), (20), (7), (9), where $E_i \in L^r(\Omega)$ $(r = n \geqq 3$ or $r > 2 = n)$ satisfies $-D_i E_i \geqq 0$ in $\mathcal{D}'(\Omega)$, has at least one solution. This result seems new.

ACKNOWLEDGEMENTS: This work was partially supported by research contract No. 87046 of JNICT, and was written while the author was visiting the Academy of Nancy/Metz.

REFERENCES

[A] M. Artola, Sur une classe de problèmes paraboliques quasilinéares, Boll. Un. Mat. Ital. (6) 5-B (1986), 51-70.

[B] G. Bottaro, Alcune condizioni sufficienti per l'existenza e'unicita della soluzione di una disequazione variazionale non coerciva, Annali Mat. Pura Appl. 106 (1975), 187-203.

[BM] G. Bottaro and M.E. Marina, Problema di Dirichlet per equazioni ellipticche di tipo variazionale su insiemi non limitati, Boll. Un. Mat. Ital. (4) 8 (1973), 46-56.

[BMP] L. Boccardo, F. Murat and J.P. Puel, Existence of bounded solutions for nonlinear elliptic unilateral problems, Annali Mat. Pura Appl. (1988) (in print).

[CC] J. Carrillo and M. Chipot, On nonlinear elliptic equations involving derivative of the nonlinearity, Proc. Royal. Soc. Edinburgh 100-A (1985), 281-294.

[CF] F. Chiarenza and M. Frasca, Une disequazione variazionale associata a un operatore elliptico con degenerazione di tipo A_2, Le Matematiche (Catania) 37 (1982), 239-250.

[CM] M. Chipot and G. Michaille, Uniqueness results and monotonicity properties for strongly nonlinear elliptic variational inequalities, Pre-print #347, IMA - Univ. Minnesota, 1987.

[CR] M. Chipot and J.F. Rodrigues, Comparison and stability of solutions to a class of quasilinear parabolic operators, Proc. Royal Soc. Edinburgh 110-A (1988), 275-285.

[KS] D. Kinderlehrer and G. Stampacchia, An Introduction to Variational Inequalities and their Applications, Academic Press, New York, 1980.

[L] J.L. Lions, Quelques Méthodes de Résolution des Problèmes aux Limits Non Linéares, Dunod, Paris 1969.

[RT] J.M. Rakotoson and R. Temam, Relative rearrangement in quasilinear variational inequalities, Indiana Univ. Math. J. 36 (1987), 757-810.

[R] J.F. Rodrigues, Obstacle Problems in Mathematical Physics, North-Holland, Amsterdam, 1987.

[T] N.S. Trudinger, On the comparison principle for quasilinear divergence structure equations, Arch. Rational Mech. Anal. 57, (1974), 128-133.

J.F. Rodrigues
CMAF et University of Lisbon
Av. Prof. Gama Pinto, 2
1699 Lisboa Codex
Portugal

A. VISINTIN

Generalized coarea formula

1. INTRODUCTION

Let Ω be any domain of $\mathbb{R}^N$ ($N \geq 1$), and set

$$V(u) := \int_\Omega |\nabla u| := \sup \{ \int_\Omega u \operatorname{div} \eta \, dx : \eta \in C_c^1(\Omega)^N, |\eta| \leq 1 \}, \quad \forall u \in L^1(\Omega). \tag{1.1}$$

The functional fulfils the classical Fleming-Rishel *co-area formula* [3], [4, p. 20]

$$V(u) = \int_R V(H_s(u)) \, ds \; (\leq +\infty), \quad \forall u \in L^1(\Omega), \tag{1.2}$$

where $H_s(y) = 0$ if $y < s$, $H_s(y) = 1$ if $y \geq s$.

This paper announces and illustrates the results of [7], [8]. In Section 2 we introduce the class $GC(\Omega)$ of all the functionals $\Lambda: L^1(\Omega) \to [0, +\infty]$ which fulfil (1.2), and give some examples. In Section 3 we present some results on the minima of a class of *nonconvex functionals*. Then in Section 4 we show that the class $GC(\Omega)$ is strictly related to that of set applications $\mathcal{P}(\Omega) \to [0, +\infty]$. This geometrical interpretation of $GC(\Omega)$ is the starting point for introducing a new concept of *fractional dimension* for set boundaries, which is better behaved than the standard ones (Hausdorff measure, for example) for certain problems of the calculus of variations. In Section 5 we show that certain functionals of $GC(\Omega)$ can be used for modelling two-phase systems with *surface tension* effects [6], [7].

2. GENERALIZED CO-AREA FORMULA

The way we generalize (1.2) is quite different from other extensions of (1.2), also known as co-area formulae [2].

DEFINITION 1: We shall denote by $GC(\Omega)$ the class of functionals $\Lambda: L^1(\Omega) \to [0, +\infty]$ which are proper (i.e. $\Lambda \not\equiv +\infty$) and which fulfil the following *generalized co-area formula*

$$\Lambda(u) = \int_{\mathbf{R}} \Lambda(H_s(u))\, ds \ (\leq +\infty), \quad \forall u \in L^1(\Omega), \tag{2.1}$$

with the convention that the integral is replaced by the superior integral if the function $\mathbf{R} \to [0, +\infty]: s \to \Lambda(H_s(u))$ is not measurable.

For any $\Lambda \in GC(\Omega)$, we also set $Dom(\Lambda) := \{u \in L^1(\Omega): \Lambda(u) \neq +\infty\}$ and $\hat{\Lambda}(u) := \Lambda(-u)$, $\forall u \in L^1(\Omega)$.

PROPOSITION 1 ([8]): For any $\Lambda \in GC(\Omega)$,

$$\Lambda(u+c) = \Lambda(u), \quad \forall u \in L^1(\Omega), \forall c \in \mathbf{R}, \tag{2.2}$$

$$\Lambda(cu) = c\Lambda(u), \quad \forall u \in L^1(\Omega), \forall c > 0, \tag{2.3}$$

$$\Lambda(c) = 0, \quad \forall c \in \mathbf{R}, \tag{2.4}$$

$$\hat{\Lambda} \in GC(\Omega), \tag{2.5}$$

if Λ is convex, then $Dom(\Lambda)$ is a convex cone, $Dom(\Lambda+\hat{\Lambda})$ is a linear subspace of $L^1(\Omega)$ and $\Lambda + \hat{\Lambda}$ is a seminorm. (2.6)

The proof of these properties is straightforward.

Examples of functionals of $GC(\Omega)$:

(i) Trivial cases: $\Lambda^0 \equiv 0$, $\Lambda_c(u) = 0$ if u = const. a.e. in Ω, $\Lambda_c(u) = +\infty$ otherwise.

(ii) After (1.2), $V \in GC(\Omega)$. Moreover, $Dom(V) = BV(\Omega)$ and $V = V^{**}$, namely, V is convex and lower semicontinuous in $L^1(\Omega)$.

(iii) The functional

$$\Lambda_{osc}(u) := \operatorname*{ess\,osc}_{\Omega} u \ (:= \operatorname*{ess\,sup}_{\Omega} u - \operatorname*{ess\,inf}_{\Omega} u) \ (\leq +\infty), \ \forall u \in L^1(\Omega). \tag{2.7}$$

(iv) For any measurable function $g: \Omega^2 \to \mathbf{R}^+$, we set

$$\Lambda_g(u) := \iint_{\Omega^2} |u(x)-u(y)| g(x,y)\, dx\, dy \ (\leqq +\infty), \ \forall u \in L^1(\Omega). \qquad (2.8)$$

It is easy to check that $\Lambda_g \in GC(\Omega)$, by the identity

$$|\xi-\eta| = \int_{\mathbf{R}} |H_s(\xi) - H_s(\eta)| ds, \ \forall \xi, \eta \in \mathbf{R}. \qquad (2.9)$$

(here applied with $\xi = u(x)$, $\eta = u(y)$), and by Fubini's theorem. Λ_g is also convex and, by Fatou's lemma, lower semicontinuous in $L^1(\Omega)$. That is, $\Lambda_g = \Lambda_g^{**}$.

(v) As a particular case of example (iv), we take

$$g_r(x,y) := |x-y|^{-(N+r)}, \ \forall x,y \in \Omega (x \neq y), \ \forall r \in]0,1[, \qquad (2.10)$$

and set $\Lambda_r := \Lambda_{g_r}$. This is the standard seminorm of the fractional Sobolev space $W^{r,1}(\Omega) (= \mathrm{Dom}(\Lambda_r))$.

(vi) For any measurable function $f: \Omega \times \mathbf{R}^+ \to \mathbf{R}^+$, we set

$$\tilde{\Lambda}_f(u) := \iint_{\Omega\times\mathbf{R}^+} (\operatorname*{ess\,osc}_{\Omega\cap B_h(x)} u) \cdot f(x,h)\, dx\, dh \ (\leqq +\infty), \ \forall u \in L^1(\Omega), \qquad (2.11)$$

where $B_h(x) := \{y \in \mathbf{R}^N : |x-y| < h\}$. It is easy to check that $\tilde{\Lambda}_f \in GC(\Omega)$, by example (iii) and by Fubini's theorem. $\tilde{\Lambda}_f$ is convex and, by Fatou's lemma, also lower semicontinuous in $L^1(\Omega)$, as is Λ_{osc}. That is $\tilde{\Lambda}_f = \tilde{\Lambda}_f^{**}$.

(vii) As a particular case of example (vi), we take

$$f_r(x,h) := h^{-(1+r)} \text{ a.e. for } (x,h) \in \Omega \times \mathbf{R}^+, \ \forall r \in]0,1[, \qquad (2.12)$$

and set $\tilde{\Lambda}_r := \tilde{\Lambda}_{f_r}$. Then

$$\Lambda_r(u) \leqq \text{const.} \cdot \tilde{\Lambda}_r(u), \ \forall u \in L^1(\Omega), \ \forall r \in]0,1[, \qquad (2.13)$$

whence $\mathrm{Dom}(\tilde{\Lambda}_r) \subset \mathrm{Dom}(\Lambda_r)$.

PROPOSITION 2 ([8]): If either $\Lambda = \Lambda_{osc}$, or $\Lambda = \Lambda_g$, or $\Lambda = \tilde{\Lambda}_f$, then $\Lambda \in GC(\Omega)$ and $\Lambda = \Lambda^{**}$. Moreover, if either $\Lambda = V$, or $\Lambda = \Lambda_r$, or $\Lambda = \tilde{\Lambda}_r$ $(0 < r < 1)$, then the injection of the Banach space $\mathrm{Dom}(\Lambda)$ into $L^1(\Omega)$ is compact, provided that Ω fulfils the regularity assumptions of the classical Rellich compactness theorem.

3. ON A CLASS OF NONCONVEX FUNCTIONALS

Let Ω be a domain of $\mathbf{R}^N$ $(N \geqq 1)$ and

$$\phi: \mathbf{R} \to \mathbf{R} \cup \{+\infty\} \text{ be lower semicontinuous, } \phi \not\equiv +\infty, \tag{3.1}$$

$$\phi(y) \geqq -C_1|y| - C_2 (C_1, C_2 \in \mathbf{R}^+;\ C_2 = 0 \text{ if } \mu(\Omega) = +\infty); \tag{3.2}$$

set

$$(-\infty <)\ \Phi(u(x)) := \int_\Omega \phi(u(x))\, dx\ (\leqq +\infty),\ \forall u \in L^1(\Omega). \tag{3.3}$$

THEOREM 1 ([7]): Let $\Lambda \in GC(\Omega)$ and $\Lambda = \Lambda^{**}$. Assume that (3.1), (3.2) hold and that

$$\text{any connected component of } \{y \in \mathbf{R}: \phi^{**}(y) < \phi(y)\} \text{ is bounded.} \tag{3.4}$$

Then for any $u \in L^1(\Omega)$

$$\partial(\Phi + \Lambda)(u) = \partial\Phi(u) + \partial\Lambda(u) \text{ in } L^\infty(\Omega), \tag{3.5}$$

$$(\Phi + \Lambda)^{**}(u) = \Phi^{**}(u) + \Lambda(u). \tag{3.6}$$

COROLLARY ([7]): Under the previous assumptions,

$$\text{if } \partial(\Phi + \Lambda)(u) \neq \emptyset, \text{ then } \partial\phi(u(x)) \neq \emptyset \text{ a.e. in } \Omega. \tag{3.7}$$

Otherwise stated: For any $\xi \in L^\infty(\Omega)$, if $u \in L^1(\Omega)$ is an *absolute* minimum of $v \to (\Phi + \Lambda)(v) - \int_\Omega \xi(x)v(x)\, dx$, then $\partial\phi(u(x)) \neq \emptyset$ a.e. in Ω. A similar statement holds for *relative* minima in the strong topology of $L^1(\Omega)$. In both cases, if ϕ is nonconvex, then certain values are a priori excluded from range of u, for any $\xi \in L^\infty(\Omega)$. These results can be used to model *pattern formation* phenomena, see Section 5; they also play a crucial role in the construction of a model of the *evolution of non-Cartesian surfaces* of prescribed mean curvature [9].

4. FRACTAL BOUNDARIES

We shall denote by $\mathcal{A}$ the σ-algebra of (equivalence classes of) Lebesgue measurable subsets of Ω, and by χ_A the characteristic function of any $A \in \mathcal{A}$. We also set

$$SA(\Omega) := \{F: \mathcal{A} \to [0, +\infty] : F(\emptyset) = F(\Omega) = 0\}.$$

There is a natural correspondence between the set applications of SA(Ω) and the functionals of GC(Ω):

PROPOSITION 3 ([8]): (i) For any $\Lambda \in GC(\Omega)$,

$$F_\Lambda \in SA(\Omega), \text{ where } F_\Lambda(A) := \Lambda(\chi_A), \quad \forall A \in \mathcal{A}. \tag{4.1}$$

(ii) For any $F \in SA(\Omega)$,

$$\mathcal{L}_F \in GC(\Omega), \text{ where } \mathcal{L}_F(u) := \int_{\mathbf{R}} F(\{u > s\})\, ds, \ \forall u \in L^1(\Omega). \tag{4.2}$$

(iii) The transformations

$$F: GC(\Omega) \to SA(\Omega): \ \Lambda \to F_\Lambda, \tag{4.3}$$

$$\mathcal{L}: SA(\Omega) \to GC(\Omega): \ F \to \mathcal{L}_F, \tag{4.4}$$

and each one the inverse of the other one.

In particular, for any $\psi: P(\Omega) \to [0, +\infty]$ such that $\psi(\emptyset) = 0$, we set

$$F_\psi(A) := \begin{cases} \psi(\partial_e A) & \text{if } A \in \mathcal{A}, \\ +\infty & \text{otherwise,} \end{cases} \tag{4.5}$$

where $\partial_e A := \{x \in \Omega: \mu(B_h(x) \cap A) > 0, \mu(B_h(x) \cap (\Omega \setminus A)) > 0, \forall h > 0\}$ is the *essential boundary* of A. Then $\mathcal{L}_{F_\psi} \in GC(\Omega)$.

For any $d \in]0,N[$, let us denote by H_d the d-dimensional Hausdorff measure. Note that for any set A of class C^2, $V(\chi_A) = H_{N-1}(\partial_e A)$. However, for fractional dimensions the following occurs:

PROPOSITION 4 ([8]): For any $r \in]0,1[$, let $M_r \in GC(\Omega)$ be such that

$$M_r(\chi_A) = H_{N-r}(\partial_e A), \quad \forall A \in \mathfrak{a}. \tag{4.6}$$

Then

$$M_r \text{ is not strongly lower semicontinuous in } L^1(\Omega), \tag{4.7}$$

$$\text{the injection } \mathrm{Dom}(M_r) \to L^1(\Omega) \text{ is not compact.} \tag{4.8}$$

Because of this negative result, the most standard tools of the calculus of variations cannot be applied to the functionals M_r, with $0 < r < 1$. On the other hand, the functionals Λ_r and $\tilde{\Lambda}_r$ have the useful properties of lower semicontinuity and compactness pointed out in Section 2. Moreover, by means of the latter functionals it is possible to mimic the concepts of the set of fractional dimension, via a construction different from the standard one [1], [5].

DEFINITION ([8]): For any $A \in \mathfrak{a}$ such that $\partial_e A \neq \emptyset$ we set

$$\mathrm{Dim}_{\{\Lambda_r\}}(\partial_e A) := N - \sup\{r \in]0,1[: \Lambda_r(\chi_A) < +\infty\}; \tag{4.9}$$

this number can be called the *dimension of the essential boundary* $\partial_e A$ *relative to the functionals* $\{\Lambda_r\}_{0<r<1}$. Similarly, one can introduce $\mathrm{Dim}_{\{\tilde{\Lambda}_r\}}(\partial_e A)$, that is, the *dimension of* $\partial_e A$ *relative to the functionals* $\{\tilde{\Lambda}_r\}_{0<r<1}$.

This latter concept is strictly related to the *Minkowski-Bouligand dimension* [5, p. 287]. Also note that by (2.13)

$$\mathrm{Dim}_{\{\Lambda_r\}}(\partial_e A) \leq \mathrm{Dim}_{\{\tilde{\Lambda}_r\}}(\partial_e A), \quad \forall A \in \mathfrak{a}, \ \partial_e A \neq \emptyset. \tag{4.10}$$

5. SURFACE TENSION EFFECTS IN TWO PHASE SYSTEMS

Consider an ice-water system occupying a "smooth" bounded domain $\Omega \subset \mathbf{R}^3$, and subject to a given relative temperature field $\theta \in L^1(\Omega)$. Let $c(x)$ denote the local water concentration and set $u(x) := 2c(x)-1$; so $u = 1$ in water, $u = -1$ in ice, $-1 < u < 1$ in so-called *mushy regions*. Following [6], [7], we propose to represent the free enthalpy by

$$\Psi_\theta(u) := \begin{cases} -\dfrac{L}{2\tau_E}\displaystyle\int_\Omega (\theta u + \frac{\alpha}{2}u^2)dx + \frac{\sigma}{2}V(u) \text{ (+ boundary terms) if } |u| \leq 1 \\ \qquad\qquad\qquad\qquad\qquad\qquad\qquad\qquad \text{a.e. in } \Omega, \\ +\infty \qquad\qquad\qquad\qquad\qquad\qquad \text{otherwise;} \end{cases} \tag{5.1}$$

here L, τ_E, σ and α are physical constants. This functional has at least one absolute minimum, here interpreted as a state of *stable equilibrium*. As Ψ_θ is nonconvex, it may also have one or more relative minima with respect to the strong topology of $L^1(\Omega)$, interpreted as states of *metastable equilibrium*. Note that if θ = const. in Ω, then $\theta \equiv \alpha$ and $\theta \equiv -\alpha$ are the maximal *superheating* of ice and *supercooling* of water, respectively.

By Theorem 1 we have:

PROPOSITION 5 ([6], [7]): For any either absolute or relative minimum u of Ψ_θ, $|u| = 1$ a.e. in Ω.

So u corresponds to a two-phase system with no mushy region. Moreover, denoting by S the ice-water interface and by H the mean curvature (assumed positive for an ice ball, for example), if θ is continuous (on S), then

$$\theta = -\frac{2\sigma\tau_E}{L} H \text{ on } S. \tag{5.2}$$

An alternative model is obtained by replacing V by either Λ_r or $\tilde{\Lambda}_r$, for any $0 < r < 1$. Then the above results, with the exception of (5.2), still hold. This allows us to represent very irregular interfaces S, of fractional dimension in the sense of (4.9), as appear in dendritic formations and in snowflakes, for instance, [8].

REFERENCES

[1] K.J. Falconer, The Geometry of Fractal Sets, Cambridge University Press, Cambridge, 1985.

[2] H. Federer, Geometric Measure Theory, Springer-Verlag, Berlin, 1969.

[3] W.H. Fleming and R. Rishel, An integral formula for total gradient variation, Arch. Math. 11 (1960), 218-222.

[4] E. Giusti, Minimal Surfaces and Functions of Bounded Variation, Birkhäuser, Boston, 1984.

[5] B.B. Mandelbrot, Fractals. Form, Chance and Dimension, Freeman, San Franciso, 1977.

[6] A. Visintin, Surface tension effects in phase transitions, in Material Instabilities in Continuum Mechanics and Related Mathematical Problems (J.M. Ball, Ed.). Claredon Press, Oxford, 1988, pp. 505-537.

[7] A. Visintin, Non-convex functionals related to multiphase systems (Preprint, 1987).

[8] A. Visintin, Generalized co-area formula and fractal sets, Japan J. Appl. Math. (to appear).

[9] A. Visintin; Pattern evolution (Preprint, 1988).

A. Visintin
Dipartimento di matematica
Università di Trento
I-38050-Povo
Trento
Italy